A Textbook of

COMPUTER ARCHITECTURE

By

Amit Kumar Mishra

B.E. (I.T.) with Hons.; M.Tech (C.S.), Ph.D. (Cont.)

Anand Engg. College

Agra

S.K. KATARIA & SONS®

Publisher of Engineering Books

4885/109, Prakash Mahal, Dr. Subhash Bhargav Lane,

Opposite Delhi Medical Association, Daryaganj, New Delhi-110002 (INDIA)

Phone: +91-11-43551243, Mobile: +91-9871775858, 7678425209

e-mail: katariabook@gmail.com, katariabooks@yahoo.com

Website: www.skkatariaandsons.com

(e-books, subscriptions and bulk buying also available on our website)

Published by:

S.K. Kataria & Sons®

4885/109, Prakash Mahal, Dr. Subhash Bhargav Lane,
Opposite Delhi Medical Association, Daryaganj, New Delhi-110002 (INDIA)
Phone: +91-11-43551243, *Mobile:* +91-9871775858, 7678425209
e-mail: katariabook@gmail.com, katariabooks@yahoo.com
Website: www.skkatariaandsons.com

First Edition: July 2012, Reprint: 2025

Price: ₹ 395/-

Laser Typesetting by : AISA Enterprises, Delhi

Printed at : Al-Aziz Books, Sahibabad

Preface

This book is designed to serve the class of students, researchers and faculty members who are interested in the subject of "A Textbook of Computer Architecture". There is a surge of interest today in the field of Computer Architecture. This book is about the functions and structure of typical computer system.

It contains all the fundamental features that need to be help field of Computer Architecture. The idea and the scope emerged from my own experience in attempting to acquire good understanding of computer architecture. In many university this course has the one of the subject at different level. I feel that a reading of book Computer Architecture will motivate the reader to organised their thought and make them perfect. In order to gain through knowledge of this subject one has normally to refer a number of text books on different topics such as Register Transfer and Microoperations, Computer Organization and Design, Input-Output Organization, Memory Organization and Advanced Concepts of Computer Architecture.

Keeping this need in mind I initiated the writing of book Computer Architecture that provide to the student will all the information required by him/her at one place. So for that each topic is developed through discussion of the core concepts, maximum topics explain with examples and program where it is needed. After each chapter Summary, and Review Questions as Exercise has given. At last last of this book appendices has been added with RISC/CISC and Performance Evaluation.

For systematic study, the book has been divided as per the requirements of the syllabus and consists of **Seven Chapters** in all.

In the end the author wishes to thank M/s S.K. Kataria & Sons (Publisher) for taking a lot of pains and initiative for bringing out the book in a short space of time.

AMIT KUMAR MISHRA

Acknowledgement

It is obvious that writing a textbook of this nature is really a tough task and in this endeavour many people have lent their helping hands for the success of this book. First of all, I want to thanks numerous student from B.Tech. and MCA courses who pushed me towards writing this book by requesting more Information regarding "A Textbook of Computer Architecture" course in systematic way. The student's Questions and Answer were a great help to shaping the contents. I would like to acknowledge the contribution and support provided by my friends and colleagues towards completion of this book.

I am grateful to my grandfathers Lt. Sh. Chandrabhan Mishra and Sh. Triguna Nand Mishra for providing innovative and encouraging environment to work on this project. I am also grateful to my father Sh. Radheyshyam Mishra and mother Smt. J. Mishra without their blessing this book would not come into existence.

I am deeply beholden to my dearest wife Mrs. Aparna Mishra and Son Krishna for always being with me. Their moral support, in moments of despair and doubt has given me the necessary backing and confidence. I specially recognize her contribution in the language improvement and proof reading of this book.

I would like to thanks my younger brother Piyush Kumar Mishra for supporting me to complete this book.

Dr. B. D. Gupta, Honourable Director Anand Engineering College, Agra and Dr. A.K Saxena, Professor of Deptt. of Electrical Engineering (D.E.I, Agra) deserve special thanks for their constant encouragement and motivation. I wish to thank Dr. S.P. Tripathi, Head of Deptt. of Computer Science (I.E.T. Lucknow), for constant cooperation, direct inspiration and technical suggestion. I am indeed lucky to we associated with Sharda Group of Institutions where lots of support from Mr. P. K. Gupta (Chairman) and Mr. Y. K. Gupta (Vice Chairman) and their nice cooperation.

I am also very grateful to Mr. Sachin and Anoop for supporting me complete this book in such a beautiful manner and well in time.

Lastly, I deeply express my thanks to the M/s S.K. Kataria & Sons for publishing this book in a such nice getup.

Amit Kumar Mishra

AUTHOR

This book is Dedicated to

Lord Shiva

His Blessings & Loving Care

Contents

1

Chapter

Introduction

INSIDE THIS CHAPTER

1.1. Introduction to Computer **1.2.** Computer Architecture and Organization **1.3.** System Definition **1.4.** System Representation **1.5.** Design Process **1.6.** Gate Level Design **1.7.** Register Level **1.8.** Processor Level.

1.1. INTRODUCTION TO COMPUTER

Computer is the most powerful tool man has ever created. Computers have made a great impact on our everyday life. There presence is felt at almost every working place *viz.* Homes, colleges, offices, industries, hospital, banks, retail stores, railway, research, and design organizations and so on. Computers, large and small, are used nowadays by all kinds of people for a variety of takes in a modern and industrialized society. A computer is a machine for manipulating data according to a list of instructions known as a program. Computers are extremely versatile. In fact, they are universal information processing machines. Computer programming (often shortened to programming or coding) is the process of writing, testing, debugging/troubleshooting, and maintaining the source code of computer programs. This source code is written in a programming language (C/C++).

TIPS

A computer system is the synthesis of hardware and software. A typical computer system employs a computer that uses programmable devices to store, retrieve, and process data. The personal computer or PC exemplifies one example of a computer system".

A computer system is shown in Fig. 1.1 with various functional blocks such as :

1. CPU: Central Processing Unit
2. Main Memory: Stores Program and Data
3. Input Device-Keyboard etc.

4. Output Device-VDU, Printer etc.
5. Auxiliary Memory Unit-Hard-disk etc.

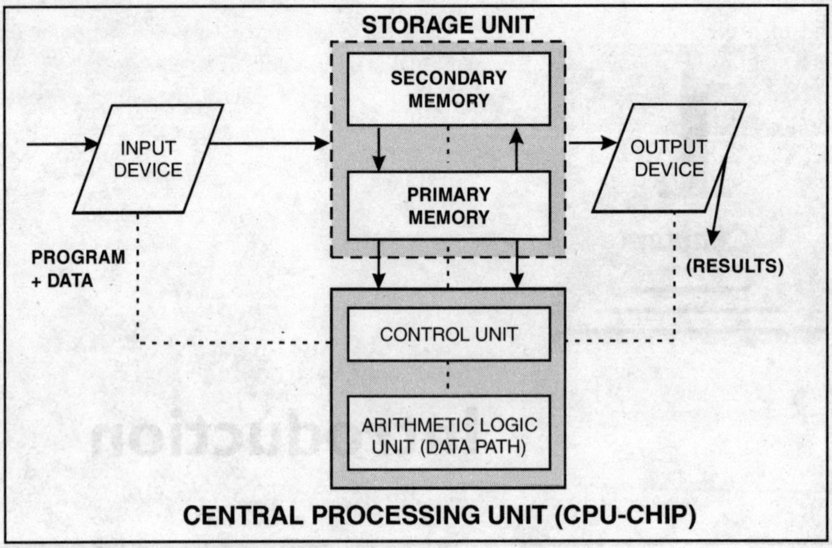

Fig. 1.1. Computer System Organisation.

(i) CPU—The Central Processing Unit

The main processing unit in a computer system, consisting of digital electronics circuits on a Silicon Chip or any other chip that executes set of user written instructions (program) to process user data as per user logistics. This chip is called micro-processor and hence the name is Micro-computer. The user program and data is stored in same memory and is executed under the control of control unit.

(ii) Main-Memory—Program And Data Storage

In the context of computing, the program is based upon logistics and the work the user wants to do with the computer and data refers to the symbols that a computer uses to represent facts and figures. The computer system processes the Binary Data such as 01010000111 patterns. The user data and programs are stored here.

(iii) Input Device

It is used for instant input to the system, such as keyboard, mouse and scanner etc.

(iv) Output Device

It is used for output from the system such as VDU-Visual Display unit or printer.

(v) Auxiliary Memory Unit

For large storage of program and data we use Hard-disks of various capacities or other media like CDROM and DVD etc.

The study of computer architecture, on the other hand, focuses on the structure and behavior of the computer system and refers to the logical aspects of system implementation as seen by the programmer. Computer architecture includes many elements such as instruction sets and formats, operation codes, data types, the number and types of registers, addressing modes, main memory access methods, and various I/O mechanisms. The architecture of a system directly affects the logical execution of programs.

Computers are extremely versatile. In fact, they are *universal* information-processing machines. According to the Church–Turing thesis, a computer with a certain minimum threshold capability is in principle capable of performing the tasks of *any* other computer, from those of a personal digital assistant to a supercomputer, as long as time and memory capacity are not considerations. Therefore, the same computer designs may be adapted for tasks ranging from processing company payrolls to controlling unmanned spaceflights. Due to technological advancement, modern electronic computers are experientially more capable than those of preceding generations (a phenomenon partially described by Moore's Law).

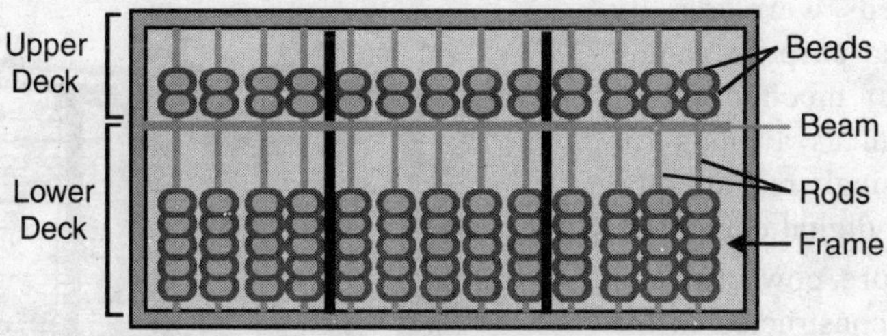

Fig. 1.2. Chinese Abacus.

ALAN TURING MACHINE

TIPS

A computer is a machine for manipulating data according to a list of instructions known as a program.

Computers take numerous physical forms. Early electronic computers were the size of a large room, and such enormous computing facilities still exist for specialized scientific computation—supercomputers —and for the transaction processing requirements of large companies, generally called mainframes. Smaller computers for individual use, called personal computers, and their portable equivalent, the laptop computer, are ubiquitous information processing and communication tools and are perhaps what most non-experts think of as "a computer". However, the most common form of computer in use today is the embedded computer, small computers used to control another device. Embedded computers control machines from fighter aircraft to digital cameras. So we can say that, A computer is basically a programmable computing, machine. Earlier computer were use for complex computations and use by only scientist and engineers. These computers being very fast can be used not only for computation but also to store and retrieve information, to control certain processes and machine, to measure and display certain physical and electrical quantities.

Do You Know

British mathematician Alan Turing proposed the idea of a machine that could process equations without human direction in 1936 machine (Turing machine).

1.1.1. History of Computing

ENIAC was a milestone in computing history. Originally, the term "computer" referred to a person who performed numerical calculations, often with the aid of a mechanical calculating device or analog computer. Examples of these early devices, the ancestors of the computer, included the abacus and the Antikythera mechanism, an ancient Greek device for calculating the movements of planets, dating from about 87 BC. The end of the Middle Ages saw a reinvigoration of European mathematics and engineering, and Wilhelm Schickard's 1623 device was the first of a number of European engineers to construct a mechanical calculator. The abacus has been noted as being an early computer, as it was like a calculator in the past.

VON NEUMANN

Charles Babbage was the first to conceptualize and design a fully programmable computer as early as 1820, but due to a combination of the limits of the technology of the time, limited finance, and an inability to resist tinkering with his design, the device was never actually constructed in his lifetime. A number of technologies that would later prove useful in computing, such as the punch card and the vacuum tube had appeared by the end of the 19th century, and large-scale automated data processing using punch cards was performed by tabulating machines designed by Hermann Hollerith. During the first half of the 20th century, many scientific computing needs were met by increasingly sophisticated, special-purpose analog computers, which used a direct mechanical or electrical model of the problem as a basis for computation. These became increasingly rare after the development of the programmable digital computer.

CHARLES BABBAGE

A succession of steadily more powerful and flexible computing devices were constructed in the 1930s and 1940s, gradually adding the key features of modern computers, such as the use of digital electronics (largely invented by Claude Shannon in 1937) and more flexible programmability. Defining one point along this road as "the first digital electronic computer" is exceedingly difficult. Notable achievements include the Atanasoff-Berry Computer (1937), a special-purpose machine that used valve-driven (vacuum tube) computation, binary numbers, and regenerative memory.

Fig. 1.3.

The team who developed ENIAC, recognizing its flaws, came up with a far more flexible and elegant design, which has become known as the Von Neumann architecture (or "stored program architecture"). This stored program architecture became the basis for virtually all modern computers. A number of projects to develop computers based on the stored program architecture commenced in the mid to late-1940s; the first of these were completed in Britain. The first to be up and running was the Small-Scale Experimental Machine, but the EDSAC was perhaps the first practical version that was developed.

Valve (tube) driven computer designs were in use throughout the 1950s, but were eventually replaced with transistor-based computers, which were smaller, faster, cheaper, and much more reliable, thus allowing them to be commercially produced, in the 1960s. By the 1970s, the adoption of integrated circuit technology had enabled computers to be produced at a low enough cost to allow individuals to own a personal computer.

Do You Know

The American mathematician Howard Aiken developed the mark I calculating machine which was built by IBM in the 1930.

1.1.2. How Computers Work

While the technologies used in computers have changed dramatically since the first electronic, general-purpose computers of the 1940s, most still use the stored program architecture (sometimes called the Von Neumann architecture). The design made the universal computer a practical reality. The architecture describes a computer with four main sections : the Arithmetic and Logic Unit (ALU), the control circuitry, the memory, and the input and output devices (collectively termed I/O). These parts are interconnected by bundles of wires (called "buses" when the same bundle supports more than one data path) and are usually driven by a timer or clock (although other events could drive the control circuitry).

Conceptually, a computer's memory can be viewed as a list of cells. Each cell has a numbered "address" and can store a small, fixed amount of information. This information can either be an instruction, telling the computer what to do, or data, the information which the computer is to process using the instructions that have been placed in the memory. In principle, any cell can be used to store either instructions or data. The ALU is in many senses the heart of the computer. It is capable of performing two classes of basic operations. The first is arithmetic operations; for instance, adding or subtracting two numbers together. The set of arithmetic operations may be very limited; indeed, some designs do not directly support multiplication and division operations (instead, users support multiplication and division through programs that perform multiple additions, subtractions, and other digit manipulations). The second class of ALU operations involves *comparison* operations : given two numbers, determining if they are equal, or if not equal which is larger.

The I/O systems are the means by which the computer receives information from the outside world, and reports its results back to that world. On a typical personal computer, input devices include objects like the keyboard and mouse, and output devices include computer monitors, printers and the like, but as will be discussed later a huge variety of devices can be connected to a computer and serve as I/O devices.

The control system ties this all together. Its job is to read instructions and data from memory or the I/O devices, decode the instructions, providing the ALU with the correct inputs according to the instructions, "tell" the ALU what operation to perform on those inputs, and send the results back to the memory or to the I/O devices. One key component of the control system is a counter that keeps track of what the address of the current instruction is; typically, this is incremented each time an instruction is executed, unless the instruction itself indicates that the next instruction should be at some other location (allowing the computer to repeatedly execute the same instructions).

The ALU and control unit (collectively called a central processing unit or CPU) have typically been located on a single integrated circuit called a microprocessor. Since the 1980, the functioning of such a computer is in principle quite straightforward. Typically, on each clock cycle, the computer fetches instructions and data from its memory. The instructions are executed, the results are stored, and the next instruction is fetched. This procedure repeats until a *halt* instruction is encountered.

Instructions, like data, are represented within the computer as binary code—a base two system of counting. The particular instruction set that a specific computer supports is known as that computer's machine language. Using an already-popular machine language makes it much easier to run existing software on a new machine; consequently, in markets where commercial software availability is important suppliers have converged on one or a very small number of distinct machine languages.

Larger computers, such as some minicomputers, mainframe computers, servers, differ from the model above in one significant aspect; rather than one CPU they often have a number of them. Supercomputers often have highly unusual architectures significantly different from the basic stored-program architecture, sometimes featuring thousands of CPUs, but such designs tend to be useful only for specialized tasks.

1.1.3. Computer Evolution

(*i*) 1623, 1642: Wilhelm Strickland/Blaise Pascal built a mechanical counter with carry.

(*ii*) 1823-34: Charles Babbage built the difference engine.

(*iii*) 1943-44: John Mauchly (Professor) and J. Presper Eckert (graduate student) built ENIAC (Electronic Numerical Integrator and Computer) at U. Penn.

(*iv*) 1944: Howard Aiken used "separate data and program memories" in MARK I computer – *Harvard Architecture*.

(*v*) 1945-52: John Von Neumann proposed a "*stored program computer*" EDVAC (Electronic Discrete Variable Computer) – *Von Neumann Architecture* – use the same memory for program and data.

First General-Purpose Computer (1943-44)

Electronic Numerical Integrator and Calculator (ENIAC) built in World War II was the first general purpose computer (Fig. 1.4)

It was used for computing artillery firing tables. It was 80 feet long by 8.5 feet high and several feet wide. It has used twenty 10 digit registers, each 2 feet long. It has used 18,000 vacuum tubes. The machine could perform 5,000 additions/second. Its weight : 30 tons. It was consuming (140kW) of power.

These were the basic limitations in terms of performance, size and cost.

Fig. 1.4. ENIAC At Moore School

Von-Neumann Computing Model

- It works on **"Stored Program Concept"** or we call it Von Neumann Model.
- Main memory stores User programs and Data.
- ALU operates on binary data fetched from main memory.
- Control unit interprets instructions from main memory and executes.
- Input and output unit is operated by control unit.
- Started at Princeton Institute for Advanced Studies-IAS (Fig. 1.5)

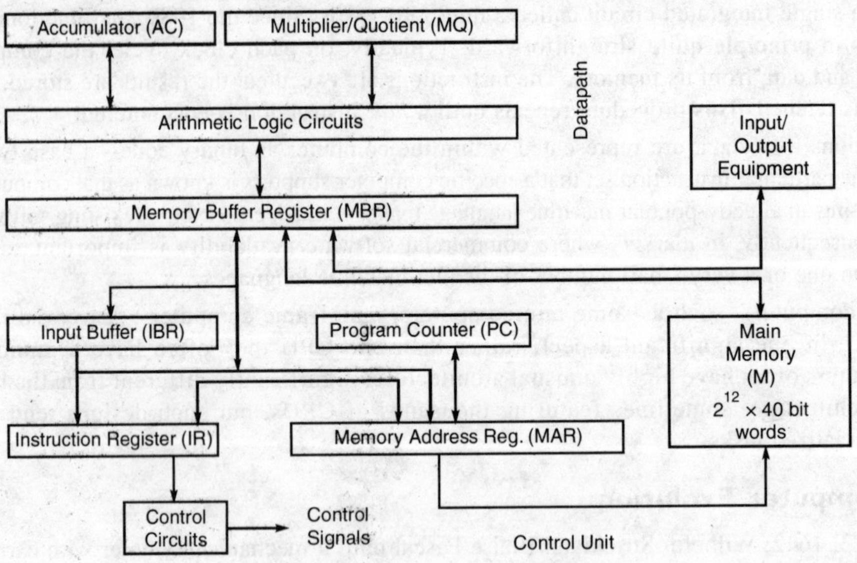

Fig. 1.5. ISA–Detailed Diagram

IAS-TECHNICAL DETAILS

Set of Registers (Storage In CPU)

- Memory Buffer Register (MBR).
- Memory Address Register (MAR).

- Instruction Register (IR).
- Instruction Buffer Register (IBR).
- Program Counter.
- Accumulator.
- Multiplier Quotient.

1000 × 40 Bit Words

- Binary number
- 2 × 20 bit instructions.

The Computer was Completed 1952

IAS computer had 4096 memory locations of 40 bits each. Each word could hold either one piece of data or two instructions (20 bits each). Instruction length is related to the word length in a simple way.

Instructions Format in IAS Machine

8-Bit Opcode	12-Bit-Address

Working Principle of IAS Computer

1. The IAS functions by fetching instructions from memory and executing them one at a time.
2. Since there are two instructions per word, it actually fetches two instructions at once.
3. The second instruction is held in the IBR: instruction buffer register.
4. The control unit and the ALU both contain registers that can hold data temporarily.
5. Memory Buffer Register [MBR] - used to store or receive a word from memory.
6. Memory Address Register [MAR] - indicates where in memory to store or receive a word.
7. Instruction Register [IR] - contains the current instruction.
8. Instruction Buffer Register [IBR] - holds the right-hand instruction from memory (not always used).
9. Program Counter [PC] - Indicates the memory address of the next instruction to fetch.
10. Multiplier/Quotient [MQ] - holds the least significant bits of some ALU operators.

Computers take numerous physical forms. Early electronic computers were the size of a large room, and such enormous computing facilities still exist for specialized **scientific** computation **supercomputers** and for the **transaction processing** requirements of large companies, generally called mainframes. Smaller computers for individual use, called **personal computers** or **Microcomputers** and their portable equivalent, the **laptop computer**, are ubiquitous information-processing and **communication** tools and are perhaps what most non-experts think of as "a computer". However, the most common form of computer in use today is the **embedded computer,** small computers used to control another device.

Computers were classified into **different generations** based on revolutions in the technology used in the development. By convention, commercial electronic computers are taking to be the first generation rather than the electromechanical machines that preceded them. Today computer generations are not commonly referred to due to the long standing of the VLSI **(Very Large Scale Integrated Circuits)** technology and the lack of revolutionary technology in sight.

Generations of Computer

Early modern computers are typically grouped into four "generations." Each generation is marked by improvements in basic technology. These improvements in technology have been extraordinary and each advance has resulted in computers of lower cost, higher speed, greater memory capacity, and smaller size. This grouping into generations is not clear-cut nor is it without debate. Many of the inventions and discoveries that contributed to the modern computer era do not neatly fit into these strict categories. The reader should not interpret these dates as strict historical boundaries.

First Generation : Approximately 1945-1955

Characteristics : vacuum tubes

Second Generation : Approximately 1955-1965

Characteristics : transistors

Third Generation : Approximately 1965-1975

Characteristics : integrated circuits

Fourth Generation : Approx. 1975-1995

Characteristics : large-scale integration

Fifth Generation : 1995's and future

Characteristics : very large integration

Do You Know

Modern Computers inherently follow the ideas of stored program laid out by John Von Neumann in 1945.

First Generation Computers (1945 to 1955)

This was maked by appearance of UNIVAC-I in 1950, in which vacuum tuber were used. Their memory were made of thin tuber of liquid memory and magnetic drums. The **vacuum tube** was invented in 1906, by an electrical engineer named Lee De Forest (1873–1961). During the first half of the twentieth century, it was the fundamental technology that was used to construct radios, televisions, radar, X-ray machines, and a wide variety of other electronic devices. It is also the primary technology associated with the first generation of computing machines.

Fig. 1.6. Hollerith's tabulator for punched card.

The first operational electronic general-purpose computer, named the ENIAC (Electronic Numerical Integrator and Computer), was built in 1943 and used 18,000 vacuum tubes. It was constructed with government funding at the University of Pennsylvania's Moore School of Engineering, and its chief designers were J. Presper Eckert, Jr. (1919–1995) and John W. Mauchly (1907–1980). It was almost 30.5 meters (100 feet) long and had twenty 10-digit registers for temporary calculations. It used **punched cards** for input and output and was programmed with plug board wiring. The ENIAC was able to compute at the rate of 1,900 additions per second. It was used primarily for war-related computations such as the construction of **ballistic** firing tables and calculations to aid in the building of the atomic bomb. The Colossus was another machine that was built during these years to help fight World War II. A British machine, it was used to help decode secret enemy messages. Using 1,500 vacuum tubes, the machine, like the ENIAC, was programmed using plug board wiring.

Fig. 1.7. Electronic Vacuum Tubes.

These early machines were typically controlled by plug board wiring or by a series of directions encoded on paper tape. Certain computations would require one wiring while other computations would require another. So, while these machines were clearly programmable, their programs were not stored internally. This would change with the development of the stored program computer.

Fig. 1.8. ENIAC.

The team working on the ENIAC was probably the first to recognize the importance of the stored program concept. Some of the people involved in the early developments of this concept were J. Presper Eckert Jr. (1919–1955) and John W. Mauchly (1907–1980), and John Von Neumann (1903–1957). During the summer of 1946, a seminar was held at the Moore School that focused great attention on the design of a stored program computer. About thirty scientists from both sides of the Atlantic Ocean attended these discussions and several stored programmed machines were soon built.

Do You Know

ENIAC was produced by partnership between the U.S. government and the university of Pennylvania. It consisting of 18000 vaccum tubes, 70000 resistors, 1000 capacitors, 6000 switches and 10 lacks soldered joints.

Another important stored program machine of this generation was the UNIVAC (UNIVersal Automatic Computer). It was the first successful commercially available machine. The UNIVAC was designed by Eckert and Mauchly. It used more than 5,000 vacuum tubes and employed magnetic tape for bulk storage. The machine was used for tasks such as accounting, actuarial table computation, and election prediction. Forty-six of these machines were eventually installed. The UNIVAC, which ran its first program in 1949, was able to execute ten times as many additions per second as the ENIAC. In modern dollars, the UNIVAC was priced at $4,996,000. Also, during this period, the first IBM computer was shipped. It was called the IBM 701 and nineteen of these machines were sold.

For example, IBM 701, IBM704, IBM 709. During this period concept of stored program was introduced by JOHN VON NEUMAN. But due to Leigh cost of memory and vacuum tuber, it could not be much popular instead, electrostatic memories were used.

Second Generation Computers (1955 to 1965)

As commercial interest in computer technology intensified during the late 1950s and 1960s, the second generation of computer technology was introduced—based not on vacuum tubes but on transistors.

John Bardeen (1908–1991), William B. Shockley (1910–1989), and Walter H. Brattain (1902–1987) invented the transistor at Bell Telephone Laboratories in the mid-1940s. By 1948, it was obvious to many that the transistor would probably replace the vacuum tube in devices such as radios, television sets, and computers. One of the first computing machines based on the transistor was the Philco Corporation's Transac S-2000 in 1958. IBM soon followed with the transistor-based IBM 7090. These second generation machines were programmed in languages such as COBOL (Common Business Oriented Language) and FORTRAN (Formula Translator) and were used for a wide variety of business and scientific tasks. Magnetic disks and tape were often used for data storage.

Third Generation Computers (1965 to 1975)

The third generation of computer technology was based on integrated circuit technology and extended from approximately 1964 to 1970. Jack Kilby (1923–) of Texas Instruments and Robert Noyce (1927–

1990) of Fairchild Semiconductor were the first to develop the idea of the integrated circuit in 1959. The integrated circuit is a single device that contains many transistors. Arguably the most important machine built during this period was the IBM System/360. Some say that this machine single handedly introduced the third generation. It was not simply a new computer but a new approach to computer design. It introduced a single computer architecture over a range or family of devices. In other words, a program designed to run on one machine in the family could also run on all of the others. IBM spent approximately $5 billion to develop the System/360. One member of the family, the IBM System/360 Model 50, was able to execute 500,000 additions per second at a price in today's dollars of $4,140,257. This computer was about 263 times as fast as the ENIAC.

> **Do You Know**
>
> *Transister is a device composed of semiconductor material that amplifies a signal or opens or closes a circuit invented in 1947 at Bell labs.*

During the third generation of computers, the central processor was constructed by using many integrated circuits. It was not until the fourth generation that an entire processor would be placed on a single **silicon** chip—smaller than a postage stamp.

Fourth Generation Computers (1975 to 1995)

The fourth generation of computer technology is based on the microprocessor. Microprocessors employ Large Scale Integration (LSI) and Very Large Scale Integration (VLSI) techniques to pack thousands or millions of transistors on a single chip.

> **Do You Know**
>
> *John Napier invented logs in 1614. It allows multiplication and division to be reduced to addition and subtraction.*

The Intel 4004 was the first processor to be built on a single silicon chip. It contained 2,300 transistors. Built in 1971, it marked the beginning of a generation of computers whose lineage would stretch to the current day. In 1981, IBM selected the Intel Corporation as the builder of the microprocessor (the Intel 8086) for its new machine, the IBM-PC. This new computer was able to execute 240,000 additions per second. Although much slower than the computers in the IBM 360 family, this computer cost only $4,000 in today's dollars! This price/performance ratio caused a boom in the personal computer market. In 1996, the Intel Corporation's Pentium Pro PC was able to execute 400,000,000 additions per second. This was about 210,000 times as fast as the ENIAC–the workhorse of World War II. The machine cost only $4,400 in inflation-adjusted dollars.

Microprocessor technology is now found in all modern computers. The chips themselves can be made inexpensively and in large quantities. Processor chips are used as central processors and memory chips are used for dynamic Random Access Memory (RAM). Both types of chips make use of the millions of transistors etched on their silicon surface. The future could bring chips that combine the processor and the memory on a single silicon die.

During the late 1980s and into the 1990s cached, pipelined, and superscaler microprocessors became common place. Because many transistors could be concentrated in a very small space, scientists were able to design these single chip processors with on-board memory (called a cache) and were able to exploit instruction level parallelism by using instruction pipelines along with designs that permitted more than one instruction to be executed at a time (called superscaler). The Intel Pentium Pro PC was a cached, superscaler, pipelined microprocessor.

Also, during this period, an increase in the use of parallel processors has occurred. These machines combine many processors, linked in various ways, to compute results in parallel. They have been used for scientific computations and are now being used for database and file servers as well. They are not as ubiquitous as uniprocessors because, after many years of research, they are still very hard to program and many problems may not lend themselves to a parallel solution.

> **Do You Know**
>
> *In 1965, Jack Kilby, an engineer with Texas instruments developed the integrated circuits (IC).*

Fifth Generation Computers (1995 Onwards)

The fifth generation is extremely large scale integration. It uses the concept of parallel processing. This generation aim to have computers with some knowledge of real word with logical reasoning power.

Quick View of Generations of Computers

First Generation : 1945-1955

Characteristics : (not all first generation computers had all these characteristics)

- vacuum tube based
- punched tape input or output
- about 1,000 circuits per cubic foot

Examples :

- Harvard Mark I (electromechanical)
- Whirlwind
- ENIAC
- EDSAC
- UNIVAC I, UNIVAC II, UNIVAC 1101
- RCA BIZMAC
- NCR CRC 102A, NCR CRC 102D
- Honeywell Datamatic 1000
- Burroughs E101, Burroughs 220
- IBM models 604, 650 (drum memory), 701, 702, 704, 705, 709

Second Generation : 1955-1965

Characteristics :

- used transistors
- about 100,000 circuits per foot

Examples :

- UNIVAC 1107, UNIVAC III
- RCA 501
- Philco Transact S-2000
- NCR 300 series
- IBM 7030 Stretch
- IBM 7070, 7080, 7090, 1400 series, 1600 series
- Honeywell 800, 400 series
- General Electric GE 635, 645, GE 200
- Control Data Corp. CDC 1604, 3600, 160A
- LARC
- Burroughs B5000, 200 series

Third Generation : 1965-1975

Characteristics :

- large scale integrated circuits
- 10 million circuits per square foot

Examples :

- Burroughs 6700
- Control Data 3300, 6600, 7600
- Honeywell 200
- **IBM System/360**, System 3, System 7
- NCR Century Series
- RCA Spectra 70 series
- UNIVAC 9000 series
- General Electric GE 600 series, GE 235

Fourth Generation : 1975-1995

Characteristics :

- very large scale integration
- consisting of thousand of transistors in a chip
- billions of circuits per cubic foot

Examples :

- IBM System 3090, IBM RISC 6000, IBM RT
- ILLIAC IV
- Cray 2 XMP
- HP 9000

Fifth Generation : 1995 onwards

Characteristics :

Combinations of some or all of the following technologies :

- extremely large scale integration
- parallel processing
- high speed logic and memory chips
- high performance, micro-miniaturization
- voice/data integration; knowledge-based platforms
- artificial intelligence, expert systems
- virtual reality generation
- satellite links

1.1.4. Characteristics of Computer

There are some characteristics of computer that are :

(*i*) **High Speed :** A computer is fast information processing electronic device. It carries out all sorts of computations within a fraction of a second. It executes millions of instructions per second.

(*ii*) **Accuracy :** It gives accurate results for correct input data. Here accuracy means correctness of the processed data. If the input data is erroneous, then the output will not be correct.

(*iii*) **Versatility :** Its role is versatile. It is used for scientific calculations, business processing, computer games, teaching, training etc.

(*iv*) **Diligence :** It does not feel tired. It can be used for hours and hours. It can also be used for a number of days or months continuously. It will work satisfactorily without fatigue.

(*v*) **No IQ and Feelings :** It has no IQ and feelings like human beings.

1.1.5. Application of Computers

The new technology is being developed day-by-day due to needs of people are increasing regularly. The ability of computer has benefited the human life a lot. The applications of computer are as follows :

(*i*) Weather forecasting
(*ii*) Management of organization
(*iii*) Entertainment
(*iv*) CAD/CAM
(*v*) E-mail
(*vi*) Scientific and Education works
(*vii*) Advertisement
(*viii*) Medical
(*ix*) Database processing and word
(*x*) Sports activity

Do You Know

The first calculating machine which was produced in large number was invented by Frenchman Thomas de colmor in 1820.

1.1.6. Digital Computers

Computer which are use today are digital computer. They manipulate numbers. They operate on binary digits 0 and 1. They understand information composed of only 0s and 1s. In the case of alphabetic information, the alphabets are coded in binary digits. A binary digit is called bit. A group of 8 bits is called a byte. Components which convert analog quantity to digital or digital quantity to analog quantity are connected to a digital computer as a peripheral where needed. Processing of analog quintile encountered in industrial control and instrumentation, not in general purpose Compton, text manipulation or information storage, retrieval or transmission.

Do You Know

Boole was mathematician and logistician who developed ways of expressing processes using algebraic.

The computer which can process analog quantities is called an analog computer. Today, analog computers are rarely used. Earlier, analog computers were used to simulate certain systems. They were used to solve differential equations used to simulate certain systems. They were used to solve differential equation.

TIPS

A digital computer is programmable machine specially designed for making computation. The components are CPU, memory, input and output devices.

A microcomputer is a small digital computer. The CPU of a microcomputer is a microprocessor. Other components are same as those of any other digital computer. The organization of a microcomputer is shown in Fig. 1.17. The physical devices and circuitry of computer are called hardware. A physical device may be an electronic, magnetic, mechanical or an optical device etc. A sequence of instruction to perform a particular task is called program. A set of programs written for a particular computer is known as software for the computer.

The input and output devices are known as peripherals. Sometimes the term peripherals also include memory program or subroutines stored in ROM's, PROM's, EPROM's and/or EEPROM's are known as firmware the commonly available firmware are : monitors, microprograms, subroutines for input and output devices.

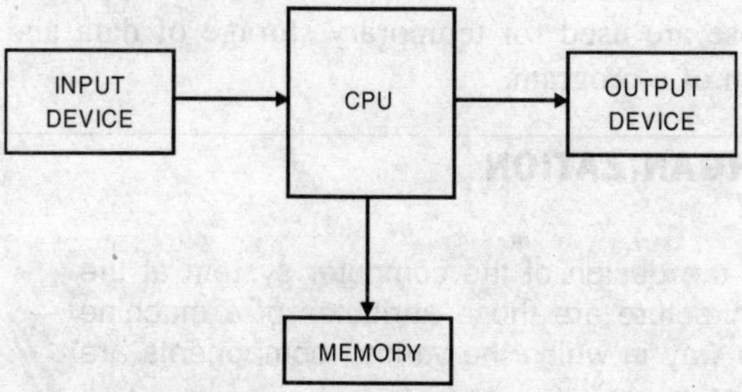

Fig. 1.9. CPU : Central Processing Unit.

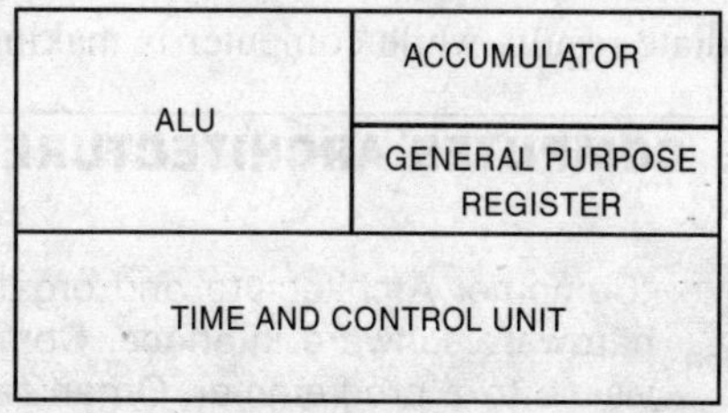

Fig. 1.10. A Central Processing Unit (CPU).

CPU

> **Do You Know**
>
> *PARAM-10000 the indian super computer is capable of performing one billion mathematical calculations per second.*

A Central Processing Unit (**CPU**), or sometimes simply **processor,** is the component in a digital computer that interprets instructions and processes data contained in computer programs. CPUs provide the fundamental digital computer trait of programmability, and are one of the necessary components found in computers of any era, along with primary storage and input/output facilities. A CPU that is manufactured using integrated circuits is known as a microprocessor. Since the mid-1970s, single-chip microprocessors have almost totally replaced all other types of CPUs, and today the term "CPU" is usually applied to some type of microprocessor.

The Central Processing Unit (CPU) fetches the instruction from the memory and performs specific tasks. It stores results in the memory or send the results to the output device according to the instruction given in the program. The CPU controls and communicates with memory and input and output devices under the controls of the CPU programs and data are stored in memory and displayed on CRT. The schematic diagram of a CPU is shown in Fig. 1.10. The CPU is large computer is implemented on any or more circuit boards. IC's are used as its components. Recent practices is to use microprocessors to perform different functions with in the CPU as a large computer.

The major sections of CPU are as :

(1) ALU (Arithmetic Logical Unit) : The arithmetic logic unit/arithmetic-logic unit (ALU) of a computer's CPU is a part of the execution unit, a core component of all CPUs. The function of ALU is to perform arithmetic operations such as addition, subtraction : and logical operation such as AND, OR and EXCLUSIVE OR ALUs are capable of calculating the results of a wide variety of basic arithmetical computations. Virtually all modern computer ALUs use the two's complement binary number representation.

> **Do You Know**
>
> *One of the first and most popular personal computers was the Apple II introduced in 1977 by Apple Computer.*

There are some operations of ALU that are following :

(*i*) Integer arithmetic operations (addition, subtraction, and sometimes multiplication, though this is more expensive)

(*ii*) Bitwise logic operations (and, not, or, xor)

(*iii*) Bit-shifting operations (shifting or rotating a word by a specified number of bits to the left or right, with or without)

Many standard ALUs do not handle integer division or any floating point operations since they can be emulated in software. However, several algorithms do exist for implementing division in hardware.

(2) Timing Control Unit : This unit controls the entire operation of a computer. It acts as a brain. It also controls all the devices the are connected to CPU. It generates timing signals that are necessary for input and output devices.

(3) Accumulator : The accumulator is a register which contains one of the operands and stores results of most arithmetic and logical operations.

(4) General Purpose Registers : General purpose are used for temporary storage of data and intermediate results while computer is making execution of a program.

1.2. COMPUTER ARCHITECTURE AND ORGANIZATION

Definition "Computer Architecture and organization is the design of the computer system at the hardware/software interface. Computer Architecture are those attributes of a machine visible to a programmer. Organization is the way in which the various components are interconnected and they perform the way user wants."

Capabilities and performance characteristics of principal functional units, for e.g., Registers, ALU, Shifters, and Logic unit. Ways in which these components CPU, Memory and I/O are interconnected. How the information flows between these components using system bus? Logic and means by which such information flow is controlled. Register Transfer Level Description using RTL language for designing. Coordination of functional units to realize the ISA-Instruction Set Architecture, the interface between hardware and software.

Definition "The overall design or structure of a computer system, including the hardware and the software required to run it, especially the internal structure of the system components such as CPU, (microprocessor) Main Memory and other I/O devices".

The Machine organization is the view of the computer that is seen by the logic designer. This includes capabilities and performance characteristics of functional units (e.g., registers, ALU, shift registers, etc.). Typically the machine organization is designed to meet a given instruction set architecture which is designed by a designer or computer architect and finally manufactured by a semi-conductor foundry. The computer architecture for a given machine is the combination of its hardware components plus its instruction set architecture (ISA). The ISA is the agreed-upon interface between all the software that runs on the machine and the hardware that executes it. The ISA allows designer to talk to the machine *i.e.* man to machine communication. A good understanding of computer architecture is important for compiler writers, operating system designers, and also to general computer programmers. The study of Computer Architecture and Organization (CAO) is very important because of following reasons;

(*i*) To understand time, space, and price tradeoffs in IT competitive world.

(*ii*) To design a better machine and optimize its performance parameters like speed of computation, memory access time and so on.

(*iii*) We need to understand the hardware environment in order to develop a simple user-friendly operating system, a compiler or a device driver.

(*iv*) We need to understand any resource constraints when working with more complex embedded systems like mobile-computing device or PDA-personal Digital Assistant or a robot system.

(*v*) We need to understand Mobile OS development and new applications for current and future usage.

(*vi*) We wish to know how floating-point arithmetic is carried out by the processor in order to model large and real-world complex systems.

Early electronic computers were the size of a large room, and such enormous computing facilities still exist for specialized scientific computation supercomputers and for the transaction processing requirements of large companies, generally called mainframes. Smaller computers for individual use, called personal computers or Micro-computers and their portable equivalent, the laptop computer, are ubiquitous information-processing and communication tools and are perhaps what most non-experts think of as "a computer". However, the most common form of computer in use today is the embedded computer, small computers used to control another device. Computers process data under the control of sets of instructions called computer programs.

1.3. SYSTEM DEFINITION

A system is an interconnection of various entities such as CPU, Input Unit etc to achieve a specific goal. A computer system is an example. The system requires inputs X1, X2, X3.... and so on. The system process these inputs based upon a program and finally produce an output,

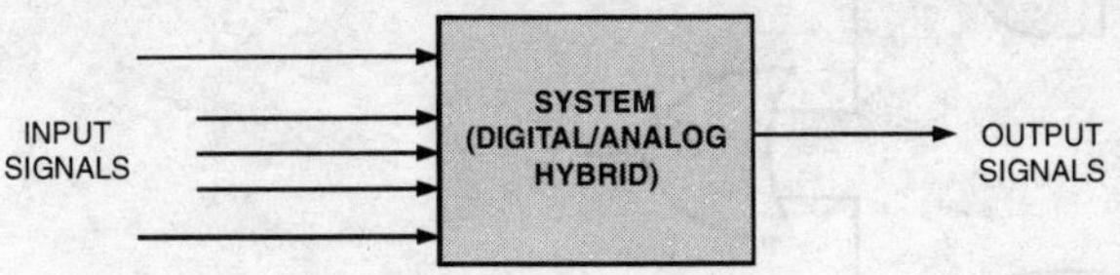

Fig. 1.11 (a). Block Diagram of System

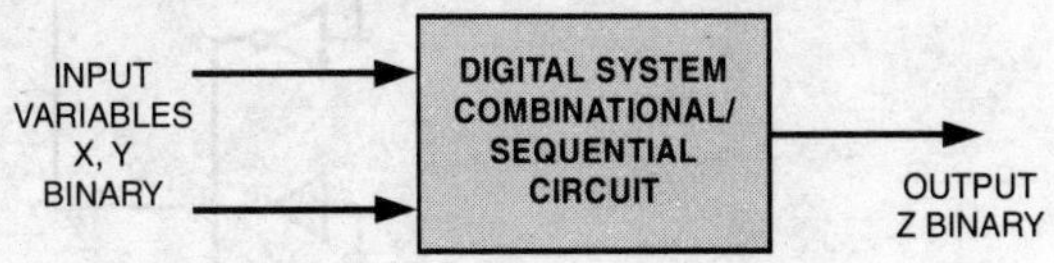

Fig. 1.11 (b). Digital System

A system may be an analog system or a digital system or hybrid. In case of analog the inputs are analog and output is also analog. An example of analog system is television system.

1.4. SYSTEM REPRESENTATION

A system is represented by a graph or a block diagram with proper inputs and outputs. A system has structure and a behavior. A computer network system such as Internet is represented by nodes and links. The structure of the internet and its behavior are two important properties of the system.

1.4.1. System Structure

It is the way how system components are organized for connection such as logic diagram to achieve a goal or function of a system.

NAND Gate Structure and Behaviour

The basic NAND Gates has the following symbol, illustrated for three inputs :

AND-Invert (NAND) : NAND represents NOT AND, *i.e.*, the AND function with a NOT applied. The symbol shown is an AND-invert. The small circle ("bubble") represents the invert function.

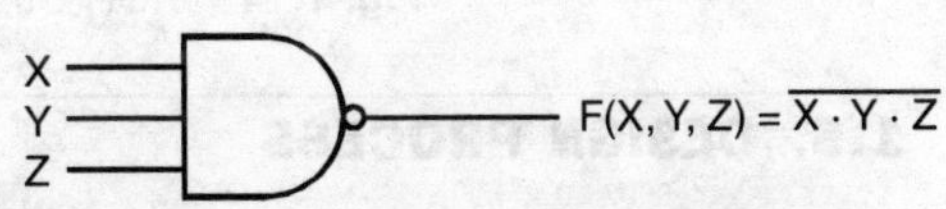

Fig. 1.12. NAND Gate Structure and Behaviour

The structure is represented by Logic Diagram and the behavior by Boolean equation F(X, Y, Z) as in Fig. 1.12. The behavior may also be represented by VHDL or Verilog Languages since these are used for design and modeling of digital systems.

Behaviour of a System

The behavior of a digital system may be represented by Boolean Function or Truth Table. The behavior of a digital system can be modeled and simulated using modern languages such as VHDL-Very High Speed Hardware description Language or Verilog. They can provide precise technology independent description of digital systems at various levels primarily at the gate and register level. Functions of a CPU in computer system may be as follows:

(*i*) Data Processing
(*ii*) Data Storage
(*iii*) Data Movement
(*iv*) Control Functions

TIPS

A system is an interconnection of system entities such as CPU, Input and output units. A system can be represented by a block diagram or a structure. The behaviour of the system depends upon the various sub-system components. How they are related?

XOR Implementations

The simple SOP implementation uses the following structure :

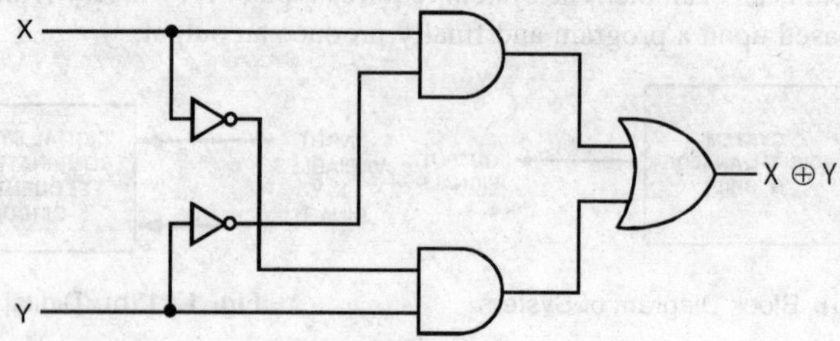

Fig. 1.13.

A NAND only implementation is

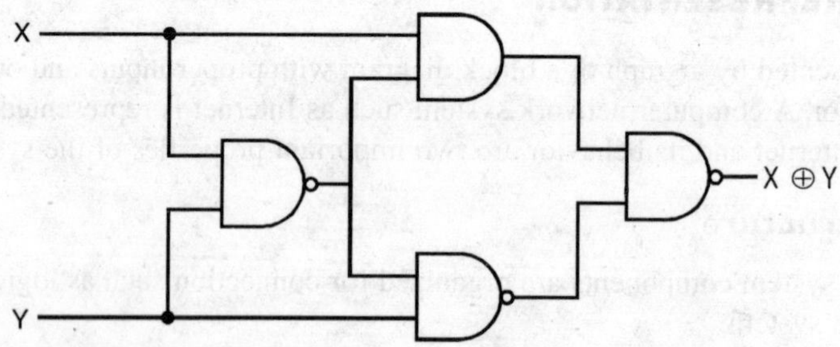

Fig. 1.14. XOR Structure Using SOP and NAND Gates

1.5. DESIGN PROCESS

A system design is a creative process, which comes from the mind of a designer. A system can be designed using system analysis and synthesis methods. Analysis is concerned with finding its behavior for a given structure. Synthesis is the determining the structure from the behavior of the system. Design process can be defined as to determine a suitable structure for a given behavior and given set of components.

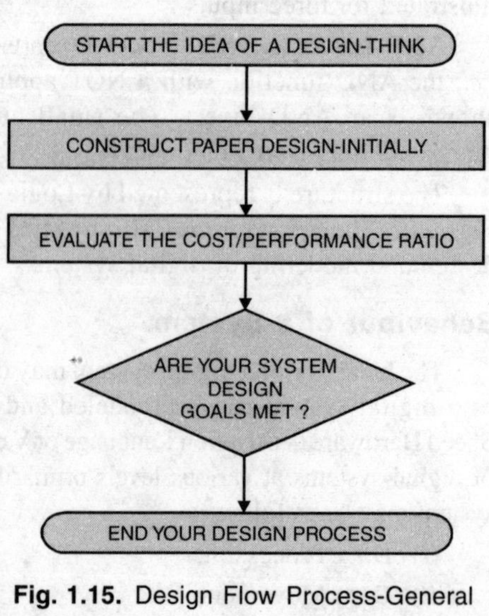

Fig. 1.15. Design Flow Process-General

Design Steps and Flow Chart

1. It starts with the initial paper design based upon the idea of a designer. He conceives a system. He constructs a system based upon initial requirements and analysis.
2. The next step is to find out the cost and performance. The system must achieve a desired behavior for a give cost and performance.

3. If the cost/performance ratio is not achieved. We do the iteration process. Here we modify and fine tune the system till we meet the goals of the system.

1.5.1. E-Cad Tools

E-CAD known as EDA—Electronic Design Automation, in particular is used for chip implementation (physical design). Computer is a complex digital system; hence we need Computer Aided Design tools for the design of CPU Chip, and main memory etc. VLSI is Very Large Scale Integration needed for the design of modern CPU and other devices. A computer system is designed as an application of digital logic design procedures. The active part of the computer (CPU), which does all the work (data manipulation and decision-making). Process Engineers are the experts who tune process that, is more circuit/physical design friendly. Lots of VLSI physical design problems can be formulated into combinatorial optimization or mathematical programming problems.

The basic steps used in chip design are given below :

1. **System Specifications :** Specifications of the size, speed, power and functionality of the VLSI system such as CPU Chip.
2. **Architectural Design :** Decisions on the architecture, e.g., RISC/CISC of ALU pipeline structure, cache size, etc. Such decisions can provide an accurate estimation of the system performance, die size, power consumption, etc.
3. **Functional Design :** Identify main functional units and their interconnections. No details of implementation.
4. **Logic Design** : Design the logic, *e.g.*, Boolean expressions, control flow, word width, registers allocation, etc. The outcome is called an RTL (***Register Transfer Level***) description. **RTL** is expressed in a **HDL** (***Hardware Description Language***), *e.g.*, **VHDL** and **Verilog** languages.
5. **Circuit Design :** Design the circuit including gates, transistors, interconnections, etc. The outcome is called a **net list.**
6. **Physical Design :** Convert the **netlist** into a geometric representation. The outcome is called a ***layout.***
7. **Fabrication :** Process includes lithography, polishing, deposition, diffusion, etc., to produce a *chip.*
8. **Packaging :** Put together the chips on a PCB (*Printed Circuit Board*) or an MCM (*Multi-Chip Module*).

1.5.2. System Design Levels

A computer system design is considered as an abstraction of several layers. Three most important levels re as follows :

(*a*) **Processor Level :** Architectural or behavioral level.

(*b*) **Register Level :** Register Transfer level, we use RTL-Register transfer Language for inner operation-known as micro-operations.

(*c*) **Gate Level :** It is also called a Logic Level.

1. PROCESSOR LEVEL : Architectural or behavioral level. The chip is designed based upon user specifications for the chip.

(*a*) We first analyze instruction set which leads to data path requirements.

(*b*) The meaning of each instruction is given by register transfers.

```
R[rd] <- R[rs] + R[rt];
```

(*c*) Data path must include storage element for ISA registers.

(*d*) Data path must support each register transfer.

(*e*) We select set of data path components and establish clocking methodology.

(*f*) We design data path to meet the requirements.

(*g*) We analyze implementation of each instruction to determine setting of control points that affects the register transfer.

(*h*) We design the control logic finally.

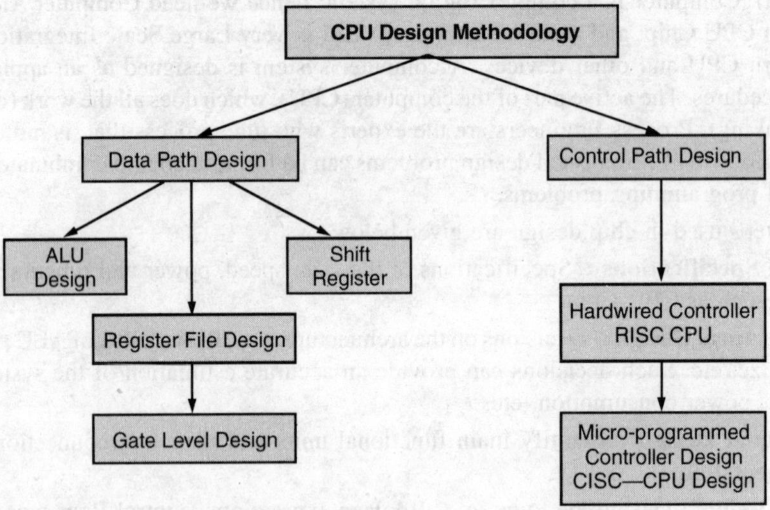

Fig. 1.16. CPU Design Methodology Diagram

1.5.3. System Hierarchy

A system hierarchy refers to design levels of a system such as sub-system or sub-sub-systems and so on.

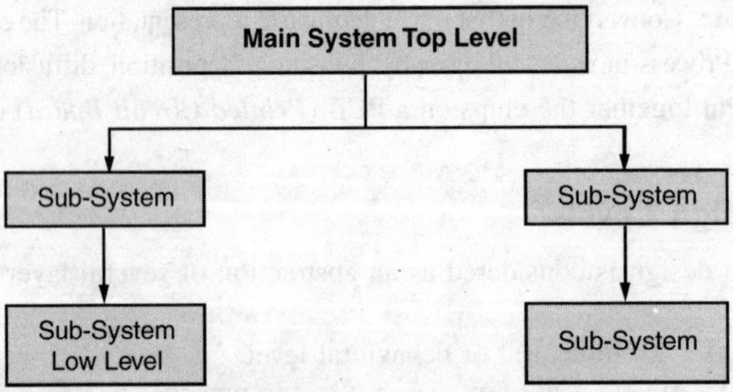

Fig. 1.17. System Hierarchy

(*a*) High level-top-level design- Main System

(*b*) Low-level design-Sub System

FULL ADDER : Let Inputs X, Y and C-Cary.

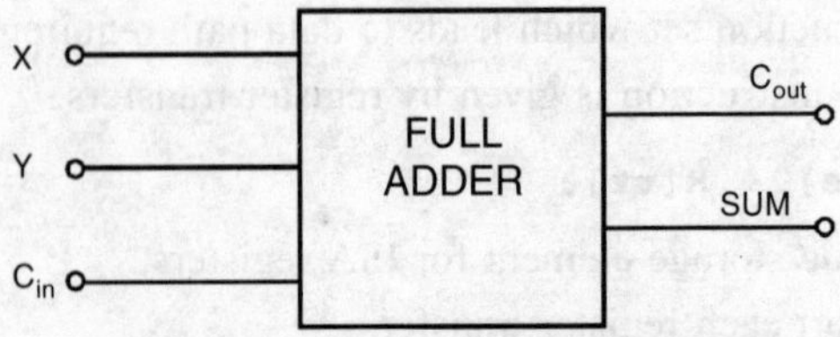

Fig. 1.18. Top-Level of a Full Adder

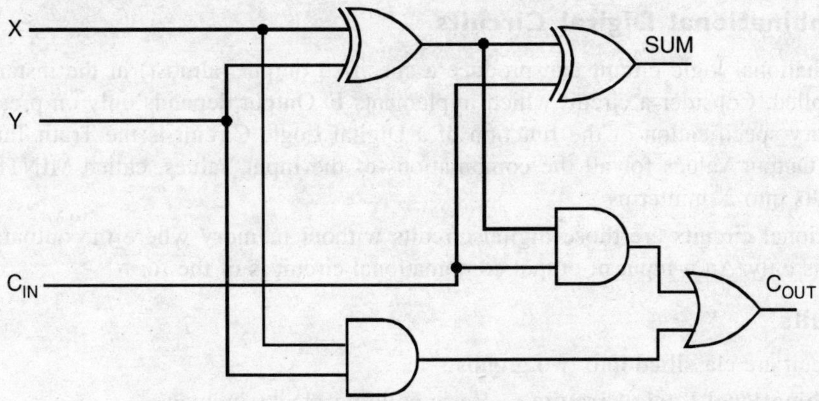

Fig. 1.19. Logic Diagram of Full Adder-Low-Level

The block diagram of a full adder is achieved using two half adders as shown below :

TABLE 1.1. System Design Levels

Level	Components	Information Units	Time in Sec (Order)	IC Density
Processor	CPU, Memory and I/O Devices	Blocks of words	10^{-3} to 10^{3} Millisec	LSI, VLSI, ULSI
Register	Registers, Counters, Decoders etc.	Words-4/8/16 Bits	Nano-Sec to Micro Seconds	MSI
Gate Level	Flips-Flops, Logic Gates	Bits	Pico to Nano Sec	SSI

1 MICRO = 10^{-6}, 1 NANO = 10^{-9} and 1 PICO = 10^{-12}

To design a complex system usually we have to follow the following steps :

1. Specifications of Processor level structure of the system
2. Specifications at register level of each component which are specified n step one.
3. Specifications at the Gate level structure of the components specified n step two.

The design approach may be top to bottom (processor to gate level) approach or bottom to top approach (gate level to processor level). The designer may use SSI, MSI or VLSI components as the case may be. System ANALYSIS is the process of determining function or behaviour of the system from a given structure. System SYNTHESIS is the process of determining structure from the behaviour of the system. CAD Tools—These are E-CAD tools used for simulation and synthesis- as Mentor Graphics Tool. Three LEVELS—Processor level, Register and Gate levels.

1.6. GATE LEVEL DESIGN

The gate level design is concerned with logic gates, (AND OR etc.) adders, decoders, counters and registers memory elements etc, where the bits 0, 1 is processed at this level by digital circuits. Their behavior is determined by truth table or K-maps.

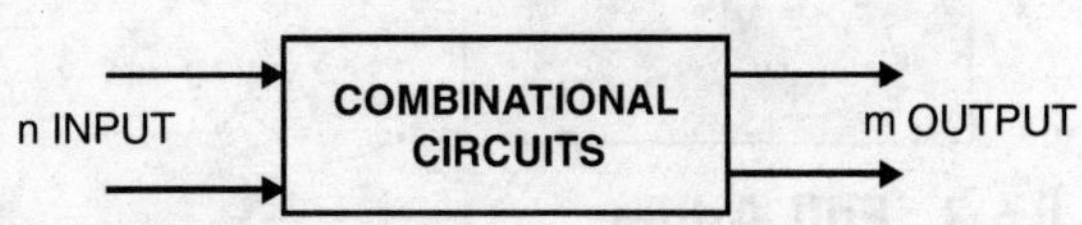

Fig. 1.20. Combinational Circuit.

1.6.1. Combinational Digital Circuits

A Combinational logic circuit can produce a specified output (almost) at the instant, when input values are applied. Consider a circuit which implements F. Output depends only on present input. The most elementary specification of the function of a Digital Logic Circuit is the Truth Table .Table that describes the Output Values for all the combinations of the Input Values, called MINTERMS n input variables results into 2^n minterms

Combinational circuits are those digital circuits without memory where the outputs are obtained from the inputs only. An n-input m-output combinational circuit is of the form.

Logic Circuits

Logic circuit are classified into two groups :

(*i*) **Combinational Logic Circuits :** Basic building blocks include :

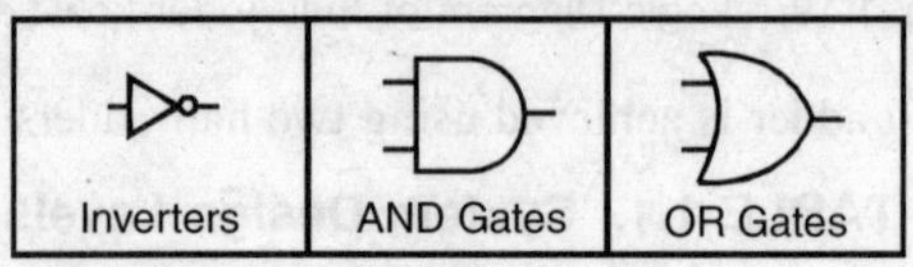

Fig. 1.21 (*a*).

(*ii*) **Sequential Logic Circuits :** Basic building blocks include flip-flops :

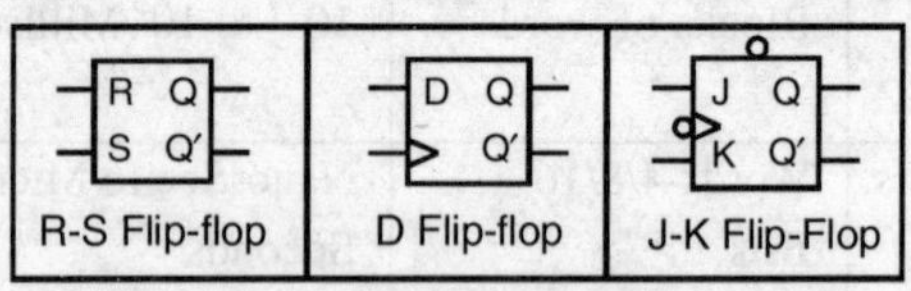

Fig. 1.21 (*b*).

Steps of Combinational Logic Design

(*i*) Start with the problem statement.
(*ii*) Determine the number of inputs variables and the required number of output variables.
(*iii*) Derive a truth table that defines the required relationship between input and output.
(*iv*) Simplify each output function (Karnaugh or K-maps).
(*v*) Draw the logic diagram.

1.6.2. Half-Adder

The digital circuit adds 2 bits and gives out carry and sum as result

Truth Table

x	*y*	*c*	*s*
0	0	0	0
0	1	0	1
1	0	0	1
1	1	1	0

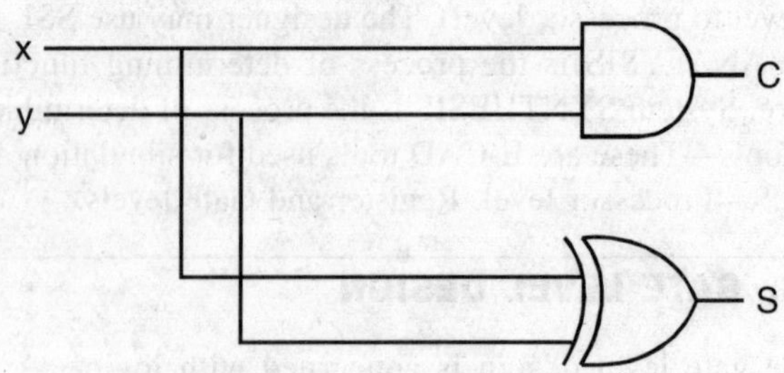

Fig. 1.22. A Half-Adder

1.6.3. Full Adder

We can change half adder of Fig. 1.23 into to a full adder by including gates for processing the carry bit. The block diagram and truth table for a full adder is shown in Fig. 1.23. Let Inputs X, Y and Cin-Carry are applied to the circuit. The output is given using truth table.

$$Cout = XY + XC_{in} + YC_{in} = XY + (X \oplus Y)\, C_{in}$$
$$Sum = X \oplus Y \oplus C_{in}$$

The term $X \cdot Y$ is called *carry generate*.

The term $X \oplus Y$ is called *carry propagate*.

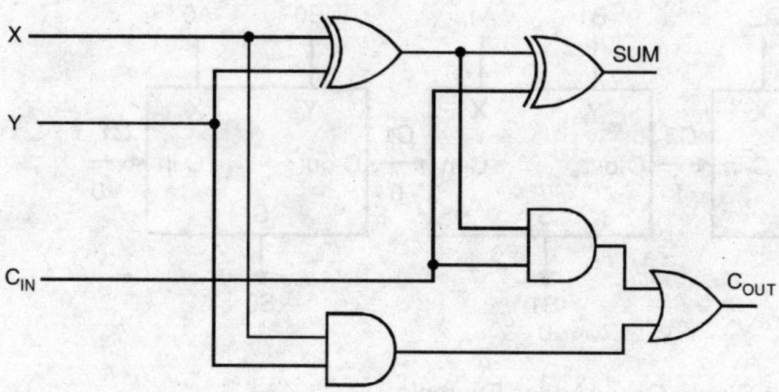

Fig. 1.23. Full-Adder Logic-Diagrams

INPUTS			OUTPUTS	
X	Y	Carry In	Sum	Carry Out
0	0	0	0	0
0	0	1	1	0
0	1	0	1	0
0	1	1	0	1
1	0	0	1	0
1	0	1	0	1
1	1	0	0	1
1	1	1	1	1

Fig. 1.24. Truth Table of Full-Adder

Additional Development

(i) Half-Adder (HA), a 2-input bit-wise addition functional block,

(ii) Full-Adder (FA), a 3-input bit-wise addition functional block,

(iii) Ripple Carry Adder, an iterative array to perform binary addition, and

(iv) Carry-Look Ahead Adder (CLA), a hierarchical structure to improve performance.

1.6.4. Four-Bit Ripple Carry Adder

If we want to do a 4 bit addition we could simply use four full adders next to each other creating a ripple carry adder. In this process four bit number X = (X3,X2,X1,X0) and Y=(Y3,Y2,Y1,Y0) are applied inputs to the FAs and the Sum=(S3,S2,S1,S0) is obtained as in Fig. 1.25 and final Carry Cout is also obtained. Here we have taken no-carry as input, but we can also get carry from previous stage, if necessary.

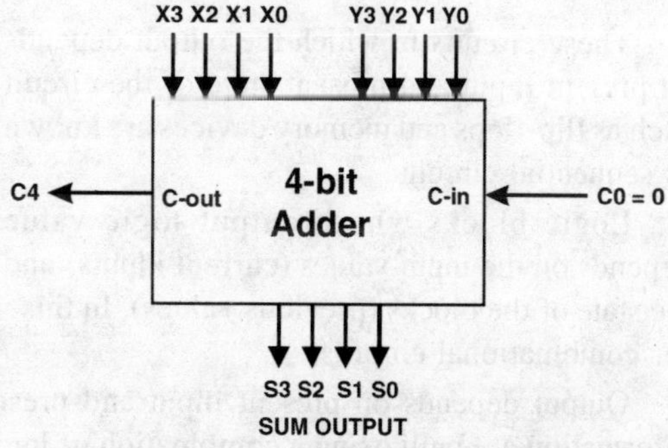

Fig. 1.25. 4-Bit Ripple Carry Adder-Top Level

Adds two 4-bit numbers :

$$X = X3\ X2\ X1\ X0$$
$$Y = Y3\ Y2\ Y1\ Y0$$

And producing the sum S = S3 S2 S1 S0,

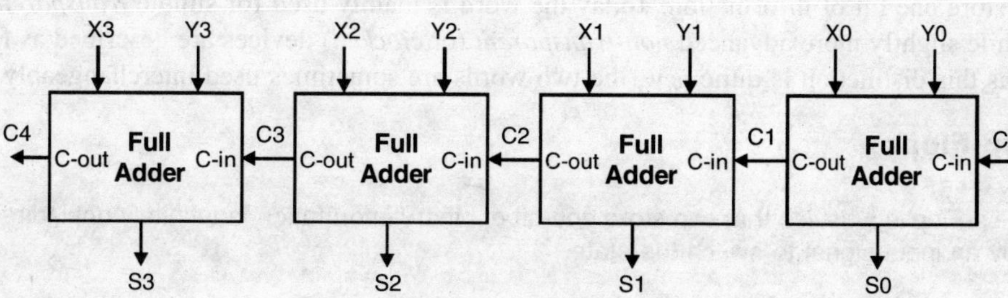

Fig. 1.26. 4-Bit Ripple Carry Adder

C-out = C4 from the most significant position j = 3

Total Propagation delay = 2 nD = 8D or 8 gate delays

An Example of 4-Bit Addition

Let's try our initial example : A = 1011 (eleven), B = 1110 (fourteen).

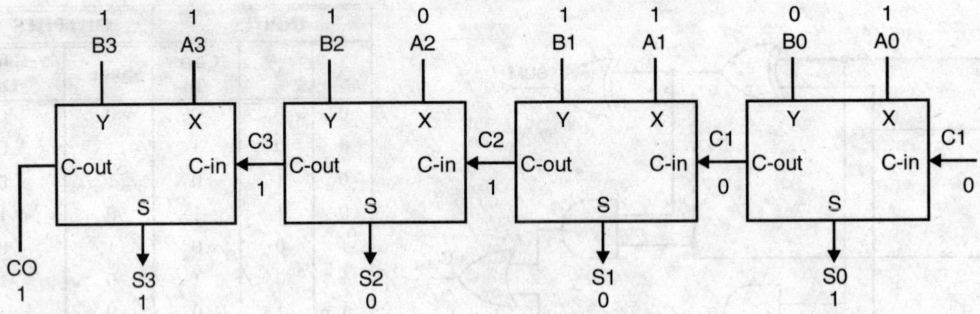

Fig. 1.27. 4-bit Ripple Carry Adder Example

1. Fill in all the inputs, including C1 = 0
2. The circuit produces C1 and S0 (1 + 0 + 0 = 01)
3. Use C1 to find C2 and S1 (1 + 1 + 0 = 10)
4. Use C2 to compute C3 and S2 (0 + 1 + 1 = 10)
5. Use C3 to compute C0 and S3 (1 + 1 + 1 = 11)

The final answer is 11001 (twenty-five).

1.6.5. Sequential Digital Circuits

Those circuits in which the output depends on present input and present state of the circuit such as flip-flops and memory devices are known as sequential circuits.

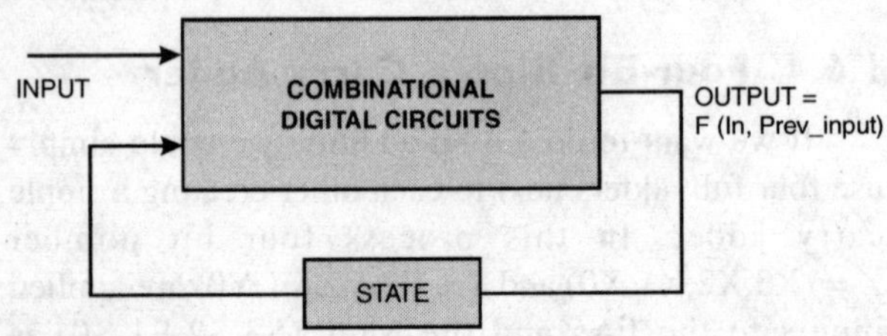

Fig. 1.28. General Sequential Circuit Diagrams

Logic blocks whose output logic value depends on the input values (current inputs) and the state of the blocks (previous values). In this we have the concept of memory which was not applicable for combinational circuits.

Output depends on present input and present state of the circuit. Digital circuits that can "store" information are built using a combination of logic gates output depends on the input, and the past history (state of the circuit) can also be represented by its truth table.

A latch is a kind of Bistable Multivibrator, an electronic circuit which has two stable states and thereby can store one bit of information. Today the word is mainly used for simple *transparent* storage elements, while slightly more advanced *non-transparent* (or *clocked*) devices are described as flip-flops. Informally, as this distinction is quite new, the two words are sometimes used interchangeably.

1.6.6. Flip-Flops

Definition Flip-Flop is a device that can store one-bit of binary information indefinitely until directed by an input signal to switch the state.

Flip-flops are the building blocks of sequential logic functions. Sequential logic has memory in it. The present output does not depend only upon the external input but also upon the present output. That means it has a feedback of present output to the input to decide the next output in combination with the next input as shown in Fig. 1.29.

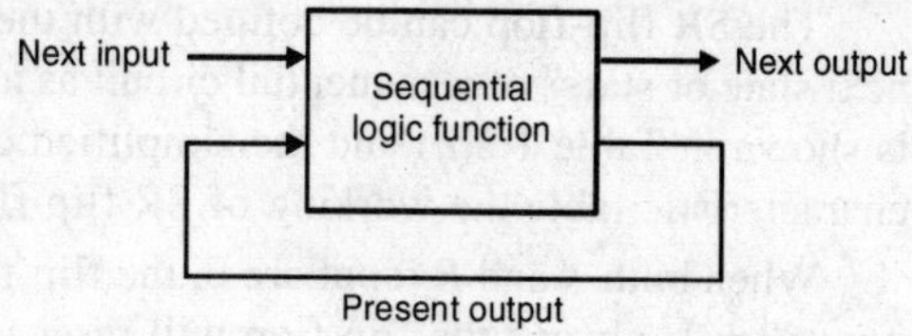

Fig. 1.29.

In combinational logic circuit, the present output is entirely dependent upon the present input, present output does not have any influence on the present input.

1-bit Memory Cell

Flip-Flop is the basic digital memory. It has two stable states known as high-state (logic-1) or low state (logic 0). Any device having two stable states is known as bi-stable. A flip-flop can be obtained by two inverters in series and connecting the output of second inverter to input of first. The 1-bit memory cell is shown in Fig. 1.30. The output of inverter *A*, and output of inverter *B* are always complementary. The circuit has two stable states, high state and low state. If the circuit is in high state, it remains in high state until its input values are changed. Similarly, if the circuit is in low state, it continues to remain in the same state until the input values are changed.

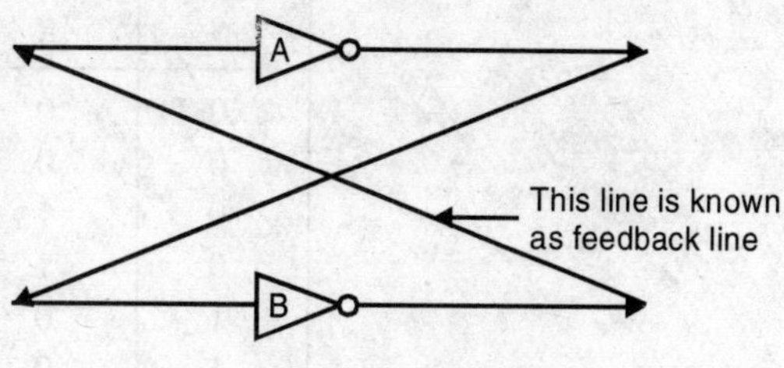

Fig. 1.30. 1-bit Memory Cell.

SR Flip-Flop

The logic diagram of SR flip-flop is shown in Fig. 1.31. The circuit has two inputs *S* (set) and *R* (reset) and two outputs *Q* and *Q′* and consists of two NOR gates where the output of one is the input to the other NOR gates. The output of the NOR gate is 0 if any input is 1, and the output is 1 only when all inputs to NOR gate is 0.

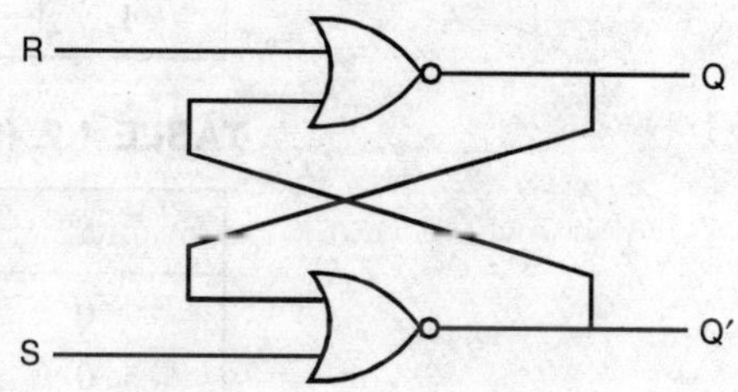

Fig. 1.31. SR Flip-Flop.

Case 1. Assume both ***S* and *R* are 0** and ***Q* is 0.**

The inputs to the lower gate are $S = 0$ and $Q = 0$ and thus, the output Q' is 1.

Thus, the input to the upper gates are $R = 0$ and $Q' = 1$ which gives output $Q = 0$.

Thus, the state of the circuit is internally consistent and remains stable as long as $R = 0 = S$.

1(*b*) : $S = 0, R = 0, Q = 1$

Input to lower gate is $S = 0$ and $Q = 1$, gives output $Q' = 0$. *i.e.* input to upper gate is $R = 0$, $Q' = 0$ gives output $Q = 1$.

Thus, for $S = 0$, $R = 0$, the flip-flop remains in previous state.

Case 2. Consider the state $Q = 0$, $Q' = 1$, $S = 0$, $R = 0$. Suppose *S* changes to value 1. Now the input to lower gate are $S = 1$ and $Q = 0$, the output of lower NOR gate will be $Q' = 0$. At this time, the input to the upper gate is $R = 0$ and $Q' = 0$, the output *Q* becomes 1. This is again a stable state.

Now the input to lower gate is $S = 1$ and $Q = 1$, and thus $Q' = 0$ and hence $Q = 1$.

Thus, as long as $S = 1$, $R = 0$, the output *Q* remains 1. Also, if *S* returns to 0, the outputs will remain unchanged.

Case 3. Consider the state $R = 1$, $S = 0$ and $Q = 0$

Input to lower NOR gate is $S = 0$, $Q = 0$ gives $Q' = 1$

Input to upper NOR gate is $R = 1$ and $Q' = 1$, gives $Q = 0$.

Case 4. When $S = 1$ and $R = 1$, both NOR gate produce an output $Q = 0$ and $Q' = 0$, which takes the flip-flop in unstable state.

The SR flip-flop can be defined with the help of a table called characteristic table, which shows the next state or states of a sequential circuit as a function of current state and inputs. The characteristic table is shown in Table 1.2(*a*) and the simplified characteristic table is shown in Table 1.2(*b*). Thus, from the characteristic table the working of SR flip-flop can be explained.

When both *S* and *R* input are 0, the flip-flop retains its previous state. When set input *S* is 0 and reset input *R* = 1, always the flip-flop will reset *i.e.* the next state is 0. When set input *S* is 1 and reset input *R* = 0, always flip-flop will set *i.e.* the next state is 1. When both set and reset input are 1, a race condition or invalid condition occurs.

TABLE 1.2. (a) Characteristic Table.

S	R	Q_n	Q_{n+1}
0	0	0	0
0	0	1	1
0	1	0	0
0	1	1	0
1	0	0	1
1	0	1	1
1	1	0	indeterminate
1	1	1	indeterminate

TABLE 1.2. (b). Simplified Characteristic Table.

S	R	Q_{n+1}
0	0	Q_n (Previous state)
0	1	0 (Reset)
1	0	1 (Set)
1	1	Invalid condition

Clocked SR Flip-Flop

The basic flip-flop as it stands is an asynchronous sequential circuit. By adding gates to the inputs of the basic circuits, the flip-flop can be made to respond to input levels during occurrence of a **clock pulse**.

The clocked SR flip-flop consists of a basic NOR flip-flop and two AND gates shown in Fig. 1.33. The outputs of two AND gates remains at 0 as long as the clock pluse is 0, regardless of the *S*, *R* input. When clock pulse goes to 1, information from *S* and *R* inputs is allowed to reach the basic flip-flop. The set state is reached with *S* = 1, *R* = 0 and *CP* = 1. For clear state, the input must be *S* = 0, *R* = 1 and *CP* = 1. The characteristic table for the flip-flop is shown in Table 1.3.

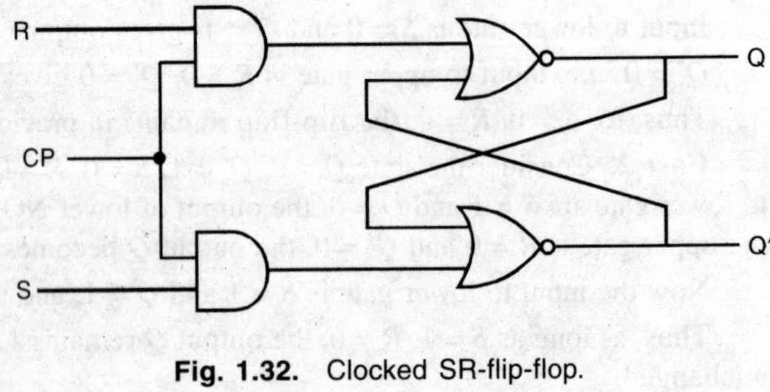

Fig. 1.32. Clocked SR-flip-flop.

TABLE 1.3. Characteristic Table of Clocked SR Flip-Flop.

Q_n	S	R	Q_{n+1}
0	0	0	0
0	0	1	0
0	1	0	1
0	1	1	indeterminate
1	0	0	1
1	0	1	0
1	1	0	1
1	1	1	indeterminate

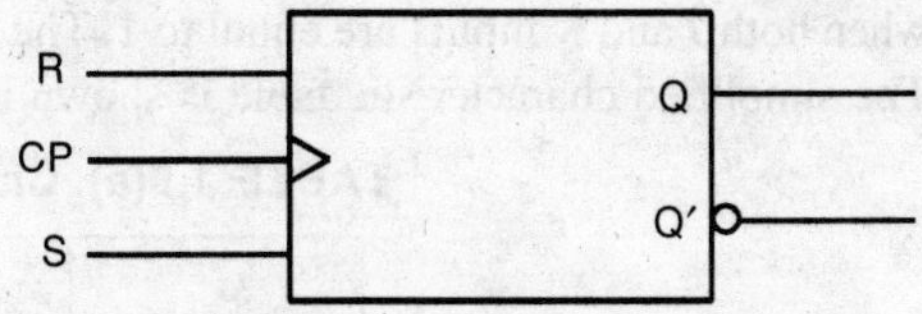

Fig. 1.33. Block Diagram of Clocked SR Flip-Flop.

Q_n is the binary state of the flip-flop at a given time *i.e.* present state. The S and R colums gives the possible values of input at that time and Q_{n+1} is state of flip-flop after the occurrence of a clock pulse *i.e.* next state. Block diagram of clocked SR flip-flop is shown in Fig. 1.33.

D Flip-Flop

The D flip-flop is a slight modification of the SR flip-flop. An SR flip is converted to D flip-flop by inserting an inverter between S and R inputs and assigning the symbol D to the single input. If $D = 1$, the output of the flip-flop goes to the 1 state and if $D = 0$, the output of the flip-flop goes to the 0-state. The logic circuit and block diagram of D flip-flop is shown in Fig. 1.34 and its characteristic table is shown in Table 1.4.

TABLE 1.4. Characteristic Table of D Flip-Flop.

D	Q_{n+1}
0	**0**
1	**1**

The D flip-flop is also known as Data flip-flop because it is storage of one bit of data. The output of the D flip-flop is always equal to the most recent value applied to the input and hence it remembers and produces the last input. It is also referred as a delay flip-flop, because it delays a 0 or 1 applied to its input for a single clock-pulse.

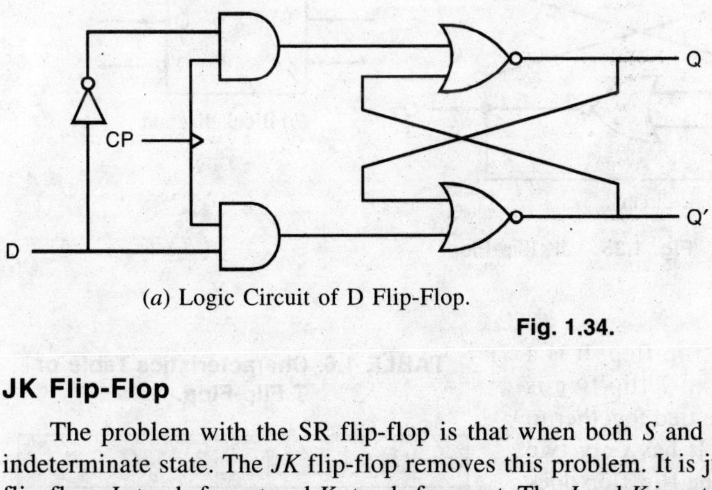

(*a*) Logic Circuit of D Flip-Flop.

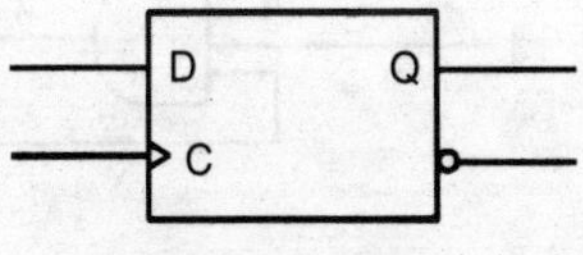

(*b*) Block Diagram of D Flip-Flop.

Fig. 1.34.

JK Flip-Flop

The problem with the SR flip-flop is that when both S and R inputs are 1, the flip-flop goes into indeterminate state. The JK flip-flop removes this problem. It is just a refinement of SR flip-flop. In JK flip-flop, J stands for set and K stands for reset. The J and K inputs behaves same as that of S and R to set and clear the flip-flop, respectively. The logic circuit and the block diagram of JK flip-flop is shown in Fig. 1.35. The output Q is ANDed with the K input and CP input and output Q' is ANDed with J and CP inputs. The JK flip-flop will behaves like SR flip-flop except for the condition when both J and K input are 1. When both J and K are 1, and $Q = 1$ the output of the upper AND gate becomes 1 upon application of clock pulse, and the flip-flop is cleared. When both J and K are 1, and $Q' = 1$, the output of the lower

AND gate becomes 1 and the flip-flop is set. Thus, in either case, the output state of the flip-flop is complemented, *i.e.* instead of the indeterminate condition, the *JK* flip-flop has a complemented condition when both *J* and *K* inputs are equal to 1. The characteristic table of the *JK* flip-flop is shown in Table 1.5(*a*). The simplified characteristic table is shown in Table 1.5(*b*).

TABLE 1.5(a). Characteristic Table of JK Flip-Flop.

J	K	Q_n	Q_{n+1}
0	0	0	0
0	0	1	1
0	1	0	0
0	1	1	0
1	0	0	1
1	0	1	1
1	1	0	1
1	1	1	0

TABLE 1.5(b). Simplified Characteristic Table of JK Flip-Flop.

J	K	Q_{n+1}
0	0	Q_n (Previous state)
0	1	0 (Reset)
1	0	1 (Set)
1	1	Q'_n (Complement of previous state)

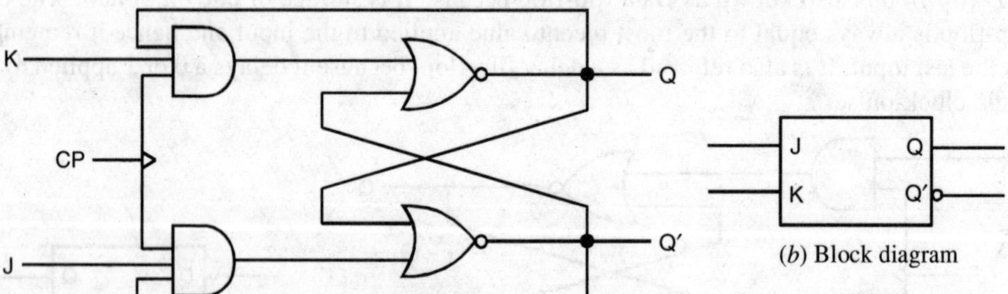

(*a*) Logic circuit (*b*) Block diagram

Fig. 1.35. JK Flip-Flop.

T Flip-Flop

T flip-flop also known as Toggle flip-flop. It is a single version input of the *JK* flip-flop. *T* flip-flop is obtained when both *J* and *K* inputs are tied together to provide a single input known as *T*. It has only two conditions. When $T = 0$, the state of the flip-flop does not change. When $T = 1$, a clock pulse complements the state of the flip-flop. The logic circuit, block diagram and characteristic table of the *T* flip-flop is shown in Fig. 1.36 and Table 1.6, respectively.

TABLE 1.6. Characteristics Table of T Flip-Flop.

Q_n	T	Q_{n+1}
0	0	0 (Q_n)
0	1	1 (Q'_n)
1	0	1 (Q_n)
1	1	0 (Q'_n)

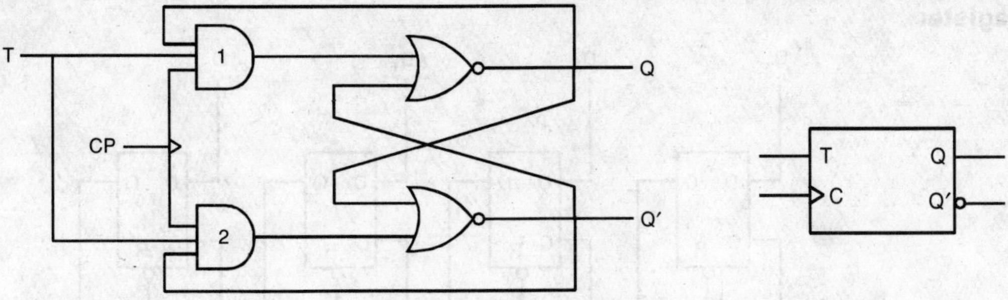

Fig. 1.36. T Flip-Flop.

1.6.7. D-Latch

Forbidden input values are forced not to occur by using an inverter between the inputs as shown in the Fig. 1.37.

These circuits are very useful in making a memory units-RAM and ROM and CPU registers and other components.

D	Q(t + 1)
0	0
1	1

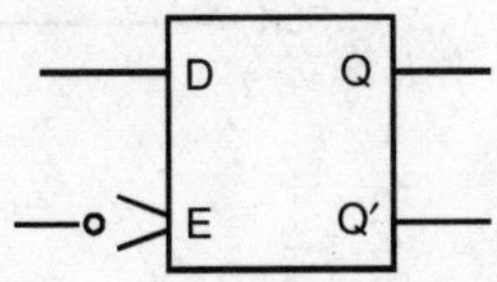

Fig. 1.37

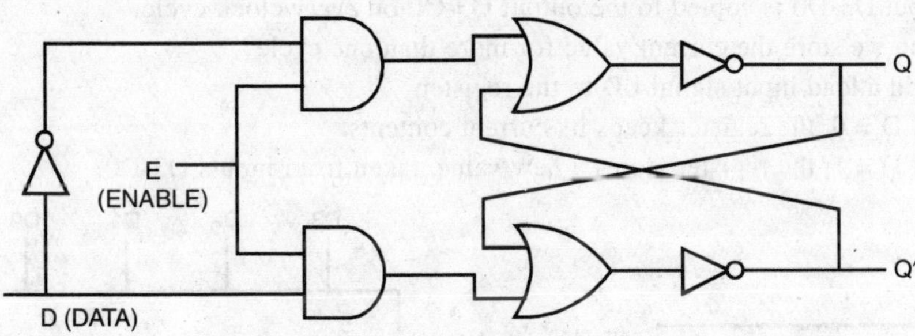

Fig. 1.38. D-Latch With Truth Table

1.6.8. Registers—A Basic Register

Basic registers are easy to build. We can store multiple bits just by putting a bunch of flip-flops together. Flip-flops are limited because they can store only one bit. We had to use two flip-flops for our two-bit counter examples. Most computers work with integers and single-precision floating-point numbers that are 32-bits long.

Definition A register capable of shifting its binary information in one or both directions is called a shift-register.

A register is an extension of a flip-flop that can store multiple bits. Registers are commonly used as temporary storage in a processor. They are faster and more convenient than main memory. More registers can help speed up complex calculations.

A register is a good example of sequential analysis and design. They are also frequently used in building larger sequential circuits. Registers hold larger quantities of data than individual flip-flops.

Registers are central to the design of modern processors. There are many different kinds of registers.

Examples : Registers inside a CPU chip.

RAM is a collection of memory cells. Each cell may be one flip-flop. This register uses D flip-flops, so it's easy to store data without worrying about flip-flop input equations. All the flip-flops share a common CLK and CLR signal.

4-Bit Register

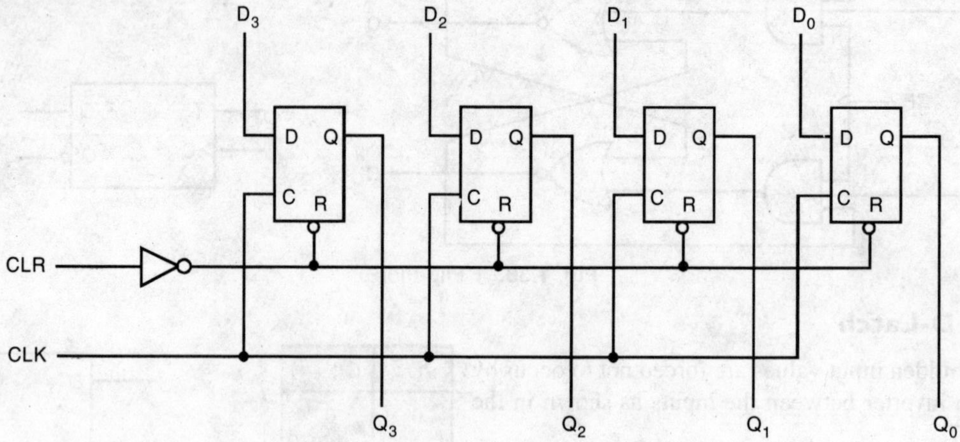

Fig. 1.39. Logic Diagram for a Basic 4-bit Register

Adding a Parallel Load Operation

- The input D3-D0 is copied to the output Q3-Q0 on *every* clock cycle.
- How can we store the current value for more than one cycle?
- Let's add a load input signal LD to the register.
  - If LD = 0, the register keeps its current contents.
  - If LD = 1, the register stores a new value, taken from inputs D3-D0.

LD	Q(t + 1)
0	Q(t)
1	$D_3 - D_0$

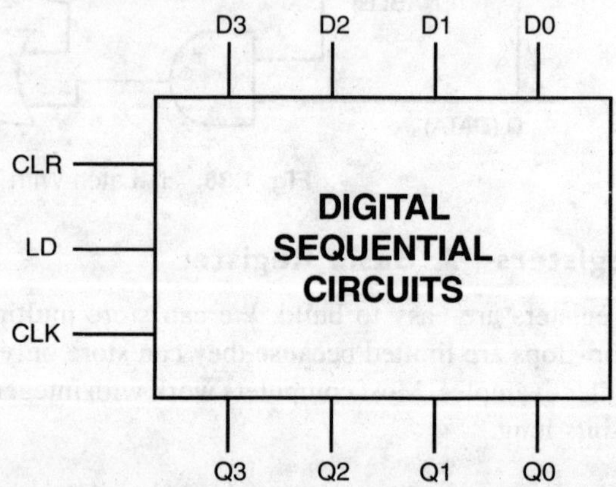

Fig. 1.40. Register With Parallel Load

Clock Gating

We could implement the load ability by playing with the CLK input, as shown below.

- When LD = 0, the flip-flop C inputs are held at 1. There is no positive clock edge, so the flip-flops keep their current values.
- When LD = 1, the CLK input passes through the OR gate, so the flip-flops can receive a positive clock edge and can load a new value from the D3-D0 inputs.
- This is called clock gating, since gates are added to the clock signal.
- There are timing problems similar to those of latches. Here, LD must be kept at 1 for the correct length of time (one clock cycle) and no longer. The clock is delayed a little bit by the OR gate.

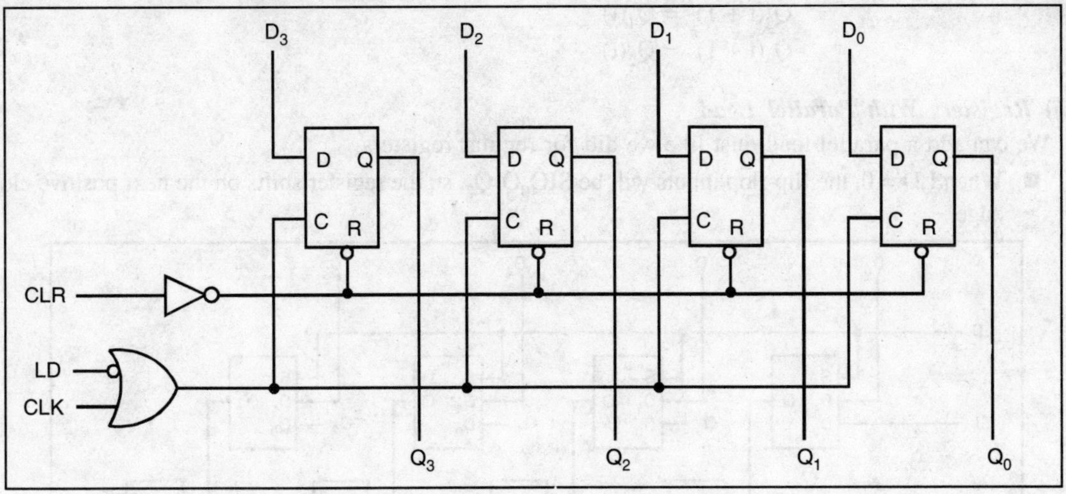

Fig. 1.41. Register With Clock Gating

- In more complex scenarios, different flip-flops in the system could receive the clock signal at slightly different times.
- This "clock skew" can lead to synchronization problems.

Shift Register

A shift register "shifts" its output once every clock cycle. SI is an input that supplies a new bit to shift "into" the register. (Fig. 1.42)

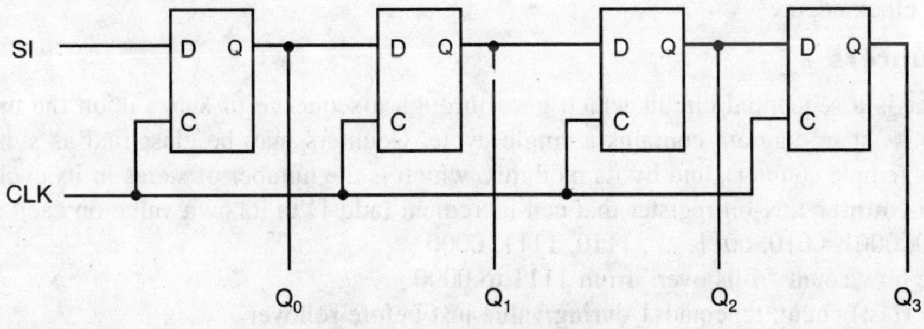

Fig. 1.42. A Shift Registers

A shift register is an n-bit register that can shift its stored data by one bit position at each clock period. A serial input (SERIN) specifies a new bit to be shifted in. A serial output (SEROUT) has the value of the bit that is shifted out. In a serial-in serial-out register only one bit is out at any clock cycle. In a serial-in parallel-out register all stored bits are out at every clock cycle.

For example, if on some positive clock edge we have:

$$SI = 1$$

$$Q_0 - Q_3 = 0110$$

Then the next state will be :

$$Q_0 - Q_3 = 1011$$

The current Q_3 (0 in this example) will be lost on the next cycle.

$$Q_0(t + 1) = SI$$

$$Q_1(t + 1) = Q_0(t)$$

$$Q_2(t + 1) = Q_1(t)$$
$$Q_3(t + 1) = Q_2(t)$$

Shift Registers With Parallel Load

We can add a parallel load, just like we did for regular registers.

- When LD = 0, the flip-flop inputs will be $SIQ_0Q_1Q_2$, so the register shifts on the next positive clock edge.

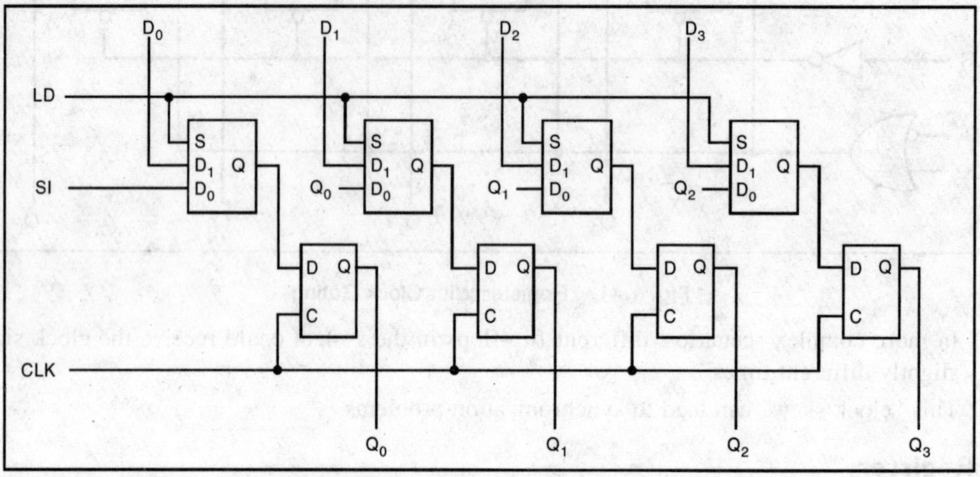

Fig. 1.43. Shift Register with Parallel Load

When LD = 1, the flip-flop inputs are D0-D3, and a new value is loaded into the shift register, on the next positive clock edge.

1.6.9. Counters

A counter is a sequential circuit which goes through a sequence of states upon the application of input pulses. Its state diagram contains a single cycle. Counters may be classified as synchronous or asynchronous (ripple counter), and by its modulus, which is the number of states in its cycle.

N-bit up-counter : N-bit register that can increment (add 1) to its own value on each clock cycle.

(*i*) 0000, 0001, 0010, 0011, ..., 1110, 1111, 0000

(*ii*) Note how count "rolls over" from 1111 to 0000

Terminal (last) count, tc, equals1 during value just before rollover

Internal Design

Register, incrementer, and N-input AND gate to detect terminal count (Fig. 1.44)

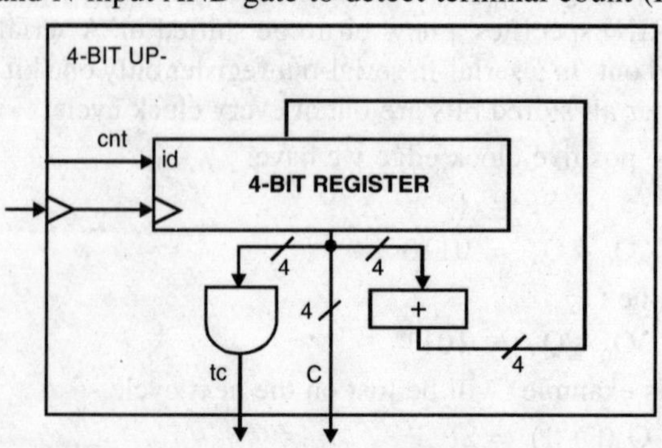

Fig. 1.44. 4-Bit Up-Counter

Counter Design Using Incrementer

Incrementer design :

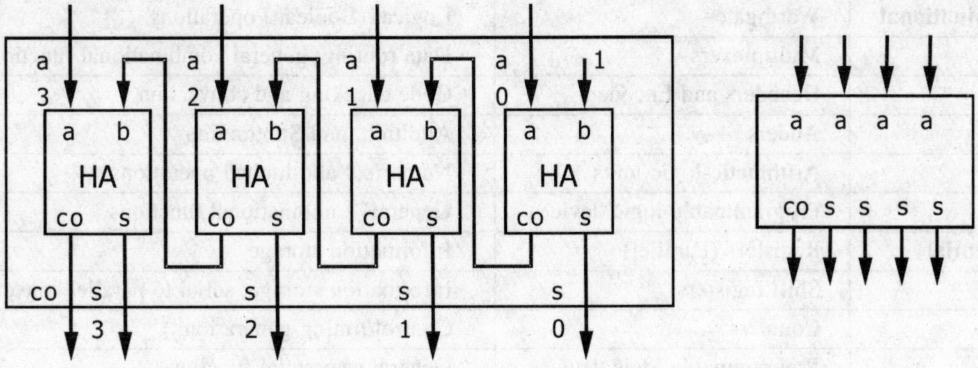

Fig. 1.45. Counter Design

(*i*) We could use carry-ripple adder with B input set to 00...001. But when adding 00...001 to another number, the leading 0's obviously don't need to be considered—so just two bits being added per column.

(*ii*) Use half-adders (adds two bits) rather than full-adders (adds three bits).

1.7. REGISTER LEVEL

At the register or register-transfer level (RTL), related information bits are grouped to form words or vectors. These words are processed by small combinational or sequential circuits.

Register-Level Components

Register-level circuits are composed of word-oriented devices *i.e.*, the devices that perform on group of bits (words). Table 1.7 shows the commonly used register transfer level components and their functions.

1.7.1. Multiplexers (MUX)

A multiplexer or data selector (abbreviated MUX) consists of a group of data inputs and group of control inputs. The control inputs are used to select exactly one data input to be outputted. Control inputs are also referred to as selection lines. The number of control inputs is directly related to the maximum number of data inputs that can be selected. A MUX with n control inputs can select from a maximum of 2^n data inputs. When $n = 2$, there are $2^2 = 4$ data inputs that can be selected.

What is a Multiplexer (MUX)?

(*i*) A MUX is a digital switch that connects one of n inputs to its single output. The Inputs can be one or more bits wide.

(*ii*) A Control signal selects the active input. An Enable signal enables the MUX to do its thing. It selects a single output from several inputs.

(*iii*) The particular input chosen for output is determined by the value of the multiplexer's control lines.

(*iv*) To be able to select among *n* inputs, $\log_2 n$ control lines are needed.

(*v*) A *DEMUX* does the opposite: it routes a signal to one of several destinations.

TABLE 1.7. Major Components Types at Register Transfer Level

Type	Component	Function
Combinational	Word gates	Logical (Boolean) operations
	Multiplexers	Data routing: general combinational functions
	Decoders and Encoders	Code checking and conversion
	Adders	Addition. and Subtraction
	Arithmetic-logic units	Numerical and logical operations
	Programmable logic devices	General' combinational functions
Sequential	Registers (Parallel)	Information storage
	Shift registers	Information storage; serial to parallel conversion.
	Counters	Control/timing generation
	Programmable logic devices	General sequential function

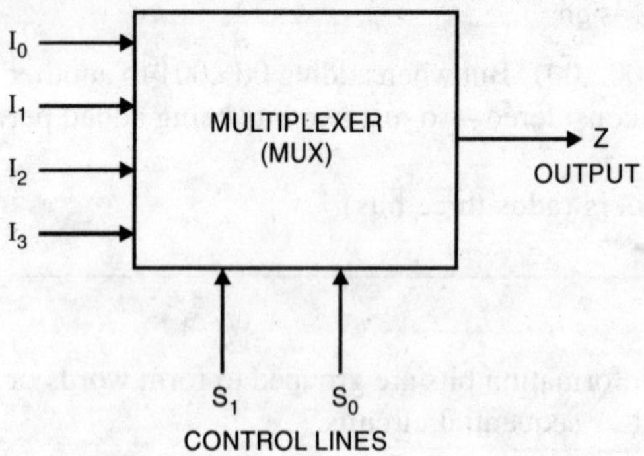

Fig. 1.46. A 4 : 1 Multiplexer Diagram

TABLE 1.8

S_1	S_0	Y
0	0	I_0
0	1	I_1
1	0	I_2
1	1	I_3

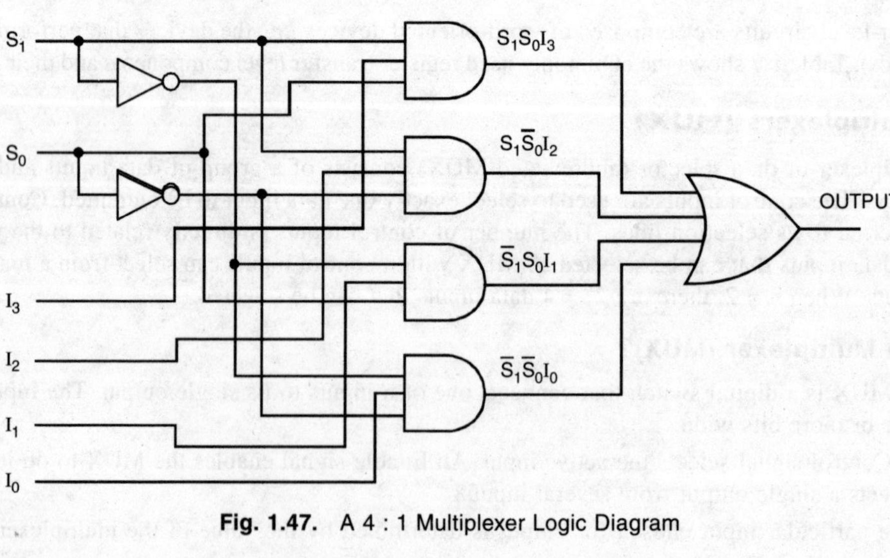

Fig. 1.47. A 4 : 1 Multiplexer Logic Diagram

1.7.2. Encoder

An encoder is a digital circuit that performs the inverse operation of a decoder. It generates a unique binary code from several input lines. Generally encoders produce 2-bit, 3-bit or 4-bit code. n bit encoder has 2^n input lines. If one of the four input lines is an active encoder produces the binary code corresponding to that line. If more than one of the input lines will be activated or all the output is undefined. We can consider don't care for these situations but in general we can solve this problem by using priority encoder.

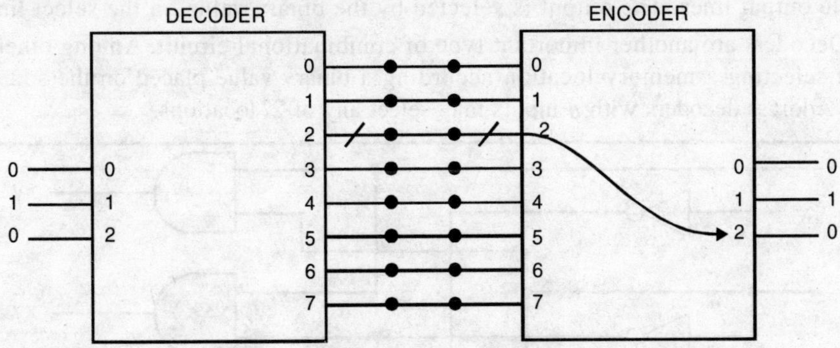

Fig. 1.48. Block-diagram Decoder and Encoder

The Input is "one hot encoding". And Output = binary representation of the "index" of the wire that is "hot" (equal to 1)

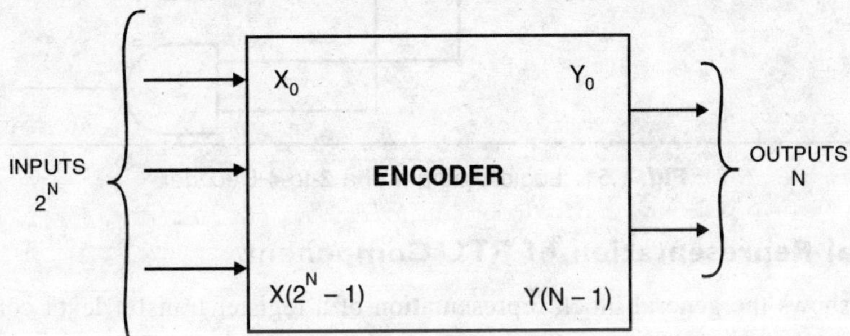

Fig. 1.49. A 2^N-to- N Binary Encoder

1.7.3. Decoder

Definition A *decoder* is a **multiple-input multiple-output logic** circuit that converts coded inputs into coded outputs, where the inputs and outputs codes are different.

Decoding the conversion of an n-bit input code to an m-bit output code with $n = m = 2n$ such that each valid code word produces a unique output code. Circuits that perform decoding are called *decoders.*

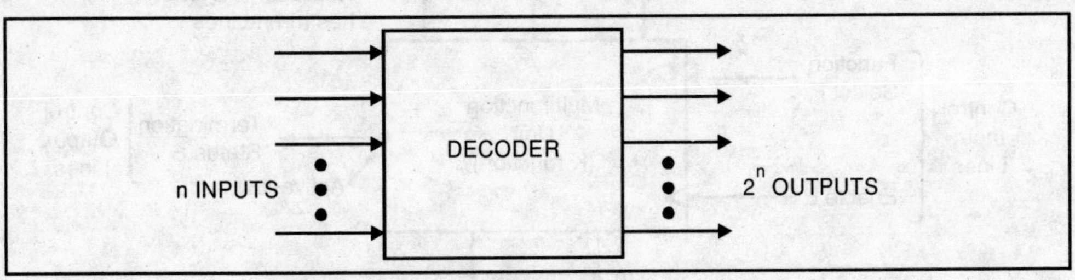

Fig. 1.50. General Decoder Diagram

Here, functional blocks for decoding are called n-to-m line decoders, where $m = 2n$, and generate $2n$ (or fewer) minterms for the n input variables. The outputs of a binary decoder provides 2^n minterms for its n inputs, and a Boolean functions can be represented as a sum of minterms, therefore a decoder and one or more OR gates can be used to realize sums of minterms of n variables. Combinational logic circuit that receives binary information on a single input and sends this information to one of many possible output lines. The output is selected by the binary value on the select lines.

Digital Decoders are another important type of combinational circuit. Among other things, they are useful in selecting a memory location according a binary value placed on the address lines of a memory bus. Address decoders with n inputs may select any of 2^n locations.

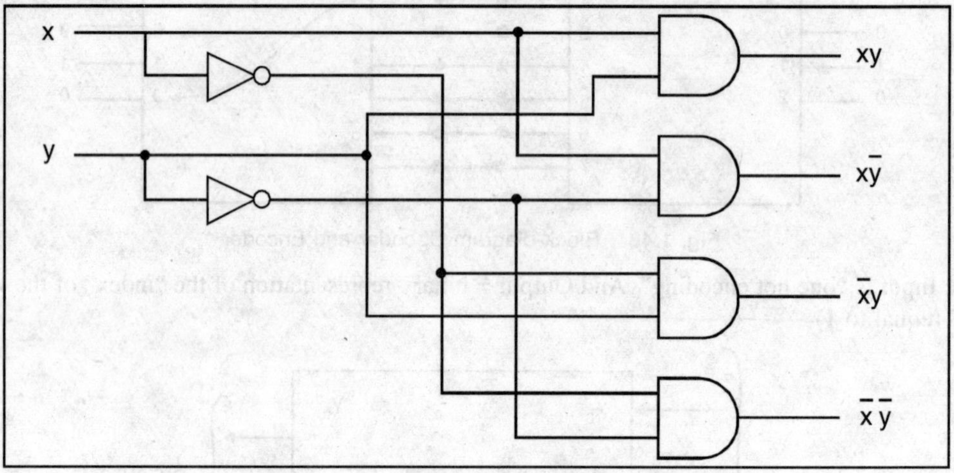

Fig. 1.51. Logic Diagram of a 2-to-4 Decoder

1.7.4. General Representation of RTL Component

Figure 1.52 shows the general block representation of a register transfer level component. The *1m* (slash *m)* on the input and output lines indicate it is a *m-bit* input and output bus. A bubble on the start or end of the line indicates an active-low signal: otherwise it is an active-high signal.

The input and output control lines are also shown separately. The input control lines associated with a multifunction block fall into two categories:

(*i*) Select lines and

(*ii*) Enable lines.

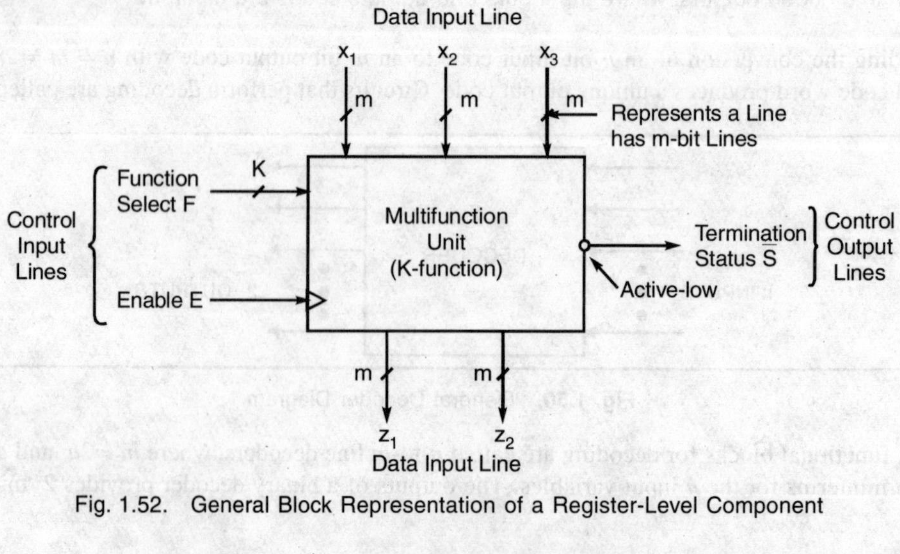

Fig. 1.52. General Block Representation of a Register-Level Component

The select lines specify one of several k possible operations. In this multifunction unit is enabled by activate the enable E control signal. The output control signal indicates when or how the unit completes its processing.

Let us see the major combinational and sequential components used in design at the register level.

Word Gate

Bit-wise logical functions can be performed on the m-bit binary words using word gate operators.

Let $$X = (X_1, X_2, X_3 \ldots\ldots\ldots X_M)$$

and $$Y = (Y_1, Y_2, Y_3 \ldots\ldots\ldots Y_M)$$

be the two *m-bit* binary words, we can perform the gate operations bitwise on X and Y to obtain the output *m*-bit word.

$$Z = (Z_1, Z_2, Z_3, \ldots\ldots, Z_m)$$

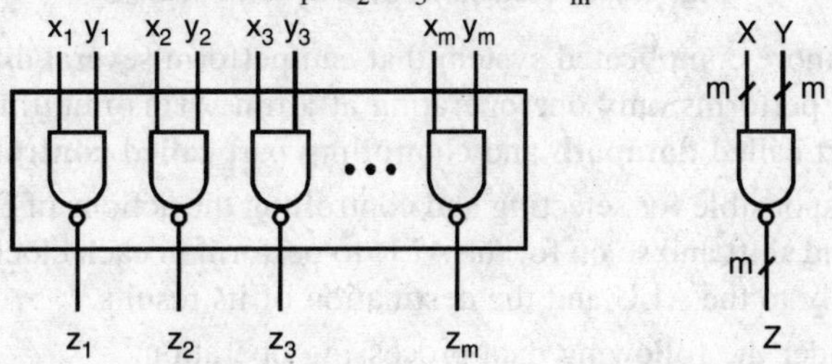

(*a*) Logic Diagram (*b*) Symbol

Fig. 1.53. Two-Input m-bit NAND Word Gate

Figure 1.53 shows the two input-m-bit NAND gate and the corresponding symbol.

In general, if is any logic operator, we write

$Z =$ ***func*** (X, Y) if $Z i =$func (x_i, yi) for $i = 1,2,3 \ldots m$.

For example of NAND gate output is specified by $Z = X.Y$ denotes the m-bit NAND operation defined by.

1.7.5. Register-level Design

A register-level system consists of a set of registers linked by combinational data transfer and data processing circuits. A block diagram can define its structure and the set of operations it performs on data words can define its behavior. Each operation is implemented by one or more elementary register transfer steps and it can be defined as

Condition : $$Z = Func(X1, X_2, \ldots X_K);$$

where, *Func is* a function to be performed or an instruction to be executed in one clock cycle.

Here $X_1, X_2 \ldots X_K$ and Z denote data words or the registers that store data. The prefix "condition" denotes a control condition that must be satisfied (if condition = l) for the indicated operation to take place. Therefore when condition = 1 the function Func is computed on variables $X_1, X_2 \ldots X_K$ and result is stored in the register Z.

$$Z = (z_1, z_2, z_3, \ldots, z_m) = (x_1{\cdot}y_1, x_2{\cdot}y_2, \ldots, x_m{\cdot}y_m)$$

Register Level Design : Digital modules are designed by the registers they contain and operations that are performed on the data stored in them. The operations executed on data stored in registers are called micro operations.

Data and Control

Figure 1.54 shows a simplest register level system. It performs addition operation between the contents of register A and B, and the result is stored in Register Z. This process is simply written as

Z = A + B

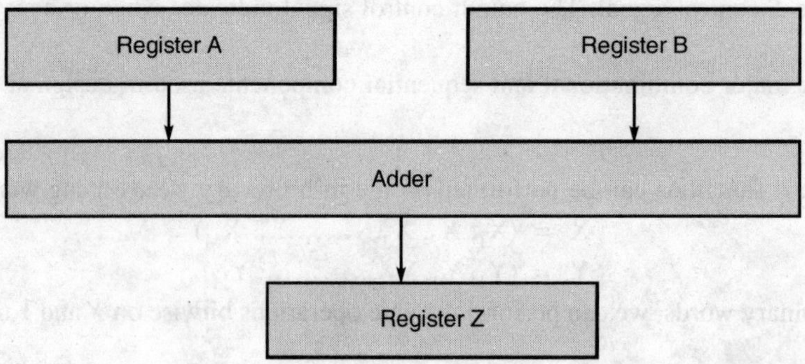

Fig. 1.54. Register Level System in CPU

Figure 1.55 shows a more complicated system that can perform several different operations. Such a multifunction system can performs only one operation at a time. This multifunction system partitioned into a data processing part called **datapath** and controlling part called **control unit.**

The control unit is responsible for selecting and controlling the actions of Datapath. It also selects an operation like add, sub, and shift and so on for the ALU to perform in each clock cycle. It also determines the input operands to apply to the ALU and the destination of its results.

For example of consider the following data processing operation.

$$Z = (A + B) \quad ...(1)$$

$$B = (A - B) \quad ...(2)$$

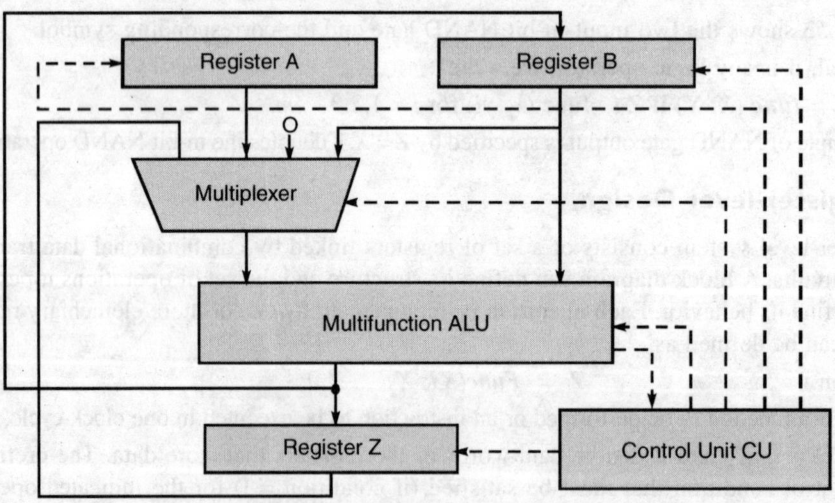

Fig. 1.55. Datapath and Control Unit

The Eq. (1) is the addition operation, in which the addition is to be performed using the contents of register *A* and *B,* and result is stored in register Z. The selection source and desist nation register is taken care of by the control unit (CU).

The Eq. (2) is the subtraction operation is to be performed using the contents of register *A* and *B* and result is stored in *B* register.

An example of large multifunction system is a computer's CPU. Its control unit is responsible for interpretation (decoding) of instructions. The CPU's Datapath—also called the E-unit (Execution unit) which is an Array File.

Description Language

A Description language provides both behavioral and structural descriptions at 'the register level. We can use VHDL for modeling the register level components. VHDL is complex and it is beyond the scope of this book, however we will use a much smaller HDL as below.

For example

```
If condition = 1 then Z = Func (X1, X2,.... XK);
```

Here **Func** can be any function.

Let us see the simple description language to describe the behavior of a complete system at the register level. This 8-bit, multiplication circuit named as multiplier *B* computes the product Z = *X* x *Y* where both the numbers are 8-bit number.

Register Level Design Techniques

The design techniques for register-level system are as follows :

1. First let us define the desired behavior of the system by a set of sequence of register transfer operations, such that each operation can be implemented directly using the available design components. This gives the desired algorithm to be executed.
2. Analyze the algorithm to determine the types of components and the number of each type required for the datapath.
3. Construct a block diagram for data path using the components identified in step 2. Make the connections between the components so that all data paths implied by algorithm are present and the given performance-cost constrains are met.
4. Analyze algorithm and datapath to identify the control signals needed. Now introduce the logic or control points necessary to apply these signals into data path.
5. Design a control unit for datapath that meets all the requirements of algorithm.
6. Verify by computer simulation, that the final design operates correctly and meets, all performance cost goals.

Algorithm design in step-l involves a creative design process analogous to writing a computer program and depends on the skill and experience of the designer. The identification of the data-processing components in step-2 is straight forward but complications arise when the possibility of sharing components exists.

For example $C: A = A + B, C = C + D;$

These relations define two addition operations. Since these additions do not involve the same operands, they can be one in parallel if two independent adders are provided. However, costs can be reduced by sharing a single adder and performing the two additions sequentially thus

$$C(t_o) : A = A + B;$$
$$C(t_o + I) : C = C + D;$$

The step 3 requires defining an interconnection structure that links the component needed by the various parts of algorithm. Identifying the control signals and design of control unit is done in step 4 and step 5 respectively.

The step-6 is the design verification by computer simulation. The simulation can be used to verify and correct functional errors before new design is committed to hardware.

(*i*) **Register** is a group of flip-flops which is used for temporary storage, transfer of data and shifting of information, etc.

(*ii*) The symbolic notation used to describe a small operation (micro operation) transfer among registers is called **REGISTER TRANSFER LEVEL** or register transfer language.

(*iii*) A group of bits (i.e. word) performed in the group of logic gates are called word gates.

(*iv*) Multiplexer is a combinational logic circuit that receives binary information from one of 2^n inputs and directs it to a single output line. Multiplexer may be used to implement any combinational logic function.

(*v*) **Decoder** is a combinational circuit that converts binary information from the n-bit coded inputs to a maximum of 2^n unique outputs.

(*vi*) An **encoder** has 2^n (or less) input lines and *n* output lines. The output lines generate the binary codes from corresponding input code.

1.8. PROCESSOR LEVEL

The processor level or system level is the highest in hierarchy of computer design. It is concerned with storage and processing of blocks of information such as programs and data files. The components: required for performing these functions are complex. Usually sequential circuits are used which are based on VLSI technology.

Processor Level Components

The following types of components which are generally used at this level can be divided mainly four groups as

1. PROCESSOR-CPU
2. MEMORIES –RAM/ROM
3. I/O DEVICES –KEYBOARD, MOUSE ETC
4. INTERCONNECTION OF NETWORKS

Central Processing Unit (CPU)

The primary function of a central processing unit is to execute sequences of instructions stored in a memory, which is external to the central processing unit. When the functions of the processors are restricted, those processors become more specialized processor such as I/O processor.

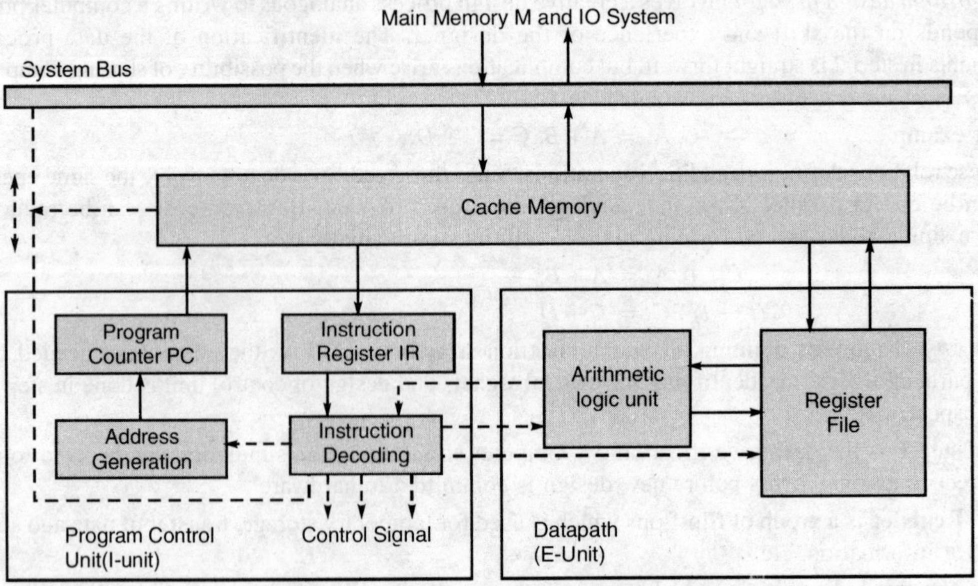

Fig. 1.56. A CPU at The Register Level

Figure 1.56 shows the internal structure of the CPU at the register level. The CPU contains different units such as Data path and control units. The control part generates the addresses of instructions and data stored in external memory. In this structure, a cache memory is placed between the main memory *M* and the CPU. The cache is a fast buffer memory designed to hold an active portion of the system's address space. Each memory request generated by the CPU is first directed to the cache. If the required information is not currently assigned to the cache, the request is redirected to *Main Memory* and the cache - automatically updated from memory.

The Control unit fetches instructions from the cache or memory *M* and decodes them to derive the control signals needed for their execution. The CPU's datapath has the arithmetic-logic circuits that execute most instructions. It also has a set of registers for temporary data storage. The CPU manages a system bus, which is the main communication between the CPU-cache subsystem, main memory, and the I/O devices.

Memory Devices—Main and Auxiliary

The memory unit is an essential component in any digital computer. It is needed for storing programs data. Most general purpose computers would run more efficiently if they were equipped with additional storage beyond the capacity of the main memory. This main memory does not have enough space e in one memory unit to accommodate all the programs used in a typical computer. The memory unit that communicates directly with the CPU is called the main memory. Devices that provide backup storage are called auxiliary memory. Magnetic disks and tapes are used as common auxiliary memory devices. They are used for storing system programs, large data files, and other backup information. Only programs and data currently needed by the processor reside in main memory.

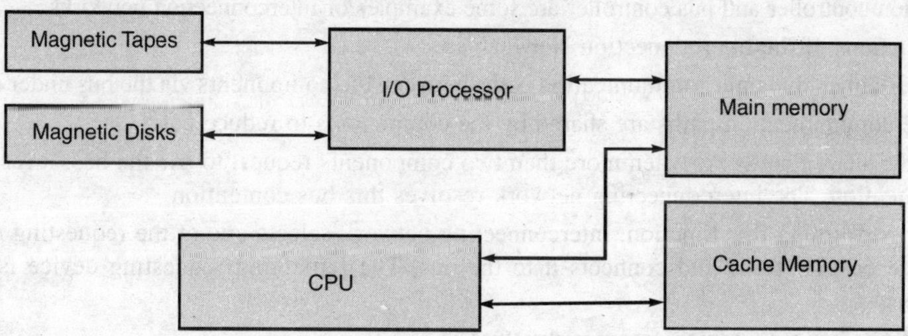

Fig. 1.57

The memory hierarchy system consists of all storage devices employed in a computer system. Figure 1.57 shows the typical memory hierarchy. The auxiliary memories are magnetic disks and tapes are removable memory devices. The main memory occupies a central position by being able to communicate directly with the CPU and with auxiliary memory devices through an I/O processor. A special very high-speed memory called a cache is sometimes used to increase the speed of processing by making current programs and data available to the CPU at a rapid rate. The cache memory is employed between CPU and main memory. CPU logic is usually faster than main memory access time, with the result that processing speed is limited primarily by the speed of main memory. A technique used to compensate for the mismatch in operating speeds is to employ an extremely fast, small cache between the CPU and main memory whose access time is close to processor logic clock cycle time. The I/O processor manages data transfers between auxiliary memory and main memory, the cache organization is concerned with the transfer of information between main memory and CPU.

I/O Devices

A computer communicates with outside world by means of input-output (I/O) systems. The main function of I/O system is to transfer information between CPU or memory and the outside world.

Interfacing : Designing logic circuits (hardware) and writing instructions (software) to enable the processor to communicate with peripherals is called interfacing, and the logic circuits used for such purposes are called I/O ports or interfacing devices. The most important constituents of computer system are CPU, memory and I/O (input/output) devices. The CPU fetches instruction (opcodes and operands) from memory, processes them and stores the results in memory. *The main function of I/O system (I/O interface) is to transfer information between CPU or memory.*

The I/O devices cannot be connected directly to the system bus due to the following reasons.

1. Variety of peripherals with different methods of operation is available. So it would be impractical to incorporate the necessary logic with in the CPU to control the range of devices.
2. The data transfer rate of peripherals is much slower than memory or CPU. So it is impractical to use the high speed system bus to communicate directly with the peripherals.
3. Generally, the peripherals used in a computer system have different data formats and word lengths than that of CPU used in it.

So to overcome all these difficulties, it is necessary to use a module or interface device in between system bus and peripherals called I/O module or I/O system or I/O interface.

Interconnection Networks

The processor level components, CPU, 10 devices communicate via system bus (address bus, data bus and control bus). In a computer system, when many components are used, communication between these components may be controlled by a subsystem called an *interconnection network.* Switching network, communication controller and bus controller are some examples of interconnection networks.

The Functions of the Interconnection Network are :

- To establish dynamic communication paths between the components via the bus under its control.
- The communication paths are shared by the components to reduce cost.
- In the shared network, when more than two components request to use the bus, it results in bus contention. The interconnecting network resolves this bus contention.
- For performing this function, interconnecting network selects one of the requesting devices on some priority basis and connects it to the bus. The remaining requesting device is kept in a queue.

Figure 1.58 shows a simple computer structure.

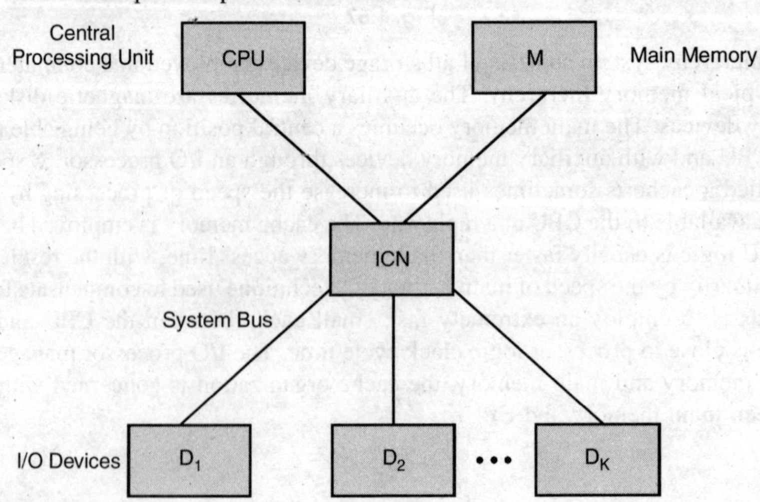

Fig. 1.58

Processor-Level Design

While designing any system, it is very much difficult to give a detailed description of the desired system behavior. Because of this reason, the processor level design is critical as compared to register level design. The common approach to design at this level is to take a prototype design of known performance. Then according to the necessity new technologies are added and new performance requirements are achieved.

Performance Characteristics

In this section we introduce some basic aspects of computer system performance characteristics. The total time needed to execute application programs is the most important measure of computer system performance. In other words we can say that the *speed of the computer system is an important characteristic to define the performance of the computer system.* The speed of the computer system depends as various factors. Let us discuss these factors.

Hardware : The speed of the processor used in the computer systems basically decides the speed of computer system. For example, system having processor Pentium-IV runs faster than system having Pentium-I processor. However, system speed is not only depends on the processor speed, but it affected by the supported hardware.

Programming Language : Nowadays, programs are usually written in high-level languages. These languages require separate compiler to translate programs into machine level language. Therefore, the computer system is affected by the performance of the compiler and hence the language used for the program.

Pipelining : The processor executes the instruction in steps or phases such as fetch decode and execute. By overlapping these phases of successive instructions we can achieve a substantial improvement in the performance of the computer system. This technique is known as pipelining.

Parallelism : It is possible to perform transfers to and from secondary memory like storage disk or tapes in parallel with program execution in the processor or with activity in other I/O devices Such technique is known as parallelism. The most of computer system use this parallelism to improve the system performance.

Types of memory and 10 devices: The performance of the computer depends on the type of memory and I/O devices supported by it.

Prototype Structures

The processor-level design using prototype-structures involves following steps in the design processes.

1. Select a prototype design as per the system requirements and adopt it to satisfy the given performance constraints.
2. Determine the performance of proposed system.
3. If the performance is unsatisfactory design is to be modified and repeat step-I.
4. The above steps (1 to 3) are to be continued until the acceptable design is obtained and the desired performance constraints are achieved.

These steps are widely followed for designing a computer system. While designing new systems, the precautions are always taken to remain compatible with existing hardware and software standards. The reason is that when these standards are to be changed, computer owners have to spend more to retain users and programmers. Also the well tested software is to be replaced by the modified software.

So in the new design of the computer system the drastic changes in the previous design are generally avoided. Because of all these reasons, there is slow evaluation of computer architecture.

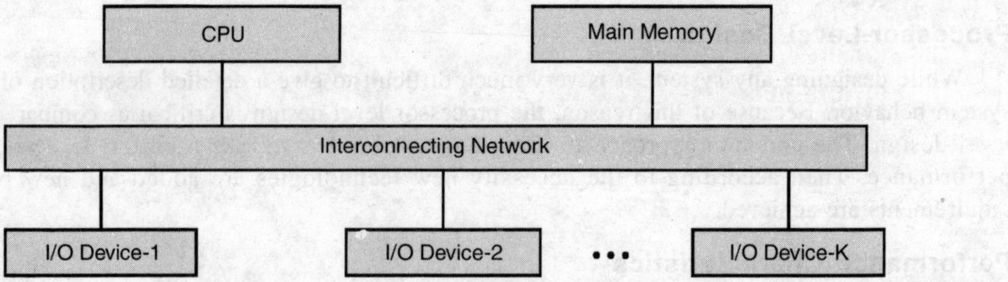

Fig 1.59. Basic Computer Structure

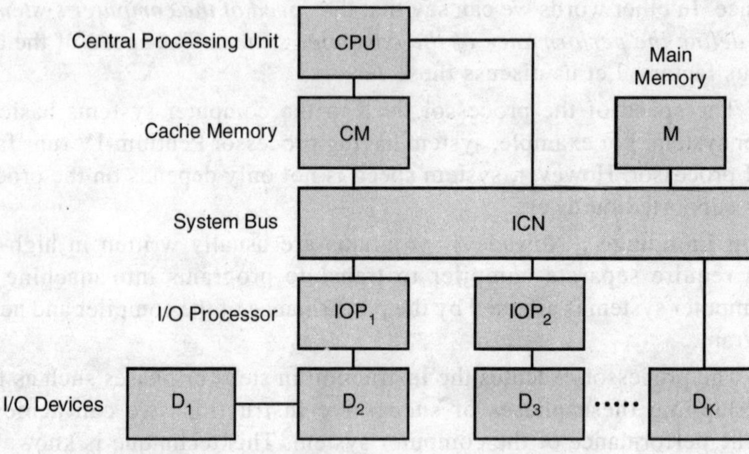

Fig. 1.60. Computer Structure With I/O Proc and Cache

Figure 1.61 shows the structure of first generation computers. The second and subsequent generations of computer involve special-purpose I/O processors and cache memory in addition to basic components ed within the basic system.

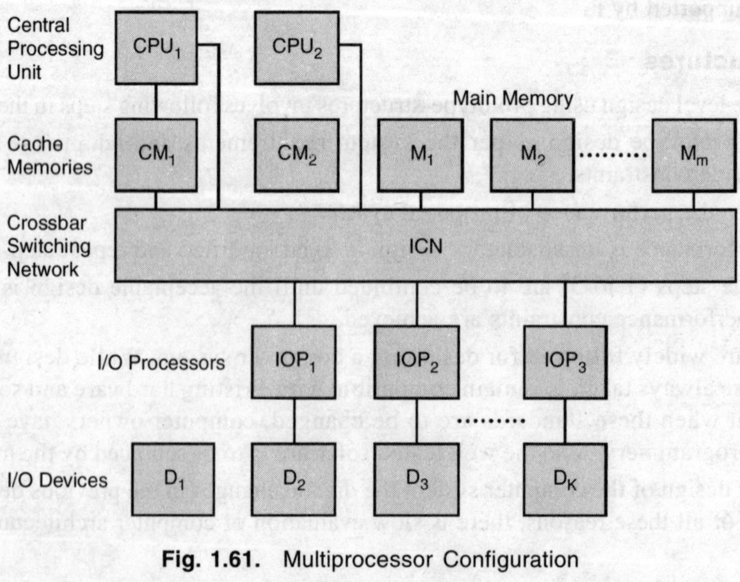

Fig. 1.61. Multiprocessor Configuration

Computer System Performance

The over all performance of the system will depend upon its sub-systems and components.

(*i*) CPU Execution Time

(*ii*) CACHE Bandwidth

(*iii*) Main Memory Bandwidth

(*iv*) I/O Performance

Performance measures can be used to guide design decisions. They can also be used to compare different architectures, different implementations of the same architecture, or different compilers for a given architecture.

We will use the following definition for the performance of ***machine X*** on *a given program*:

$$\textbf{PERFORMANCE} = \textbf{1/(Execution Time)} \quad ...(i)$$

MIPS : MIPS is a popular performance metric often used by computer vendors. MIPS stand for million instructions executed per second by cpu and are defined by the following equation:

$$\text{MIPS} = \frac{\text{Instruction Count}}{\text{Execution Time}}$$

Note that the above definition only holds for a given program.

$$\text{MIPS} = \frac{\text{Clock Rate}}{\text{(CPI)}} \quad ...(ii)$$

Clock-rate is the reciprocal of **CPU Clock-frequency.**

MFLOPS : MFLOPS stands for million floating point operations performed by CPU per second and is defined by :

$$\text{Execution Time} = \frac{\text{Number of Floating Point Operations}}{\text{Time-taken for Execution}}$$

The two main measures of performance for a computing machine can be studied under :

(*i*) **Execution Time :** Total time taken by a task to execute, from its submission to its completion.

(*ii*) **Throughput :** Number of tasks completed per unit time.

The **time parameter** can be defined in different ways, depending on what we are measuring;

System CPU Execution Time

Total time operating systems spends executing tasks for a given program.

We define the ***Speedup* of machine** X over Y as :

$$\frac{\text{Performance (X)}}{\text{Performance (Y)}}$$

The text uses the term "faster than" for this.

Measuring Performance

CPU Time : CPU time spent on the given process. This is broken down into :

(*i*) User CPU time—the CPU time used up by the user process.

(*ii*) System CPU time (this is more difficult to compute).

Another measure of the performance of a computer system is the clock period or its inverse, the clock rate. Clock cycle is a physical metric which is independent of the program being run. Relation between these metrics (for one program):

CPU Clock Cycles = Instruction Count × Avg. Clock Cycles Per Instruction

CPU (Execution time) = CPU Clock Cycle × Clock Cycle Time ...(*iii*)

PERFORMANCE = CPU CLOCK RATE/(TOTAL INSTRUCTIONS × CPI)
WHERE CPU EXECUTION TIME = INSTRUCTIONS × CPI/(CLOCK RATE)

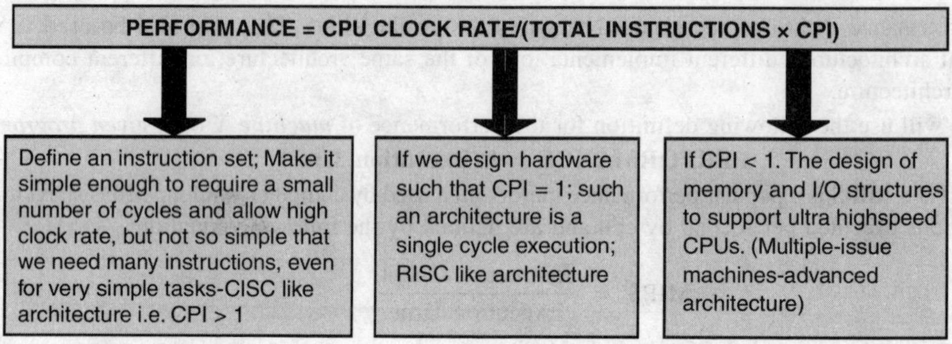

Fig. 1.62.

We can define a new term : **Clock Cycles Per Instruction (CPI),** and that gives us:

CPU Time = Instruction Count × CPI × Clock Cycle Time ...(*iv*)

Therefore, if you design an architecture that reduces the number of instructions in a program you have to be careful that it does not increase the other two factors; else you could end up with an *increase* in the overall execution time of a program. *The final performance criterion is always* **Execution time.**

- A computer clock runs at a constant rate and determines when events take placed in hardware.
- The clock cycle time is the amount of time for one clock period to elapse (e.g. 5 ns).
- The clock rate is the inverse of the clock cycle time.

Clock Per Instruction Plays an Important Role

- The CPI is the average number of cycles per instruction.
- If for each instruction type, we know its frequency and number of cycles need to execute it, we can compute the overall CPI as follows :

$$\text{CPI} = \sum_{i=1}^{i=N} \text{CPI} \times \text{F}_i \qquad ...(v)$$

where F_i is the frequency for i^{th} instruction.

Performance is determined by

1. Program Size
2. CPU Clock Speed
3. Architecture Parameter (CPI)

The primary function of a CPU is to execute sequences of instructions stored in memory. A hardware circuit which performs data processing operation is called as data path. I/O systems are used to interface the CPU with other processing units. In a computer system, many components are used, communication between these components may be controlled by a subsystem called interconnection network. Speed of the computer system is an important characteristic to define the performance of the computer system.

Summary

- A computer is a machine for manipulating data according to a list of instructions known as a program.
- A computer system is the synthesis of hardware and software. A typical computer system employs a computer that uses programmable devices to store, retrieve, and process data.
- A computer is a machine for manipulating data according to a list of instructions known as a program.
- The American mathematician Howard Aiken developed the mark I calculating machine which was built by IBM in the 1930.
- IAS computer had 4096 memory locations of 40 bits each. Each word could hold either one piece of data or two instructions (20 bits each).
- This was maked by appearance of UNIVAC-I in 1950, in which vacuum tuber were used. Their memory were made of thin tuber of liquid memory and magnetic drums.
- As commercial interest in computer technology intensified during the late 1950s and 1960s, the second generation of computer technology was introduced—based not on vacuum tubes but on transistors.
- The third generation of computer technology was based on integrated circuit technology and extended from approximately 1964 to 1970.
- The fourth generation of computer technology is based on the microprocessor.
- A computer is fast information processing electronic device. It carries out all sorts of computations within a fraction of a second.
- Computer which are use today are digital computer. They manipulate numbers. They operate on binary digits 0 and 1.
- A digital computer is programmable machine specially designed for making computation. The components are CPU, memory, input and output devices.
- "Computer Architecture and organization is the design of the computer system at the hardware/ software interface.
- "The overall design or structure of a computer system, including the hardware and the software required to run it, especially the internal structure of the system components such as CPU, (microprocessor) Main Memory and other I/O devices".
- A system is an interconnection of various entities such as CPU, Input Unit etc to achieve a specific goal.
- A system is represented by a graph or a block diagram with proper inputs and outputs. A system has structure and a behavior.
- A system is an interconnection of system entities such as CPU, Input and output units.
- A system design is a creative process, which comes from the mind of a designer. A system can be designed using system analysis and synthesis methods.
- A computer system design is considered as an abstraction of several layers.
- Those circuits in which the output depends on present input and present state of the circuit such as flip-flops and memory devices are known as sequential circuits.
- Flip-Flop is a device that can store one-bit of binary information indefinitely until directed by an input signal to switch the state.
- *T* flip-flop also known as Toggle flip-flop. It is a single version input of the *JK* flip-flop.
- A register capable of shifting its binary information in one or both directions is called a shift-register.
- A shift register is an n-bit register that can shift its stored data by one bit position at each clock period.

- A multiplexer or data selector (abbreviated MUX) consists of a group of data inputs and group of control inputs. The control inputs are used to select exactly one data input to be outputted.
- A *decoder* is a **multiple-input multiple-output logic** circuit that converts coded inputs into coded outputs, where the inputs and outputs codes are different.
- Digital modules are designed by the registers they contain and operations that are performed on the data stored in them. The operations executed on data stored in registers are called micro operations.
- The processor level or system level is the highest in hierarchy of computer design. It is concerned with storage and processing of blocks of information such as programs and data files. The components: required for performing these functions are complex. Usually sequential circuits are used which are based on VLSI technology.
- The primary function of a central processing unit is to execute sequences of instructions stored in a memory, which is external to the central processing unit.
- The memory unit is an essential component in any digital computer. It is needed for storing programs data.

Review Questions

1. What are Gate Level and Register Level Design ?
2. Add 22 + 15 in binary.
3. What is Computer Architecture and Organization ?
4. Explain –E-CAD steps.
5. Explain Processor Level Design
6. Subtract 15 from 22 in binary.
7. Convert 234_{10} into Binary Format.
8. Contrast computer architecture and computer organization.
9. What is the hexadecimal notation for 22 and 15 ?
10. What are gates? Give a brief idea of Boolean algebra
11. Discuss De-Morgan's Theorem.
12. Why NAND and NOR gates are said to be Universal Gates.
13. Simplify the following Boolean functions :
    $F(A, B, C) = \Sigma(0, 2, 3, 4, 6)$
    $F(A, B, C) = \Sigma(3, 7, 11, 15)$
14. Also discuss in brief, don't case conditions while dealing with K-map.
15. Simplify the following Boolean function and draw the logic diagram from the simplified expression.
    $$F = xy'z + x'y'z + xyz$$
16. A majority function is generated in a combinational circuit when the output is equal to 1 if the input variable has more 1's than 0's. The output is zero otherwise design a three input majority function.
17. Show that a JK flip/flop can be converted to a D flip/flop with an inverter between the J and K inputs.
18. Can you construct a 3×8 decoder, using two 2×4 decodes? If so draw it.
19. A sequential circuit has two D flip-flop A and B, two input x and y and one Output z .The flip-flop input equations and the circuit output are follows :
    $$D_A = x'y + xA$$
    $$D_B = x'B + xA$$
    $$Z = B$$
    Draw the logic diagram of circuit and show the state table also.
20. The content of a 4 bit register is initially 1101. The register is shifted six times to the right with the serial input being 101101. What are the content and the register after each shift ?

❑❑❑

2 Chapter

Register Transfer and Microoperations

INSIDE THIS CHAPTER

2.1. REGISTER TRANSFER

The registers in a digital computer are designated by capital letters (and if required are also followed by numerals) to denote the function of the register. For example, the register that contains the address of the next instruction to be executed is program counter and is designated by PC. Other registers are IR (for instruction register), MAR (memory address register), $R1$ (for processor register) etc.

The most common way of representing a register is with the help of a rectangular box with the name of the register written inside the box as in Fig. 2.1(*a*) we can also distinguish the individual bits of a register as in Fig. 2.1(*b*) with each bit assigned a letter with a subscribt number. The numbering of bits from right to left can be marked on the top of the box as in Fig. 2.1(*c*). Also, a 16-bit register is partitioned into two parts as is Fig. 2.1(*d*), where bits 0 through 7 are symbol L (for low byte) and bits 8 through 15 are assigned the symbol H (for high byte).

Definition Register transfer is defined as the transfer of information between registers with the help of common bus.

Definition The symbolic notation used to describe the microoperation transfer among registers is called a register transfer language.

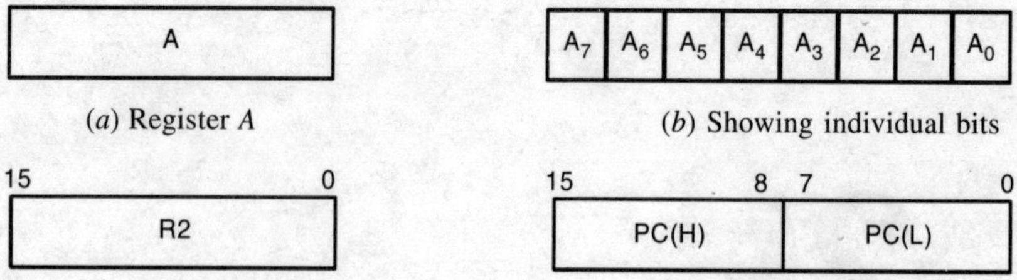

(*a*) Register *A* (*b*) Showing individual bits

(*c*) Numbering of bits (*d*) Divided into two parts

Fig. 2.1. Block Diagram of Register.

The symbolic notation used to describe the microoperation transfer among registers is called a *register transfer language*. This is a way of expressing the different microoperations sequence in the symbolic form among registers of a computer.

For example, the statement.

$$R2 \longleftarrow R1$$

simply denotes that the content of register $R1$ is to be transferred into register $R2$. After this transfer operation takes place the previous content of $R2$ will be replaced by the content of $R1$ but the content of register $R1$ does not change after the transfer. $R1$ is the source register and $R2$ is the destination register. If we want that the transfer can take place only when certain control condition satisfies, this can be shown by means of an if then statement as

$$\text{if } (P = 1) \text{ then } (R2 \leftarrow R1)$$

where P is a control signal, generated in the control section. Also, some time we use control function to separate the control variables from the register transfer operation. A control function is a Boolean variable that is equal to 1 or 0. Thus, the above statement is equivalent as

$$P : R2 \longleftarrow R1$$

where colon is used to terminate the control condition.

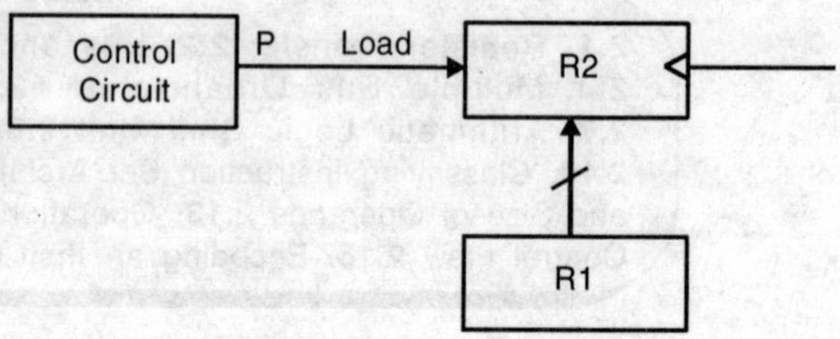

Fig. 2.2. Hardware Implementation of P : R2 ⟵ R1

Every statement written in a register transfer notation implies a hardware implementation. The block diagram that depicts the transfer from $R1$ to $R2$ is shown in Fig. 2.2.

The basic symbols used for the register transfer notation is shown in Table 2.1.

TABLE 2.1. Basic Symbols for Register-Transfer Logic.

Symbol	Description	Examples
Letters (and numerals)	Denotes a register	$R1$, $R2$, PC, IR
Parentheses ( )	Denotes a part of register	PC (0 – 7), PC (L)
Arrow ←	Denotes a transfer of information	$R2 \leftarrow R1$
Comma,	Separates two Microoperation	$R2 \leftarrow R1$, $R3 \leftarrow R1$
Subscript	Denotes a bit of a register	A_2, B_6
Colon :	Terminates a control function	$x' T_0$
Square brackets [ ]	Specifies an address for memory transfer	MBR ← M[MAR]

2.2. BUS AND MEMORY TRANSFER

Definition When a data is read from the memory or is stored in memory is referred to as memory transfer.

A digital computer consists of many processor registers and the transfer of information from one register to another is often requierd. Hence, there must be paths provided so that the transfer operations among registers can take place. Consider a Fig. 2.3, which shows the transfer among three registers $R1$, $R2$ and $R3$. The figure shows that there are six data paths.

Definition A shared communication path consisting of one or more connections lines is known as bus and the transfer of data through this bus is known as bus transfer.

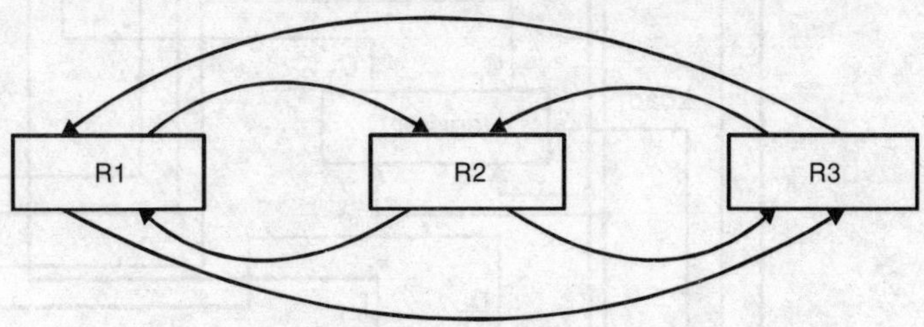

Fig. 2.3. Transfer Among Three Registers.

If the separate lines are used between each registers and all other registers, the number of wires will increase considerably. Hence, a set of common lines, one line for each bit of register, is used for the transfer. This set of common lines through which the binary information is transferred one at a time among registers is known as *bus*. A common-bus system can be constructed with the help of multiplexers and decoder where the decoder selects the destination register for the bus transfer. The multiplexer select the source register whose binary information is then placed on the bus and decoder selects one destination register to transfer the information from the bus. The construction of bus system for four registers is shown in Fig. 2.4. Two multiplexer are shown in the figure one for the low-order significant bit and one for the high-order significant bit. If the register is of n bits, n multiplexer is required to produce n-bus line. These n lines in the bus are connected to the n inputs of all registers.

TABLE 2.2. Function Table for Bus to Fig. 2.4.

S_1	S_0	*Register selected*
0	0	*A*
0	1	*B*
1	0	*C*
1	1	*D*

S_1 and S_0 are selection lines connected to selection inputs of all n multiplexers. The selection lines choose n bits of one registers and transfers those n bits into the n-line common bus. When S_1S_0 = 00, the 0 data inputs of all n multiplexers are selected and causes the n-bits from register A to transfer into the n-line common bus, since the output of this register are connected to 0 data inputs of each multiplexers. Similarly, when $S_1S_0 = 01$, the content of register B is transferred into the n-line common bus and so on. The register that is selected for the four binary possible binary value is shown in Table 2.2.

The transfer of information from bus to one of the destination register is done with the help of load control of that register. The particular load control of the register is activated by the outputs of the decoder when enabled. If the decoder is not enabled, no information from the bus will be transferred into the register, although the multiplexers places the information of the source register onto the bus.

In general, for registers of n bits, n multiplexers are needed to construct the bus of n-lines. The size of multiplexer depends upon the number of registers in the system. If there are K registers, the multiplexers size will be of $K \times 1$ since it multiplexes K data lines. For example, a common bus of 16 registers of

16 bits each requires 16 multiplexers of size 16 × 1. Four selection lines are required. Also, the size of destination decoder will be 4 × 16.

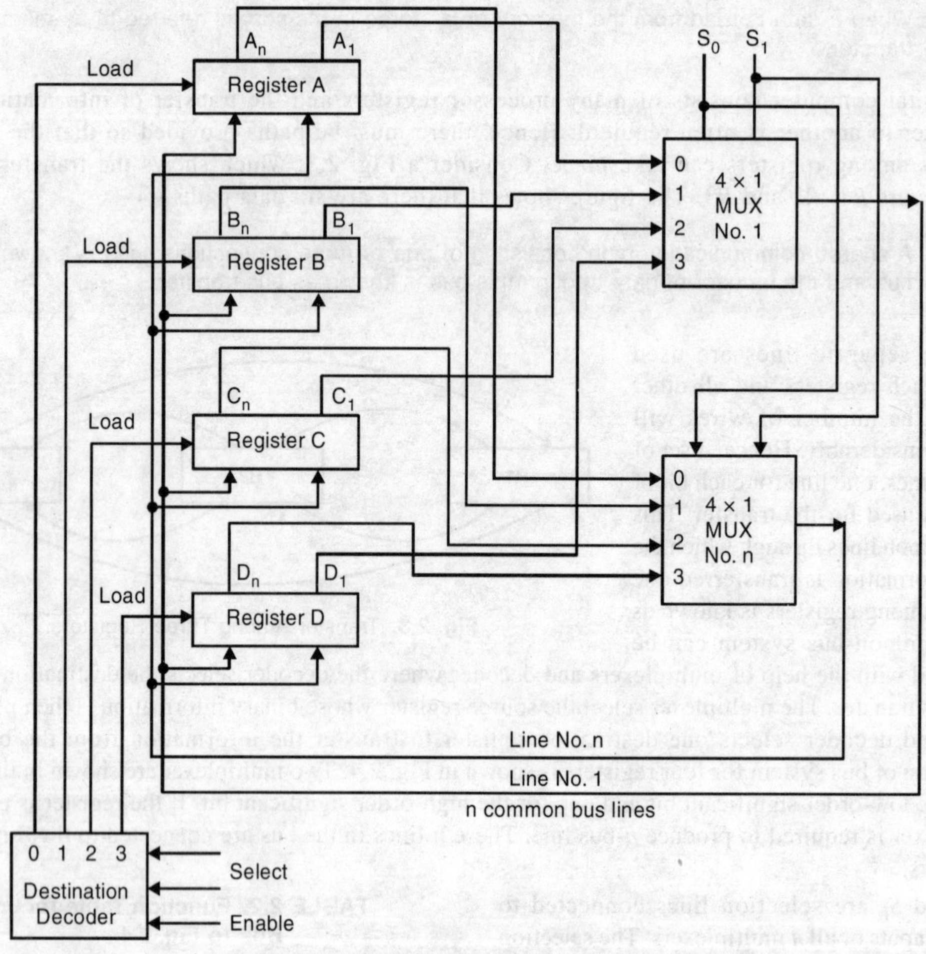

Fig. 2.4. Bus System for Four Registers.

Consider the statement

$$C \leftarrow B$$

The control function that enables this transfer must select register *B* as source and register *C* as destination register. The content of register *B* is placed on the bus and the content of bus is then transferred into the register *C* by activating its load control input.

Memory Transfer

When the information is transferred from memory word it is called a read operation and when the information is stored into a memory word it is called write operation. In both operation the memory word is specified by an address. This memory word is designated by the symbol *M*. Consider a memory unit that has single address register AR as shown in Fig. 2.5. MBR is the

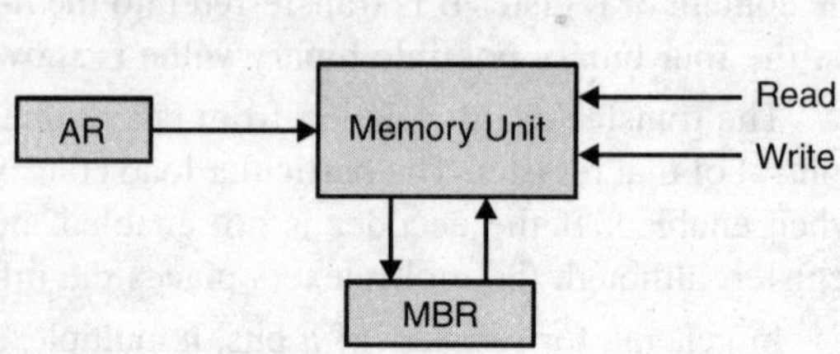

Fig. 2.5. Memory Unit that Communicates with the External Registers.

single memory buffer register used to transfer data into or out of the memory. Read and write are two memory operation.

Consider a memory unit that receives the address from the register AR, called address register and the data is to be transferred into register DR, called data register. The read operation can be

Read : $DR \leftarrow M[AR]$

The above statement causes, the information from the memory word M selected by the address in AR to be transferred into the destination register, DR.

Let the information is to be transferred from register $R1$ into the memory word M at location given by the address register AR. This operation can be stated as

Write : $M[AR] \leftarrow R1$

2.3. BUS ORGANIZATION

Definition A unidirectional bus used to carry memory addresses is called memory bus.

Definition The way in which different bus are connected to form common bus, so that CPU, memory and I/O devices can used common bus when required is called bus organization.

A basic computer consists of a memory unit, a control unit and registers. There must be a path that can be used to transfer information between memory and registers or among registers. A most efficient way of transferring information from source to destination in a system with many registers is to use a common bus. Fig. 2.6 shows the connection of eight registers and memory unit of 4096 × 16 to a common bus system. The eight registers are Address Register (AR), Program Counter (PC), Data Register (DR), Accumulator (AC), Instruction register (IR), Temporary Register (TR), Input Register (INPR) and Output Register (OUTR). The Fig. 2.6 consist of 16-bit common bus.

DO YOU KNOW

A bidirectional bus used to carry data between two units is data bus.

The outputs of seven registers and memory are connected to the common bus. The specific output that is selected for the bus lines at any given time is determined by the binary value of the selection lines S_2, S_1 and S_0 is shown in Table 2.3. The numbers along each output lines shows the decimal equivalent of the required binary selection.

TABLE 2.3. Function Table.

S_2	S_1	S_0	*Register*
0	0	1	AR
0	1	0	PC
0	1	1	DR
1	0	0	AC
1	0	1	IR
1	1	0	TR
1	1	1	Memory

For example, when $S_2S_1S_0 = 011$, the 16-bit outputs of DR are placed on the bus lines.

The lines from the common-bus are connected to the input of each registers and the data inputs of the memory. The particular register whose LD (Load) input is enabled receives the data from the bus. The memory receives the data from the bus when its write input is enabled.

The memory places its output onto the bus when the read input is activated and $S_2S_1S_0 = 111$.

The registers DR, AC, IR, TR are of 16-bits Two Register PC and AR have 12-bits since they store address.

When the contents of AR and PC are placed on the bus the four most-significant bits are set to 0's. When AR and PC receives the data from the bus only 12 least significant bits are transferred to the register.

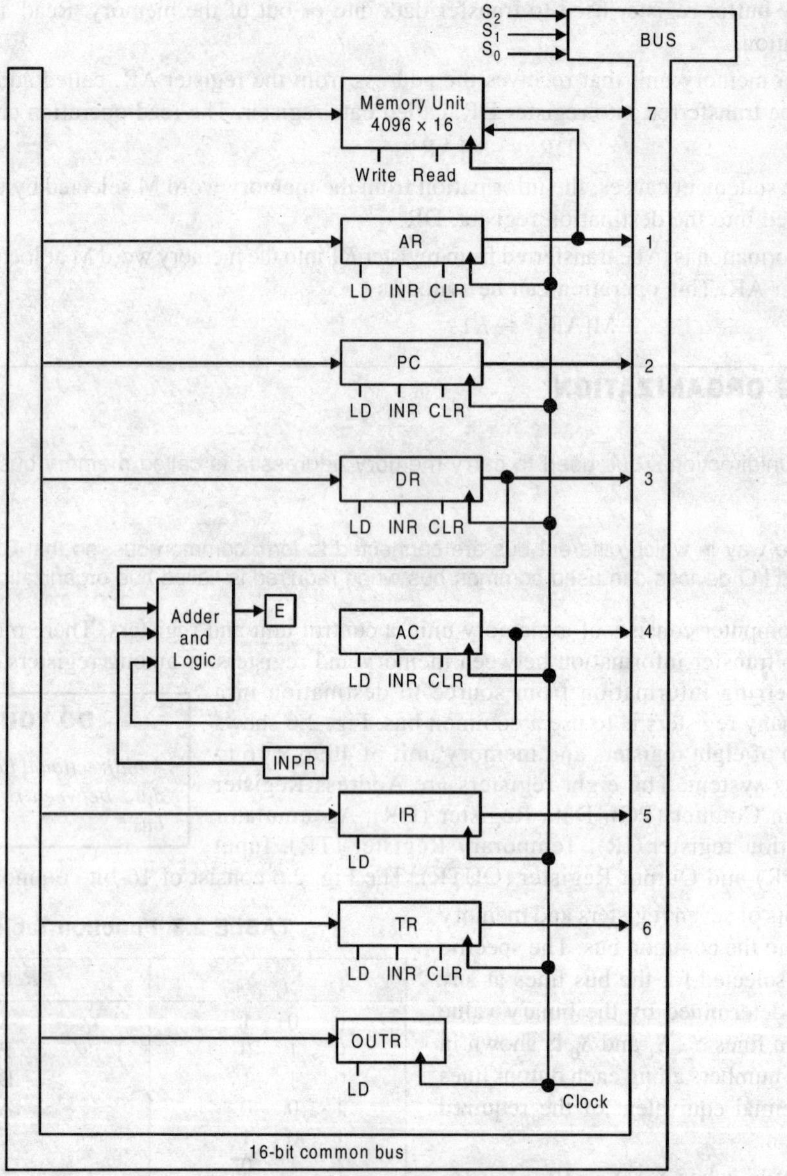

Fig. 2.6. Common Bus Organization.

The input register INPR and OUTR have 8 bits hence they communicate only with the 8 least significant bits in the bus. INPR is connected to bus to provide information but OUTR is connected to bus only to receive information from the bus. INPR receives a character from I/O device which is transferred to AC and OUTR receives a character from AC and delivers to an output device. No transfer is there from OUTR to any of the other registers.

The 16-bit common bus receives information from six registers and the memory unit. Also, the 16-bit common bus are connected to the inputs of six registers and the memory unit. The five registers have three control signals LD (Load), INR (increment) and CLR (clear). Two registers have only LD (load) control signal connected to the common bus. AR is also connected to the memory address. Thus, AR always specify the memory address. During a memory write operation the content of any register can be specified for the memory data and similarly during memory read operation any register except AC receives the memory data. The 16-bit AC receives inputs from adder and logic circuit which receives input from three registers. These three registers are 16-bit AC, 16 bit data register DR and 8-bit inputs come from input register INPR. The inputs from DR and AC are used for arithmetic and logic micro-operations. The function table shown in Table 2.3 shows the binary value of selection line $S_2S_1S_0$ that selects one of the register.

For example, in order to transfer contents of PC to Address Register, the computer requires the following instructions :

(1) Set the selection variables $S_2S_1S_0 = 010$

(2) Transfer the contents of PC to the bus

(3) Enable LD input of AR

(4) Transfer contents of bus into AR

2.4. MULTIPLE BUS ORGANIZATION

A two-bus structure used to connect the registers and the ALU of a processor is shown in Fig. 2.7. All general-purpose registers are connected to both buses *A* and *B* to form a two-bus organization. The two operand required by the ALU are routed in one clock cycle, hence instruction execution becomes faster since the ALU does not wait for the second operand as with single-bus organization. The information passed on to the bus are may be from general-purpose registers or from special purpose register. Also, the special purpose registers are divided into two groups, one group at the left of ALU and are connected to bus *A* and other group is at the right of ALU and is connected to bus *B*. The data from two special purpose registers belong to the same group cannot be transferred to the ALU at the same time.

The output of the ALU may be routed to either general purpose registers or special purpose registers. The ALU does not have any input buffer register and hence both buses will be busy in carrying the operands during binary operations. Therefore, the output of the ALU is first stored into the output register. Transfer of required operands and loading of the ALU output buffer register takes in one clock cycle. The content of ALU output register is routed to the destination with the help of either bus *A* and bus *B* in the second clock cycle.

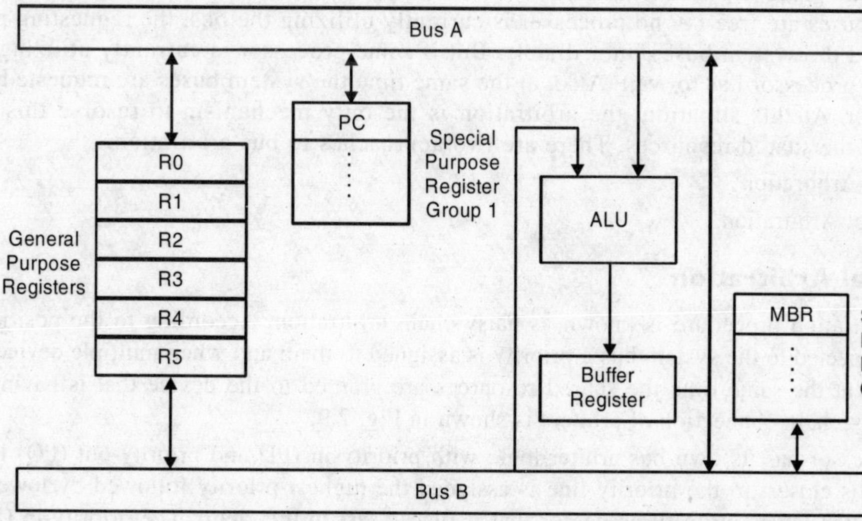

Fig. 2.7. Two-bus Organization of the Data Path.

The performance of two bus organization can be further improved by adding a third bus *C* at the output of ALU. The three-bus structure is shown in Fig. 2.8. The addition of third bus allows the system to perform the operations, such as $R3 \leftarrow R1 + R2$ in one clock cycle as there are three separate buses in the system.

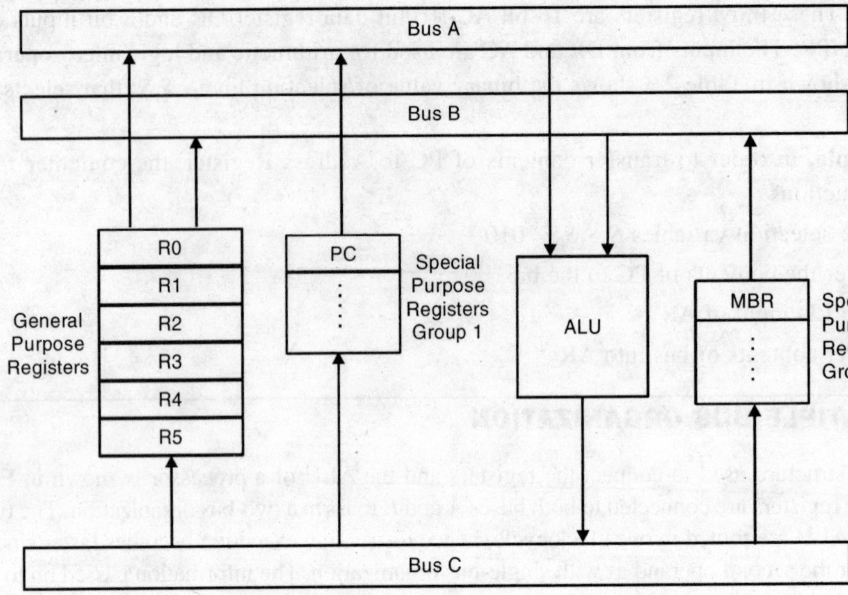

Fig. 2.8. Three-bus Organization of the Data Path.

2.5. BUS ARBITRATION

Definition Bus arbitration is defined as the process by which the requesting processor may be granted the access to the bus.

A computer system consists of several buses to facilitate the transfer of information between the components. The processors access the memory or other common resources through the system bus. If the system buses are free *i.e.* no processor is currently utilizing the bus, the requesting processor may be granted the system buses immediately. But if some processor is currently utilizing the bus, the requesting processor has to wait. Also, at the same time the system buses are requested by some other processor. At this situation, the arbitration is the only mechanism to resolve this multiple contention for the shared resources. There are two approaches to bus arbitration.

(*i*) Serial Arbitration

(*ii*) Parallel Arbitration.

2.5.1. Serial Arbitration

Serial arbitration procedure is known as daisy-chain arbitration. According to the position of the processors connected to the system bus a priority is assigned to them and when multiple devices request the use of bus at the same time the shared resources are granted to the device that is having highest priority A daisy-chain connection of arbiters is shown in Fig. 2.9.

Each processor has its own bus arbiter-logic with priority-in (PI) and priority-out (PO) lines. The processor that is closest to the priority line is assigned the highest priority followed by lower-priority processor upto the lowest priority processor that is placed last in the chain. The priority-in (PI) of the processor having the highest priority is maintained at logic 1 and the priority-out (PO) of each arbiter is connected to the PI of the next-lower priority arbiter and hence forming a chain. Whenever the highest-priority unit requests for the system bus, the unit will always receive access to the bus, if the system buses are free. Otherwise the highest-priority unit has to wait.

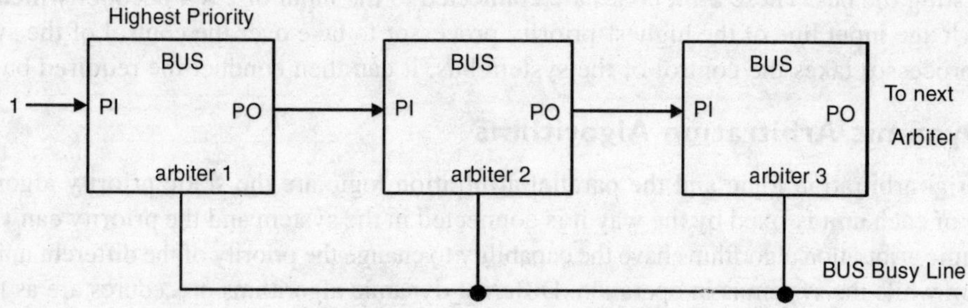

Fig. 2.9. Daisy-Chain Arbitration.

The PO output of any unit will be 1 only if its PI input is equal to 1 and the processor is not requesting for the system bus. If the processor wants to utilizes the system bus and requests for the control of bus the PI input of the corresponding arbiter must be 1 and the PO output is equal to 0. Thus, the control of system bus is with the processor whose PI input is 1 and PO output is 0.

The situation may arises when the control of the system bus is with some processor and in the mean time some higher-priority processor requests for the system bus. In this situation, the lower-priority processor must complete its bus operation, before giving the control of the bus. This situation is tested with the help of the bus busy line which is common to all arbiter as shown in Fig. 2.9. When the PI input of an arbiter is 1 and its PO output is 0, the arbiter examines the bus busy line. If the bus busy line is inactive, means no processor is using the bus, the arbiter activates the bus busy line and the control of system bus is granted to the corresponding processor. But, when some processor is using the bus, the bus busy line will be active. Then arbiter has to wait and takes the control of bus only when the bus-busy line becomes inactive.

2.5.2. Parallel Arbitration

The Fig. 2.10 shows the parallel arbitration. This technique uses an external priority encoder and decoder. Each of the bus arbiter has a bus request output line and bus acknowledge input line.

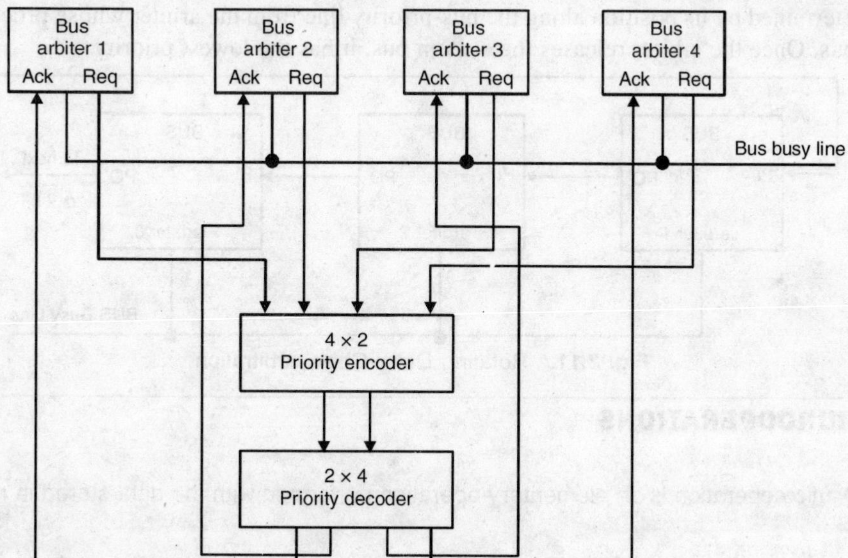

Fig. 2.10. Parallel Arbitration.

The input to the priority encoder is the bus request output lines from each arbiter and the output of the encoder generates 2-bit codes. These 2-bit codes represents the highest priority processor among these requesting the bus. These 2-bit codes are connected to the input of 2 × 4 decoder which enables the bus acknowledge input line of the highest-priority processor to take over the control of the system buses. When the processor takes the control of the system bus, it can then conduct the required bus transfer.

2.5.3. Dynamic Arbitration Algorithms

The serial arbitration logic and the parallel arbitration logic are the static priority algorithms when the priority of each unit is fixed by the way it is connected in the system and the priority can't be changed. But, a dynamic aribtration algorithms have the capability to change the priority of the different units associated in the system while the system is in operation. Different dynamic algorithms procedures are as follows :

Time-Slice

In this algorithm, the time-slice of the bus time is of fixed-length and allocates sequentially to each processor, in round-robin manner irrespective of the location of the unit in the system. Each unit is given the same amount of the time to communicated with the system bus and no preference is given to any unit.

LRU

In least recently used (LRU) algorithm, the device that has not used the system bus for longer period has been given the highest priority. No processor is given the preference as the priority are changed dynamically so that each processor can access the system bus.

FIFO

In first-in first-out algorithm, a queue is established in the same order as the request are received from the processor. Each processor has to wait in the queue for its turn to use the system bus on first-in first-out manner.

Rotating Daisy-Chain

Rotating daisy-chain is the dynamic version of the static daisy-chain algorithm. The Fig. 2.11 shows the rotating daisy-chain arbitration for three arbiters. In this algorithm, the priority line is connected to the priority out (PO) of the last processor which is connected to the priority-in of the first processor forming a loop. The processor that has the access to the system bus acts as bus controller. Each arbiter priority is determined by its position along the bus-priority line from the arbiter whose processor is using the system bus. Once the arbiter releases the system bus, it has the lowest priority.

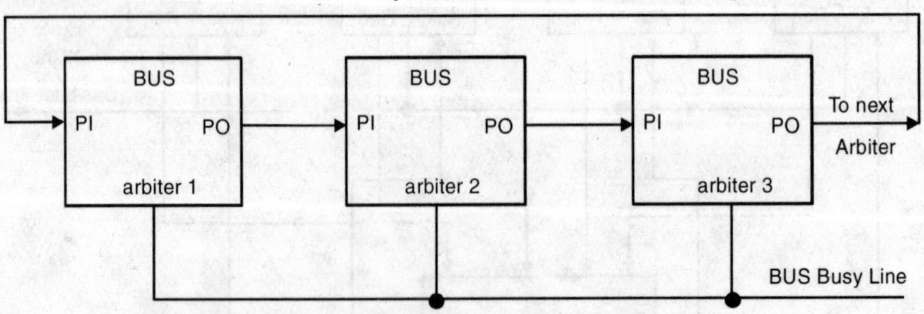

Fig. 2.11. Rotating Daisy-Chain Arbitration.

2.6. MICROOPERATIONS

Definition A microoperation is an elementary operation performed with the data stored in registers.

The operations executed on data stored in registers are called microoperations. A microoperation is an elementary operation performed on the information stored in one or more registers. The result of the operation may replace the previous binary information of a register or may be transferred to some another registers. Examples are clear, shift, count, load etc. A bidirectional shift register is capable of performing the shift right and shift left microoperations.

The microoperations most often encountered in digital computers are classified into following categories.

(1) Register transfer microoperations transfer binary information from one register to another.

(2) Arithmetic microoperations perform arithmetic operations on numeric data stored in registers.

(3) Logic microoperations perform bit manipulation operations on non numeric data stored in registers.

(4) Shift microoperations perform shift operations on data stored in registers.

The register transfer microoperation does not change the information content when the binary information moves from source register to destination register.

The arithmetic microoperation, logic microoperation and shift microoperation changes the information content during the operations.

2.6.1. Arithmetic Microoperations

The basic arithmetic microoperations are addition, subtraction, increment, decrement and shift.

Let the arithmetic microoperations defined by the statement

$$R3 \longleftarrow R1 + R2$$

specifies an add microoperation. It states that add the content of register $R1$ and that of register $R2$ and stored the result in register $R3$. Thus, to implement this statement with hardware we need three registers and the digital component that performs the addition operation. Other basic arithmetic microoperations are listed in the Table 2.4.

Definition Arithmetic microoperations are those microoperations that are used to perform arithmetic operations.

TABLE 2.4. Arithmetic Microoperations.

Symbolic notation	Description
$R3 \leftarrow R1 + R2$	Contents of $R1$ plus $R2$ transferred to $R3$
$R3 \leftarrow R1 - R2$	Contents of $R1$ minus $R2$ transferred to $R3$
$R2 \leftarrow \overline{R}2$	Complements the contents of $R2$ (1's complement)
$R2 \leftarrow \overline{R}2 + 1$	2's complement the contents of $R2$
$R3 \leftarrow R1 + \overline{R}2 + 1$	$R1$ plus the 2's complement of $R2$ (subtraction)
$R1 \leftarrow R1 + 1$	Increment the content of $R1$ by one
$R1 \leftarrow R1 - 1$	Decrement the content of $R1$ by one.

Subtraction is basically implemented through complementation and addition and can be specified by the statement

$$R \leftarrow A + \overline{B} + 1$$

where $\overline{B}$ is the 1's complement of B. When we add 1 to the 1's complement it will give 2's complement of B and adding A to the 2's complement of B gives A minus B *i.e.*, $(A - B)$. The increment and decrement microoperation are implemented with the help of combinational circuit or with a binary up-down counter. They are symbolized by plus-one or minus-one operation executed on the contents of a register. The multiplication operation and division are valid arithmetic operations. Multiplication operation is basically implemented with a sequence of add and shift microoperations and division is implemented with a sequence of subtract and shift microoperations.

1. Binary Adder

The digital circuits that generates the arithmetic addition of two binary numbers of any length is called *binary adder*. Also, the digital circuit that performs the addition of two binary bits and a previous carry is called a *full-adder*.

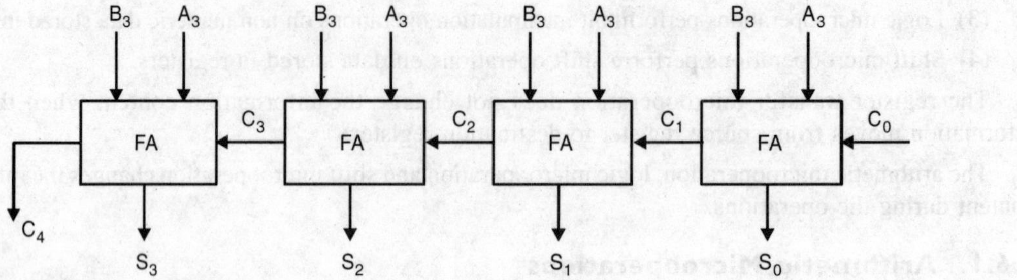

Fig. 2.12. 4-bit Binary Adder.

The binary adder circuit is constructed using full-adder circuits with the output carry from one full-adder is connected to the input carry of the next full adder. An *n*-bit binary adder requires *n* full-adders, where *n* data bits for the *A* inputs came from one register (say *R*1) and the *n* data bits for the *B* inputs came from another register (say *R*2) and the result can be stored into any of these two registers (*R*1 or *R*2) replacing its previous content or it can be stored into third register.

Figure 2.12 shows how a 4-bit binary adder can be constructed with the help of full-adders (FA).

C_0 is the input carry to the circuit and the output carry is C_4. The *S* outputs of the circuit generate the required sum bits.

2. Binary Adder-Subtractor

The subtraction of binary numbers can be done most conveniently by means of complements. The subtraction $A - B$ can be done by taking the 2's complement of *B* and adding it to *A*.

The addition and subtraction operations can be combined into one common circuit by including an exclusive-OR gate with each full adder. A 4-bit adder-subtractor circuit is shown in Fig. 2.13.

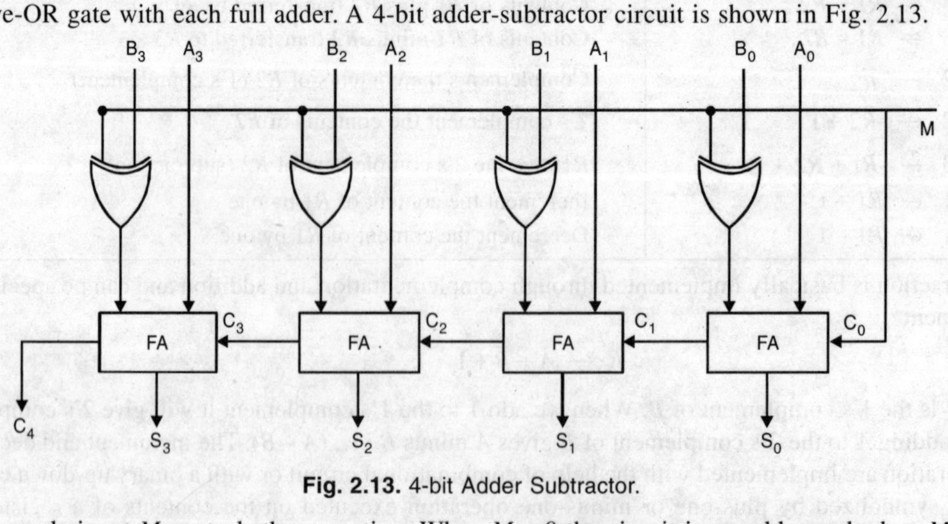

Fig. 2.13. 4-bit Adder Subtractor.

The mode input *M* controls the operation. When $M = 0$ the circuit is an adder and when $M = 1$ the circuit becomes subtractor. Each exclusive-OR gate receives input *M* and one of the inputs of *B*.

When $M = 0$, we have $B \oplus 0 = B$. The full-adders receive the value of B, the input carry is 0, and the circuit performs A plus B.

When $M = 1$, we have $B \oplus 1 = B'$ and $C_0 = 1$. The B inputs are all complemented and a 1 is added through the input carry. The circuit performs the operation A plus the 2's complement of B.

For unsigned numbers, the circuit gives $A - B$ if $A \geq B$ or the 2's complement of $(B - A)$ if $A < B$. For signed numbers, the result is $A - B$, if there is no overflow.

3. Arithmetic Circuit

The basic arithmetic microoperation listed in Table 2.4 can be implemented by means of a single arithmetic circuit where the basic component of the circuit is parallel adder. The Fig. 2.14 shows the logic circuit of 4-bit arithmetic circuit.

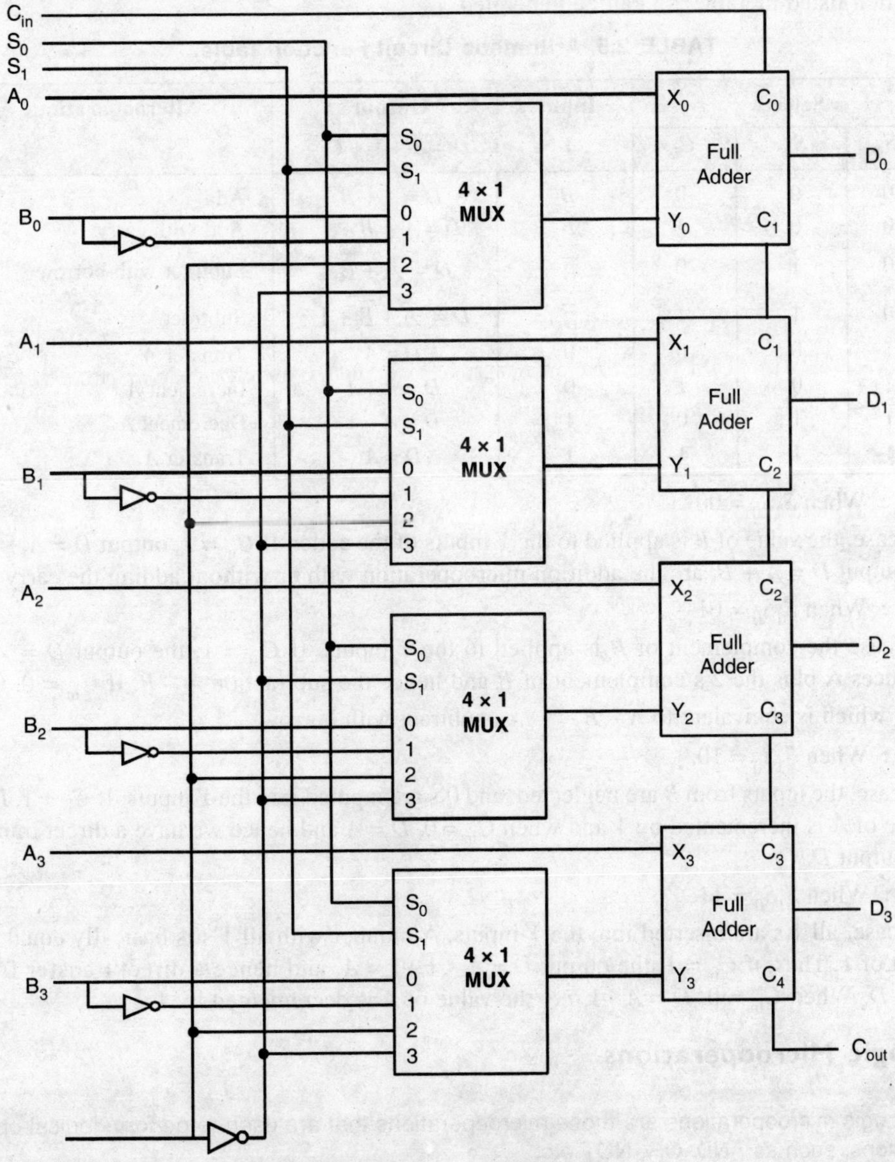

Fig. 2.14. 4-bit Arithmetic Circuit.

The arithmetic circuit consists of four full adder that makes the 4-bit adder and four multiplexers for choosing different operations. A and B are the two 4-bit inputs and D is the output. S_1 and S_0 are two selection lines. C_{in} is the input carry connected to the carry input of the full adder in the least significant position and other carries are connected to the next full adder with C_{out} the output carry. The four inputs from A goes directly to the X inputs of the binary adder and four inputs from B are connected to the data inputs of the multiplexers which also receives the complement of B. Other two data inputs of multiplexers are connected to logic 0 and logic 1 respectively.

The output of the circuit is calculated from the following arithmetic sum :

$$D = A + Y + C_{in}$$

where A is the 4-bit input at the X inputs and Y is the 4-bit binary number at the Y input of the circuit. With the help of selection lines S_1 and S_0 and by making input carry C_{in} equal to 0 or 1, the arithmetic microoperation listed in Table 2.5 can be generated.

TABLE 2.5. Arithmetic Circuit Function Table.

Select			Input	Output	Microoperation
S_1	S_0	C_{in}	Y	$D = X + Y + C_{in}$	
0	0	0	B	$D = A + B$	Add
0	0	1	B	$D = A + B + 1$	Add with carry
0	1	0	$\overline{B}$	$D = A + \overline{B}$	Subtract with borrow
0	1	1	$\overline{B}$	$D = A + \overline{B} + 1$	Subtract
1	0	0	0	$D = A$	Transfer A
1	0	1	0	$D = A + 1$	Increment A
1	1	0	1	$D = A - 1$	Decrement A
1	1	1	1	$D = A$	Transfer A

Case 1 : When $S_1S_0 = 00$

In this case, the value of B is applied to the Y inputs of the adder. If $C_{in} = 1$, output $D = A + B + 1$ and if $C_{in} = 0$, output $D = A + B$, are the addition microoperation with or without adding the carry input C_{in}.

Case 2 : When $S_1S_0 = 01$

In this case the complement of B is applied to the Y inputs. If $C_{in} = 1$, the output $D = A + \overline{B} + 1$, which produces A plus the 2's complement of B and hence the subtraction $A - B$. If $C_{in} = 0$, the output $D = A + \overline{B}$, which is equivalent to $A - B - 1$ *i.e.,* subtract with borrow.

Case 3 : When $S_1S_0 = 10$.

In this case, the inputs from B are neglected, and 0's are inputed into the Y inputs. If $C_{in} = 1$, $D = A + 1$, *i.e.* the value of A is incremented by 1 and when $C_{in} = 0$, $D = A$ and hence we have a direct transfer from input A to output D.

Case 4 : When $S_1S_0 = 11$.

In this case, all 1's are inserted into the Y inputs. A number with all 1's is basically equal to the 2's complement of 1. Thus, if $C_{in} = 1$, the output $D = A - 1 + 1 = A$, and hence a direct transfer from input A to output D. When $C_{in} = 0$, $D = A - 1$, *i.e.* the value of A is decremented by 1.

2.6.2. Logic Microoperations

Definition Logic microoperations are those microoperations that are used to perform logical operations, such as AND, OR, NOT etc.

The microoperations which perform operation such as AND, OR, NOT, XOR etc. on individual pairs of bits stored in registers are said to be logic microoperations. These operations consider each bit of the register separately and treat it as a binary variable. The XOR (exclusive OR) operation between the content of two register $R1$ and $R2$ can be represented as

$P : R1 \leftarrow R1 \oplus R2.$

The symbol $\oplus$ represents the XOR operation. Also, special symbols are used for logic microoperations OR, AND and complement operations. The symbol $\vee$ is used to denote an OR microoperation, and symbol $\wedge$ is used to denote AND microoperation.

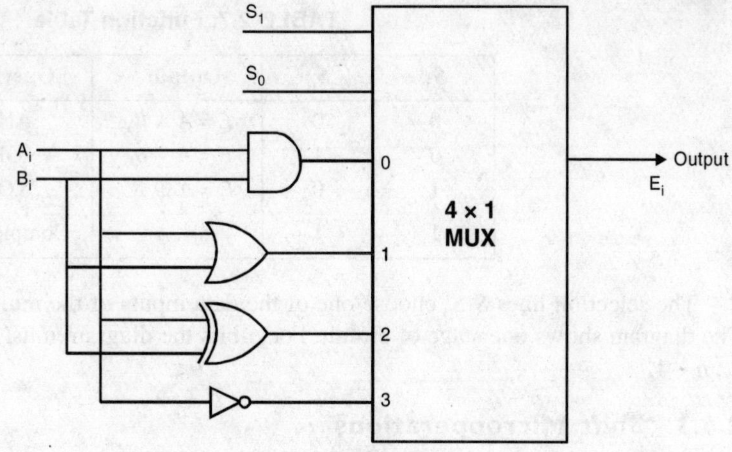

Fig. 2.15. One Stage of Logic Circuit.

There are 16 different logic microoperations that can be performed with two binary variables and are listed below in the Table 2.6.

Although there are 16 logic microoperations most of the computers uses only four— AND, OR, XOR, and complement from which all others can be derived. The Fig. 2.15 depicts the hardware implementation of four basic logic microoperations.

The output of the four gates (AND, OR, XOR, NOT) are applied to the data input of the 4 × 1 multiplexer which will give output according to the function Table 2.7.

TABLE 2.6. Sixteen Logic Microoperations.

Boolean Function	Logic Microoperation	Meaning
$F_0 = 0$	$F \leftarrow 0$	Clear
$F_1 = xy$	$F \leftarrow A \wedge B$	AND
$F_2 = xy'$	$F \leftarrow A \wedge \overline{B}$	
$F_3 = x$	$F \leftarrow A$	Transfer *A*
$F_4 = x'y$	$F \leftarrow \overline{A} \wedge B$	
$F_5 = y$	$F \leftarrow B$	Transfer *B*
$F_6 = x + y$	$F \leftarrow A \vee B$	OR
$F_7 = x + y'$	$F \leftarrow \overline{A} \vee B$	
$F_8 = x' + y$	$F \leftarrow \overline{A} \wedge B$	
$F_9 = (xy)'$	$F \leftarrow \overline{A \wedge B}$	NAND
$F_{10} = (x + y)'$	$F \leftarrow \overline{A \vee B}$	NOR
$F_{11} = x'$	$F \leftarrow \overline{A}$	Complement A
$F_{12} = y'$	$F \leftarrow \overline{B}$	Complement *B*
$F_{13} = x \oplus y$	$F \leftarrow A \oplus B$	Exclusive-OR
$F_{14} = (x \oplus y)'$	$F \leftarrow \overline{A \oplus B}$	Exclusive-NOR
$F_{15} = 1$	$F \leftarrow$ all 1's	Set to all 1's

TABLE 2.7. Function Table.

S_1	S_0	Output	Operator
0	0	$E = A \wedge B$	AND
0	1	$E = A \vee B$	OR
1	0	$E = A \oplus B$	XOR
1	1	$E = \bar{A}$	Complement

The selection lines S_1S_0 choose one of the data inputs of the multiplexer and directs it to the output. The diagram shows one stage of circuit. For n-bits the diagram must be repeated n times, for $i = 0, 1, 2, ..., n - 1$.

2.6.3. Shift Microoperations

Shift microoperations are used for serial transfer of data. The shifting of bits of register can be in both direction left or right, during which the first flip-flop receives its binary information from the serial input. When the operation is shift-left, the serial input transfers a bit into the right most position and when the operation is shift-right the bit from the serial input is transferred into the left most position.

Shift microoperation can be classified as :

(*a*) Logical, (*b*) Circular, (*c*) Arithmetic.

Definition Shift microoperations are those microoperations that are used to perform shifting of bits in either left or right direction.

1. Logical Shift

It is the easiest to perform as the bits are simply shifted to the right or to the left. If it is shift-left, a 0 bit is transferred at the right most position through serial input and the leftmost bit is lost. Similarly, if it is shift-right the right most bit is lost and a 0 bit is added at the leftmost place.

For example, $R1 \leftarrow \text{shl } R1$

$R2 \leftarrow \text{shr } R2.$

shl and shr are the symbols used for shift-left and shift-right microoperations. The above two statements specifies a 1-bit shift to the left of the content of register $R1$ and a 1-bit shift to the right of the content of register $R2$ respectively.

2. Circular Shift

Also known as rotate operation, simply circulates the bits of the register around the two ends. No bit is added or lost. If it is left circular shift, the bits are shifted in left direction and the rightmost bit takes the left most bit position. Similarly, if it is right circular shift, the bits are shift in right direction and the leftmost bit takes the rightmost bit position.

3. Arithmetic Shift

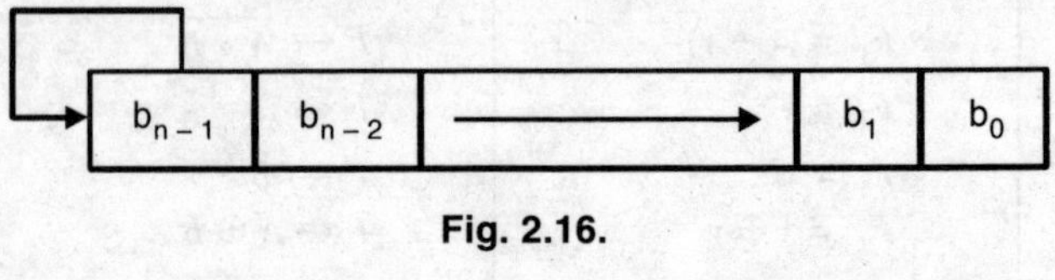

Fig. 2.16.

An arithmetic shift is a microoperation that shifts a signed binary number to the left or right. In arithmetic left or arithmetic right shift operations, we need to consider the sign bit also, because arithmetic shift requires that the sign bit (the leftmost bit) remains unchanged. 0 is the sign bit for positive and 1 is sign bit for negative numbers.

The Fig. 2.16 shows a register of *n*-bits, where bit b_{n-1} in the leftmost position contains the sign bit of a number. b_{n-2} is the most significant bit and b_0 is the least significant bit of number.

In case of arithmetic shift-right (ashr) the sign bit remains unchanged and shifts the number, including the sign bit, to the right. The bit b_0 is lost. Thus, b_{n-1} remains unchanged, b_{n-2} receives b_{n-1} and so on.

In case of arithmetic shift-left (ashl), 0 is inserted into bit position b_0 and shifts all other bits to the left. The initial bit b_{n-1} is lost and replaced by bit b_{n-2}. If the bit in b_{n-1} changes the value after the shift a sign reversal occurs.

TABLE 2.8. Function Table.

Select	Output			
S	O_0	O_1	O_2	O_3
0	S_R	I_0	I_1	I_2
1	I_1	I_2	I_3	S_L

4. 4-bit Shifter

The 4-bit combinational circuit called shifter constructed with the help of multiplexers is explained in Fig. 2.17. It consists of four data inputs I_0 to I_3 and four data output O_0 to O_3. The two serial inputs, one for shift left and other for shift right are S_L and S_R respectively. The function table for the circuit is given in Table 2.8. When the selection input $S = 0$, the input data are shifted right and when $S = 1$, the input data are shifted left. The function table explains which input goes to each output after the shift operation. In general, a shifter having n data inputs and n data outputs requires n multiplexers.

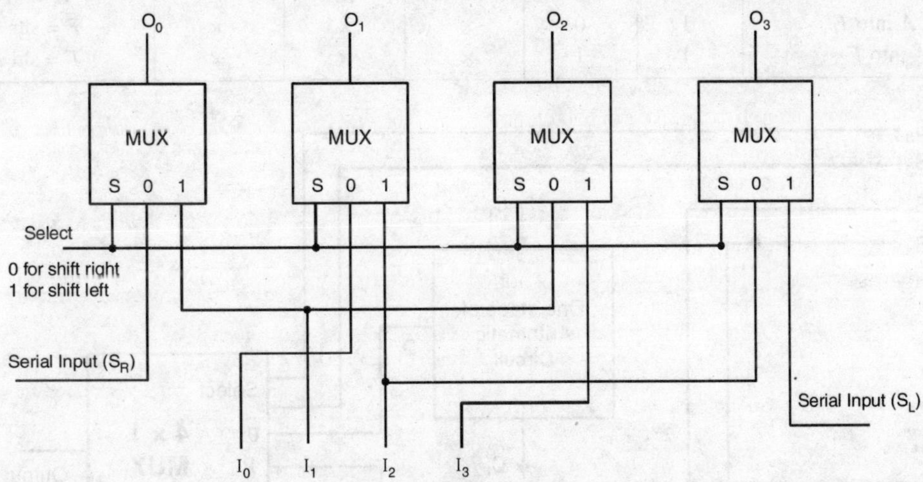

Fig. 2.17. 4-bit Shifter.

2.7. ARITHMETIC LOGIC SHIFT UNIT

In the earlier sections, we discussed the logic circuits for arithmetic operations and logic operations. These circuits can be combined to form one ALU with common selection variables. Figure 2.18 shows one stage of a complete arithmetic logic shift unit. The arithmetic, logic and shift operations are performed on the contents of two registers A and B. A_i stands for the ith bit of register A and B_i stands for the ith bit of register B. Inputs A_i and B_i are applied to both the arithmetic and shift units.

S_1 and S_0 selects a particular microoperation. A 4 × 1 multiplexer choose between the arithmetic output D_i and logic output E_i. The data input of the multiplexer are selected by the selection lines S_3 and S_2. A_{i-1} and A_{i+1} are the two data inputs of multiplexer for shift-right operation and shift-left operation, respectively. The output carry C_{i+1} is connected to the input carry C_i of the next stage in sequence.

Table 2.9 lists eight arithmetic operation, four logic operation and two shift operations. S_3, S_2, S_1, S_0 and C_{in} selects a particular operaticn. C_{in} is used to select the arithmetic operation only.

TABLE 2.9. Function Table for Arithmetic Logic Shift Unit.

Function	Operation select					Operation
	S_3	S_2	S_1	S_0	C_{in}	
Transfer *A*	0	0	0	0	0	$F = A$
Increment *A*	0	0	0	0	1	$F = A + 1$
Addition	0	0	0	1	0	$F = A + B$
Add with carry	0	0	0	1	1	$F = A + B + 1$
Subtract with borrow	0	0	1	0	0	$F = A + \overline{B}$
Subtraction	0	0	1	0	1	$F = A + \overline{B} + 1$
Decrement *A*	0	0	1	1	0	$F = A - 1$
Transfer *A*	0	0	1	1	1	$F = A$
AND	0	1	0	0	×	$F = A \wedge B$
OR	0	1	0	1	×	$F = A \vee B$
XOR	0	1	1	0	×	$F = A \oplus B$
Complement *A*	0	1	1	1	×	$F = \overline{A}$
Shift right *A* into *F*	1	0	×	×	×	$F = \text{shr } A$
Shift left *A* into *F*	1	1	×	×	×	$F = \text{shl } A$

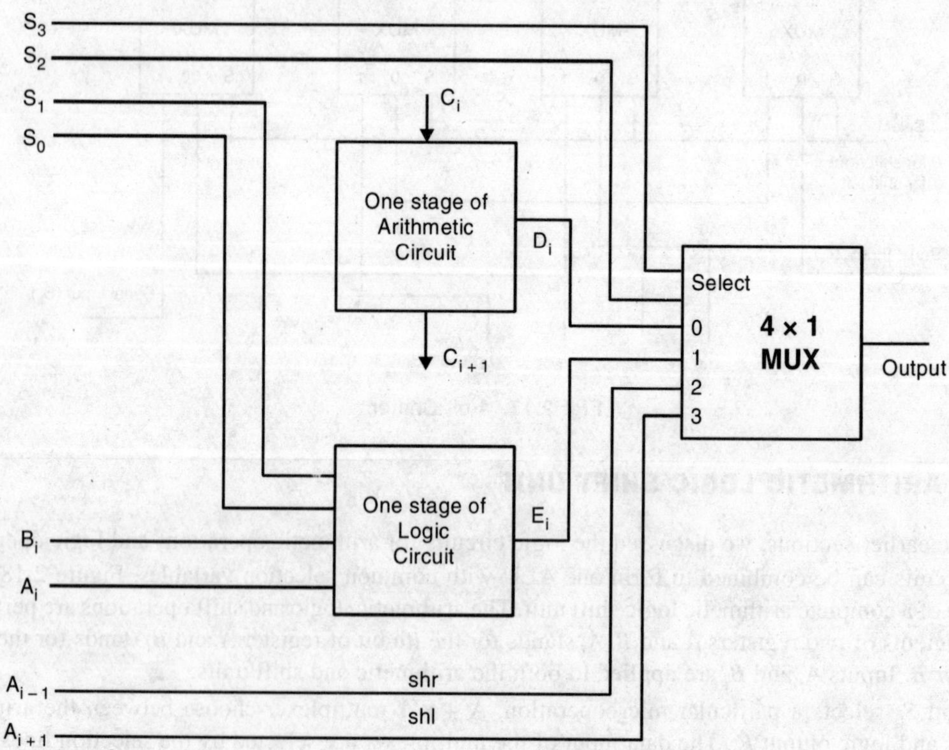

Fig. 2.18. Arithmetic Logic Shift Unit.

2.8. INSTRUCTION FORMAT

An instruction is a binary pattern designed inside a microprocessor to perform a specific function. The entire group of instructions, is called the instruction set. The instruction set determines what functions the microprocessor can perform.

➡ Each instruction basically consists of 2 parts. The 1st is called as the OPCODE and the 2nd part is called as the OPERAND.

➡ The OPCODE part of an instruction specifies the operation to be performed. The operands either provides the data or specific data. The operand can be specified in a number of ways. It includes :

1. 8-bit/16-bit internal general purpose register.
2. 8-bit/16-bit port address
3. 8-bit/16-bit memory address
4. A memory location.
5. Implicit operand : The operand is not specified, instead it is assumed in register.

Depending on the number of bytes or bit patterns required to specify on operation and the data, the instructions are of 3-type :

1. 1-byte instruction
2. 2-byte instruction
3. 3-byte instruction

1-Byte Instruction : A 1-byte instruction includes the opcode and the operand in the 8-bit only i.e. one byte. The 1-byte instruction specifies the operation to be performed and who is going to perform it. These type of instruction require 1 (single) memory location.

Format :

0   7
OPCODE

For example : MOV, B, C, whose opcode is 41 H which is 1-byte. This instruction copies the contents of register C in B.

2-Byte Instructions : The 2-byte instruction uses 1st byte to specify the operation i.e. opcode and 2nd byte to specify the operand. These instructions require 2-successive locations in the memory.

The 1st byte stored is the opcode and the 2nd byte stored is data *i.e.,* operand.

Format :

0 7	15
OPCODE	OPERAND

For example : MVI B, 57H

The 1st byte of this instruction is the opcode for instruction MVIB and second byte is the data that is to be moved to register B.

3-Byte Instructions : The 3-byte instruction use 1st byte to specify the operation *i.e.,* opcode, 2nd and 3rd bytes are used to specify the operand. Generally these instructions are used to specify 16-bit data or memory address with the instruction.

1-byte stored is opcode, 2nd byte stored is lower order 8-bits of 16-bits operand or address and 3rd-byte stored is higher order 8-bits of 16-bits operand or address.

Format :

0 7	15	23
OPCODE	OPERAND	OPERAND

For example : STA 5600H

The 1st byte of this instruction is the opcode for instruction STA 5600H and the second and 3rd bytes are the address 5600H.

OPCODE Formats : The 8085 microprocessor is on 8-bit microprocessor has 8-bit opcodes. Each instruction has a unique opcode. The opcode contains information regarding the operation, type of operation to be performed, registers to be used, flags. The opcode is fixed for each instruction.

Notations used in object code or OPCODES are as shown in the table below :

Notations	Meaning
ddd	Destination Register (S)
	Source Registers
	ddd = sss = 111 = A Register
	000 = B Register
	001 = C Register
	010 = D Register
	011 = E Register
	100 = H Register
	101 = L Register
nnn	Restart number 000 to 111
yy	An 8-bit binary data
yyyy	A 16-bit binary data unit
X	Register pair 0 = BC
	1 = DE
XX	Register pair 00 = BC
	01 = DE
	10 = HL
	11 = SP (IF Push/Pop) PSW
PPQQ	A, 16-bit memory address

2.9. INSTRUCTION SETS

Instruction set architecture is the structure of a computer that a machine language programmer must understand to write a correct (timing independent) program for that machine.

The instruction set architecture is also the machine description that a hardware designer must understand to design a correct implementation of the computer.

TIPS

The instruction set architecture serves as the interface between software and hardware.

Evolution of Instruction Sets

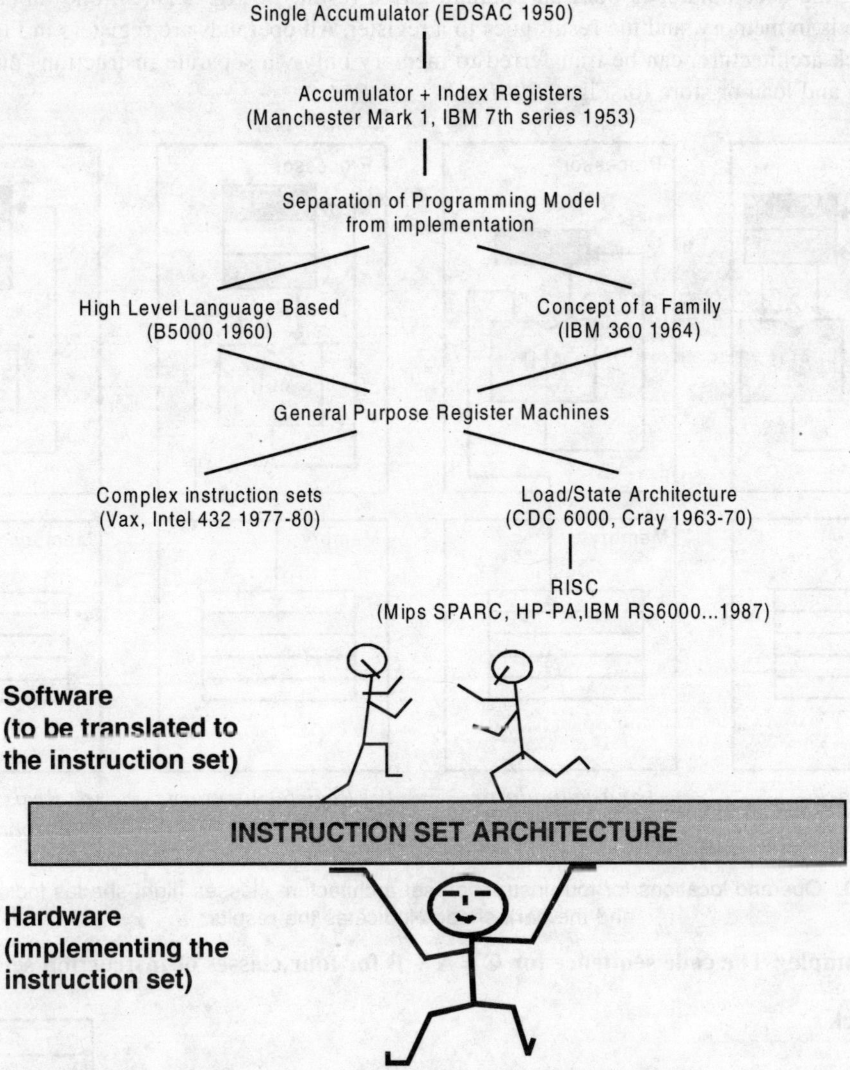

Fig. 2.19. ISA (Instruction Set Architecture)

2.10. CLASSIFYING INSTRUCTION SET ARCHITECTURE

The type of internal storage in a processor is the most basic differentiation. The major choices are :

(1) a stack

(2) an accumulator

(3) a set of registers

Operands may be named explicitly or implicitly.

The operands in a stack architecture are implicitly on the top of the stack and in an accumulator architecture one operand is implicitly the accumulator. The general-purpose register architectures have only explicit operands—either register or memory locations. In Fig. 2.20(*a*), a top of stack register (TOS), points to the top input operand, which is combined with the operand below. The first operand is removed from the stack, the result takes the place of the second operand, and TOS is updated to implicit. In Fig. 2.20(*b*), the accumulator is both an operand and a result. In Fig. 2.20(*c*), one input operand is a register, one is in memory, and the result goes to a register. All operands are registers in Fig. 2.20(*d*) and like the stack architecture, can be transferred to memory only via separate instructions-push or pop for Fig. 2.20(*a*) and load or store for Fig. 2.20(*d*).

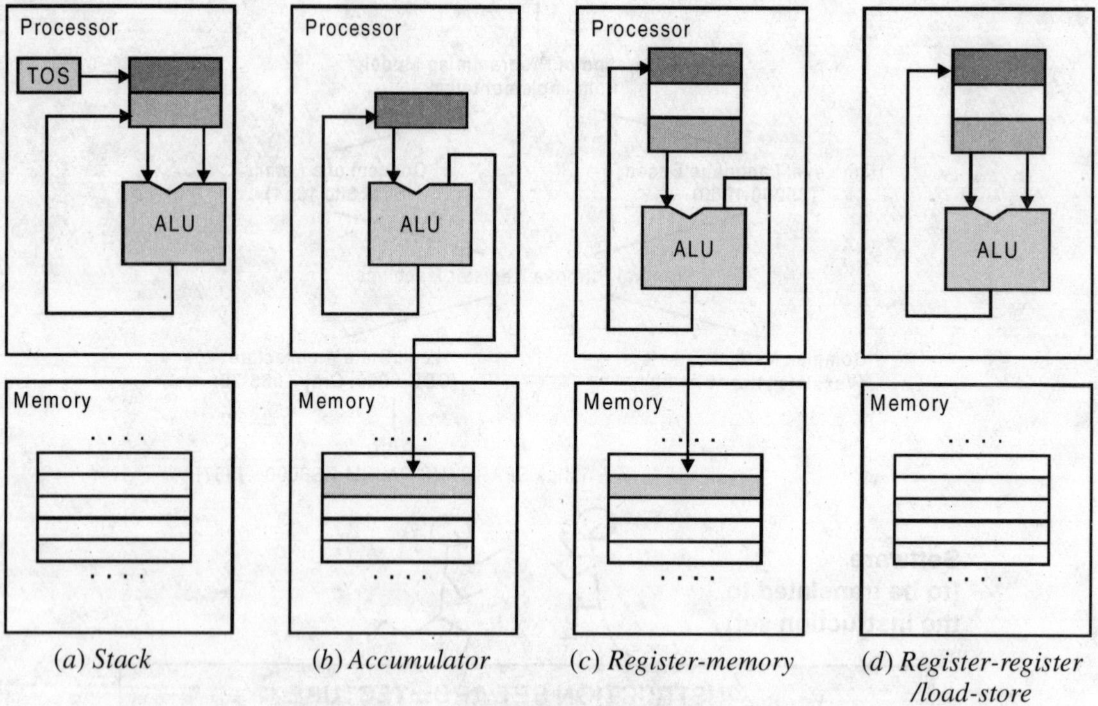

Fig. 2.20. Operand locations for four instruction set architecture classes (light shades indicate inputs and the dark shades indicates the results.

For example : The code sequence for C = A + B for four classes of instruction sets.

(*a*) **Stack**

```
Push A

Push B

Add

Pop C
```

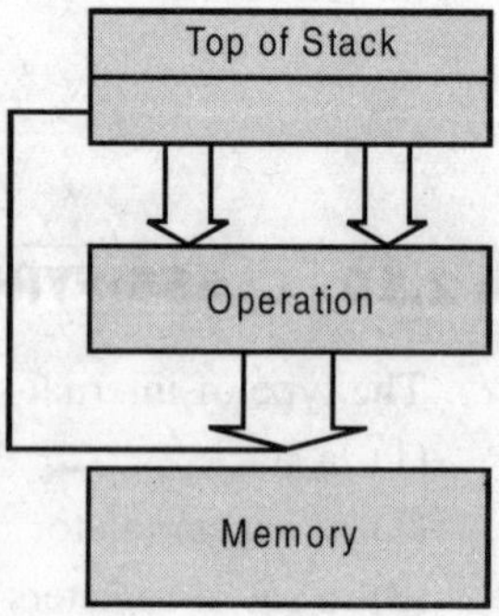

(*b*) **Accumulator**

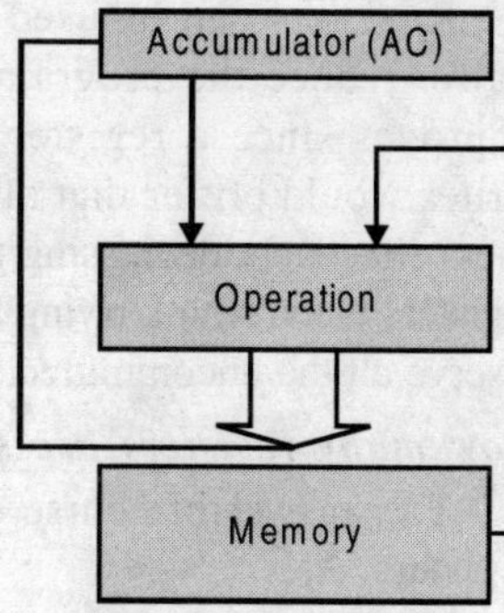


```
Load A
Add B
Store C
```

(*c*) **Register (Register-Memory)**

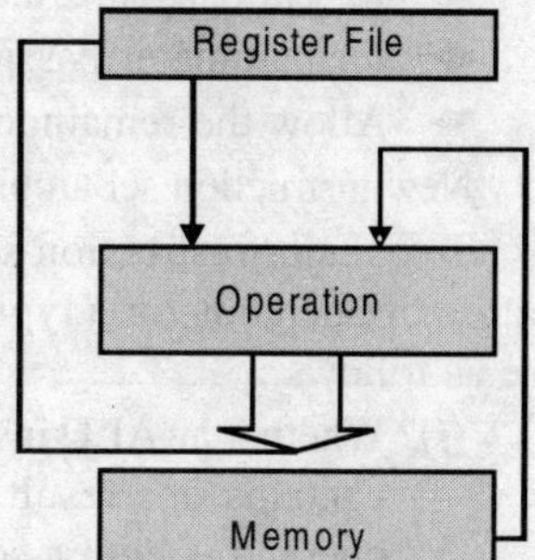

Load R_1, A

Add R_3, R_1, B

Store R_3, *C*

(*d*) **Register (Load-Store)**

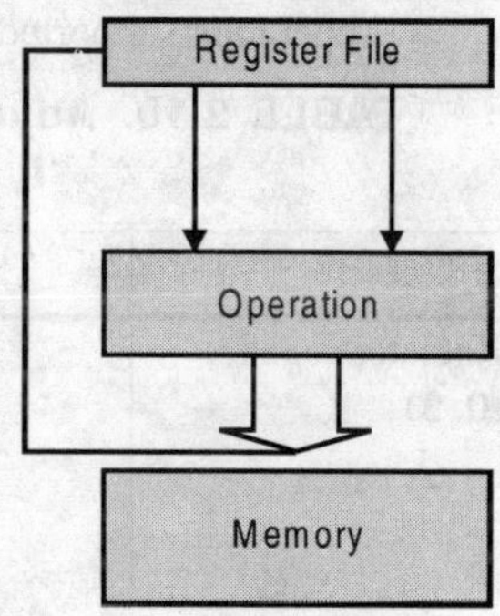


```
Load R1, A
Load R2, B
Add R3, R1, R2
Store R3, C
```

The Add instruction has implicit operands for stack and accumulator architectures, and explicit operands for register architectures. It is assumed that A, B and C all belong in memory and that the values of A and B cannot be destroyed.

There are really two classes of register computers:

(*a*) Register-memory Architecture. These can access memory as part of any instruction.

(*b*) Register-Register Architecture (load/store). These can access memory only with load and store instructions.

The third class, keeps all operands in memory is called a memory-memory architecture, but not found in computers shipping today. Some instruction set architectures have more registers than a single accumulator, but place restrictions on uses of these special registers. Such an architecture is sometimes called an extended accumulator or special-purpose register computer. The major reasons for the emergence of general-purpose register (GPR) computers are two fold :

(*i*) Registers like other forms of storage internal to the processor are faster than memory.

(*ii*) Register are more efficient for a compiler to use than other forms of internal storage.

Registers can be used to hold variables. When variables are allocated to registers, the memory traffic reduce the program speeds up since registers are faster than memory, and the code density improves since a register can be named with fewer bits than can be memory location. Compiler writes would prefer that all registers be equivalent and unreserved. By dedicating registers to special uses, effectively decreasing the number of general-purpose registers. If the number of truly general-purpose registers is too small, trying to allocate variables to registers will not be profitable. Instead, the compiler will reserve all the uncommitted registers for use in expression evaluation.

How many registers are sufficient ?

The answer of course depends on the effectiveness of the compiler. Most compilers reserve some registers.

- ➡ for expression evaluation and
- ➡ for parameter passing.
- ➡ Allow the remainder to be allocated to hold variables

New instruction set architectures tends to have more registers than their ancestors.

Two major instruction set characteristics divide GPR architectures. Both characteristics concern the nature of operands for a typical arithmetic or logical instruction or ALU instruction. These two concerns are as follows :

(*i*) Whether an ALU instruction has two or three operands. In the three-operand format, the instruction contains one result operand and two source operands. In the, two-operand format, one of the operand is both a source and a result fop the operation.

(*ii*) How many of the operands may be memory addresses in ALU instructions. The number, of memory operands supported by a typical ALU instruction may vary from none to three.

TABLE 2.10. Advantages and Disadvantages and of the three most common types of general purpose register computers.

Type	Advantages	Disadvantages
Register-register (0, 3)	• Simple. • Fixed-length instruction encoding. • Simple code generation model. • Instructions take similar numbers of clocks toexecute.	• Higher instruction count than architectures with memory references in instructions. • More instructions and lower instruction density leads to larger programs.
Register-memory (1, 2)	• Data can be accessed without a separate load instruction first. • Instruction format tends to. be easy to encode. • Yields good density.	• Operands are not equivalent as a source operand in a binary operation is destroyed. • Encoding a register number and a memory address in each instruction may restrict the number of registers. • Clocks per instruction vary by operand location.
Memory-memory (2, 2) or (3, 3)	• More compact. • Doesn't waste register for temporaries.	• Large variation in instruction size, especially for three-operand instruction. • Large variation in -work per instruction. • Memory accesses creates memory bottlenecks.

Instructions with multiple memory operands per typical ALU instruction are called register-memory or memory-memory, according to whether they have one or more than one memory operand. Table 3.1 states advantages and disadvantages of fhe three most common types of general-purpose register computers. The notation (m, n) means m is memory operands and "n is total operands. In general, computers with fewer alternatives simplify the compiler's task since there are fewer decisions for the compiler to make. Computers with a wide variety of flexible instruction formats reduce the number of bits required to encode the program. The number of registers also affects the instruction sinqe you need $\log_2$ (number of registers) for each register specifier in an instruction. Thus, doubling the number of registers takes 3 extra bits for a register-register architecture.

TIPS ▶

Computers with no memory reference per ALU instruction are called load store or register-register computers.

A GPR computer with memory-memory operations could easily be ignored by the compiler and used as a register-register computer. One of the most pervasive architectural impacts is on instruction encoding and the number of instructions needed to perform a task.

2.11. MEMORY ADDRESSING

Independent of whether the architecture is register-register or allows any operand to be a memory reference, it must define how memory addresses are interpreted and how they are specified.

2.11.1. Interpreting Memory Addressing

What object is accessed as a function of the address and the length ? All the instruction sets are byte addressed and provide access for bytes (8 bits), half words (16 bits), and words (32 bits). Most of the computers also provide access for double words (64 bits). There are two different conventions for ordering the bytes within a larger object : (*i*) Little Endian byte order, (*ii*) Big endian byte order.

(*i*) Little Endian byte order puts the byte whose address is "*x* ... *x*000" at the least significant position in the double word (the little end). The bytes are numbered.

7	6	5	4	3	2	1	0

(*ii*) Big Endian byte order puts the byte whose address is "*x*... *x*000" at the most significant position in the double word (the big end). The bytes are numbered

0	1	2	3	4	5	6	7

When operating within one computer, the byte order is often unnoticeable-only programs that access the same locations are both, say, words and bytes can notice the difference. Byte order is a problem when exchanging data among computers with different orderings. Little Endian ordering also fails to match normal ordering of words when string are compared. String appear "SDRAWKCAB" (backwards) in the registers.

A second memory issue is that in many computers, accesses to objects larger than a byte must be aligned. An access to an object of size s bytes at byte address A is aligned if A mod s = 0.

Why would someone design a computer with alignment restrictions ? Misalignment causes hardware complications, since the memory is typically aligned on a multiple of a word or double-word boundary, A misaligned memory access may, therefore, take multiple aligned memory references. Thus, even in computers that allow misaligned access, programs with aligned accesses run faster. Even if data are aligned, supporting byte, half-work, and word accesses requires an alignment network to align bytes, half words, and words in 64-bit registers.

2.11.2. Addressing Modes

Addressing modes specify constants and registers in addition to location in memory. When a memory location is used, the actual memory address specified by the addressing mode is called the effective address.

TABLE 2.11. Various Addressing Modes with Examples, Meaning and Their Usage.

Addressing Mode	Example instruction	Meaning	When used
Register	Add R_4, R_3	Regs[R_4] ← Regs[R_4] + Regs[R_3]	When a value is in a register.
Immediate	Add R_4, #3	Regs[R_4] ← Regs[R_4] + 3	For constants.
Displacement	Add R_4 100(R_1)	Regs [R_4] ← Regs[R_4] + Mem[100+Regs[R_1]]	Accessing local variables (+simulates register indirect, direct addressing modes).
Register indirect	AddR$_4$, (R_1)	Regs [R_4] ← Regs[R_4] + Mem [Regs [R_1]]	Accessing using a pointer or a computed address.
Indexed	Add R_3, R_1 + R_2	Regs [R_3] ← Regs [R_3] + Mem [Regs [R_1] + Regs [R_2]]	Sometimes useful in array addressing R1 = base of array; R2 = index amount
Direct or absolute	Add R_1, (1001)	Regs [R_1] ← Regs [R_1] + Mem [1001]	Sometimes useful for accessing static data, address constant may need to be large.
Memory indirect	Add R_1,@(R_3)	Regs [R_1] ← Regs [R_1] + Mem [Mem [Reg$_3$]]]	If R_3 is the address of a pointer *p* then mode yields *p.
Autoincrement	Add R_1,(R_2) +	Regs [R_1] ← Regs [R_2] + Mem [Regs [R_2]] Regs [R_2] ← Regs [R_2] + *d*	Useful for stepping through arrays within a loop R_2 points to start of array, each reference increments R_2 by size of an element *d*.
Autodecrement	AddR$_1$, – (R_2)	Regs [R_2] ← Regs [R_2] – *d* Regs [R_1] ← Regs [R_1] + Mem [Regs [R_2]]	Same use as auto increment. Auto decrement/increment can also act as push/pop to implement a stack.
Scaled	Add R_1 100(R_2)[R_3]	Regs [R_1] ← Regs [R_1] + Mem[100+Regs[R_2] + Regs[R_3]*d]	Used to index arrays. May be applied to any indexed addressing mode in some computers.

The most common names for the addressing modes, though the names differ among architectures. Mem array is used for main memory and Regs array is used for registers. Mem[Regs[,R_1]] refers to the contents of the memory location whose address is given by the contents of register (R_2). Table 3.7 shows the most commonly used addressing modes. In auto increment /decrement and scaled addressing modes, the variable d designates the size of the data item being accessed i.e., whether the instruction is accessing 1, 2, 4 or 8 bytes. These addressing modes are only useful when the elements being accessed are adjacent in memory.

Based on the use of various displacement sizes, a decision of what sizes to support can be made. Choosing the displacement field sizes is important because they directly affect the instruction length.

Immediate can be used in arithmetic operations, in comparisons and to-moves where a constant is wanted in a register. The last case occurs for constants written in the code-which tends to be small-and for address constants, which leads to be large. For the use of immediates it is important to know whether they need to be supported for all operations or for only a subset.

Another important instruction set measurement is the range of value for immediates. Like displacement values, the size of immediate values affects instruction length. Small immediate values are most heavily used. Large immediates are sometimes used however most likely in addressing calculations.

2.12. TYPE AND SIZE OF OPERANDS

How is the type of an operand designated ?

Encoding in the opcode designates the type of an operand, this is the method used most often, The data can be annotated with tags that are interpreted by the hardware. These tags specify the type of the operand, and the operation is chosen accordingly the type of an operand can usually be integer, single-precision floating point (1 word), character (8-bits), half word (16 bits), word (32 bits), and double precision floating point (2 words). Integers are almost universally represented as two's complement binary numbers. Characters are usually in ASCII, but the 16-bit Unicode (used in Java) is gaining popularity with the internationalization of computers.

Some architectures provide operations on character strings although such operations are usually quite limited and treat each byte in the string as a single character. Typical operations supported on character strings are comparisons and moves.

For business applications, some architectures support:

(a) *A decimal format usually called packet decimal or binary- coded decimal. It uses 4 bits to encode the values 0-9 and 2 decimal digits are packed into each byte.*

(b) *Numerical character strings usually called unpacked decimal and operations called packing and unpacking are usually provided for converting back and forth between them.*

Note

Decimal operands are used as they give result that exactly match decimal numbers. As some decimal fractious do not have an exact representation in binary. Example $(0.10)_{10}$ is a simple fraction in decimal, but the binary it requires an infinite set of repeating digits:

$$(0.000110011\overline{0011}\ldots)$$

These, calculations that are exact in decimal can be close but inexact in binary, which can be problem for financial transactions. In some architectures, objects in registers may be accessed as byte or half words.

2.13. OPERATIONS IN THE INSTRUCTION SET

The operators supported by most instructions set architecture can be categorized as shown in Table 2.12. One rule of thumb across all architectures is that the most widely executed instructions are the simple operations of an instruction set.

TABLE 2.12. Categories of instruction Operators Alongwith Examples.

Operator Type	Examples
Arithmetic and logical	Integer arithmetic and logical operations, add, subtract, and, or, multiply, divide.
Data Transfer	Loads-stores (more instructions on computers with memory addressing).
Control	Branch, jump, procedure call and return, traps.
System	Operating system call, virtual memory management instruction.
Floating Point	Floating point operations, add, multiply, divide and compare.
Decimal	Decimal add, decimal multiply, decimal-to-character conversions.
String	String move, string compare, string search.
Graphics	Pixel and vertex operations, compression /decompression operations.

All computers generally provide a full set of operations for the first three categories as mentioned in Table 2.12. The support for system functions in the instruction set varies widely among architectures, but all computers must have some instruction support for basic system functions. The amount of support in the instruction set for the last four categories may vary from none to an extensive set of special instructions. Floating point instructions will be provided in any computer that is intended for use in an application that makes much use of floating point. These instructions are sometimes part of an optimal instruction set.

Graphics instructions typically operate on many smaller date items in parallel example, performing eight 8-bit additions on two 64-bit operands.

2.14. INSTRUCTION FOR CONTROL FLOW

The measurement of branch and jump behaviour are fully independent. There is no consistent terminology for instructions that change the flow of control. They are sometimes called transfers, branch or jumps. Jump is used when the change in control is unconditional and branch is used when the change is 3 conditional.

Four different types of control flow change are:

(*i*) Conditional Branches
(*ii*) Jumps
(*iii*) Procedure calls
(*iv*) Procedure return Conditional branches dominates all other control flow changes.

2.14.1. Addressing Modes for Control Flow Instruction

The destination address of a control flow instructions must always be specified. This destination is specified explicitly in the instruction in the vast majority of cases- procedure return is the major exception, since for return the target is not known at compile time. The most common way to specify the destination is to supply a displacement that is added to the program counter (PC). Control flow instructions of this type are called PC-relative.

PC-relative branches or jumps are advantageous as the target is often near the current instruction and specifying the position relative to the current PC requires fewer bits. Using PC- relative addressing permits the code to run independently of where it is loaded. This property is called position independence. This can eliminate some work when the program is linked and is useful linking programs dynamically during execution.

To implement returns and indirect jumps when the target is not known at compile time a method other than PC-relative addressing is required. The dynamic addressing the target which changes at run time is as simple as naming a register that contains the target address. Alternatively, the jump may permit any addressing mode to be used to supply the target address.

These register indirect jumps are also useful for four other important features:

(*i*) Case or switch statements, found in most programming languages. These statements select one of the alternative among several alternatives.

(*ii*) Virtual functions or methods in object- oriented language like C++ or Java. These allow different routines to be called depending on the type of the argument.

(*iii*) High-order functions or function pointers, in languages like C or C++. This allow functions to be passed as arguments, giving some of the flavor of object- oriented programming.

(*iv*) Dynamically shared libraries allows a library to be loaded and linked at run time only when it is actually invoked by the program rather than loaded and linked statically before the program is run.

In all the four cases the target address is not known at compile time, and hence is usually loaded from memory into a register before the register indirect jump.

As branches generally use PC- relative addressing to specify their targets as important question concern how far branch targets are from branches. Knowing the distribution of these displacements will help in choosing what branch offsets to support, and thus will affect the instruction length and encoding. An architecture that require fewer instructions for the same program would have shorter branch distances Example, VAX. The number of bits needed for the displacement may increase if the computer has variable-length instructions to be aligned on any byte boundary.

2.14.2. Conditional Branch Option

Most changes in control flow are branches. Most noticeable property of branches is that a large number of the comparisons are simple tests and & large number are comparisons with Zero. Thus, some architecture choose to treat these comparisons as special cases especially if a compare and branch instruction is being used. Table 2.13 shows different branch conditions along with advantage and disadvantages.

Another looping structure is repeat instruction. It allows a single instruction or a block of instructions to be repeated upto 256 times. Three special registers are used to hold :

(*i*) Block starting address

(*ii*) Ending address

(*iii*) Repeat Counter.

The memory instructions in a repeat look will typically have autoincrement or autodecrement addressing to access a vector. The goal of such instructions is to avoid look overhead.

TABLE 2.13. Various Branch Conditions with Advantages and Disadvantages.

Name and Examples	How condition is tested	Advantages	Disadvantages
Condition Code (CC) (80×86, ARM, PowerPC, SPARC, SuperH)	Tests special bits set by ALU operations possibly under program control	Sometimes condition is set for free.	CC is extra state condition codes constrain the ordering of instructions one since they pass information from one instruction to a branch.
Condition Register (Alpha, MIPs)	Test arbitrary register with the result of a comparison	Simple	Uses up a register.
Compare and branch (PA-RISC, VAX)	Compare is part of the branch often compare is limited to subset	One instruction rather than two for a branch.	May be too much, work per instruction for pipelined execution.

A condition codes can be set by ALU operations that are needed for other purposes, measurements on programs shows that this rarely happens. The major implementation problems with condition codes arise when the condition code is set by a large or haphazardly chosen subset of the instructions, rather than being controlled by a bit in the instruction. Computers with compare and branch often limit the set of compares and use a condition register for more complex compares. Procedure calls and returns include control transfer and possibly some state saving at a minimum the return address must be saved some where some times in a special link register or just a GPR. Some older architectures provide a mechanism to save many registers, while newer architecture require the compiler to generate stores and load for each register saved and restored.

There are two basic convention in use to save registers-cither at the call site or inside the procedure being called. Caller saving means that the calling procedure must save the register that it wants preserved for access after the call, and thur the called procedure need not worry about registers. Callee saving is the opposite, the called procedure must same the registers it wants to use, leaving the caller unrestrained. There are times when caller save must be used because of access patterns to globally visible variable in two different procedures.

2.15. ENCODING AN INSTRUCTION SET

The choices mentioned above will affect how the instructions are encoded into a binary representation for execution by the processor. This representation affects not only the size of the compiled program, it affects the implementation of the processor, which must decode this representation to quickly find the operation and its operands. The operation is typically specified in one field, called the opcode.

This decision depends on the range of addressing modes and the degree of independence between opcodes and modes. Some older computers have one to five operands with 10 addressing modes for each operand. For such a large number of combinations, typically separate address specifier is needed for each operand. The address specifier tells what addressing mode is used to access the operand. At the other extreme are load-store computers with only one memory operand and only one or two addressing modes. In this case, the addressing mode can be encoded as part of the opcode.

When encoding the instructions, the number of registers and the number of addressing modes both have a significant impact on the size of instructions, as the register field and addressing mode field may appear many times in a single instruction. The architect must balance several competing forces when encoding the instruction set:

1. The desire to have as many registers and addressing modes as possible.
2. The impact of the size of the register and addressing mode fields on the average instruction size and hence on the average program size.
3. A desire to have instructions encoded into lengths that will be easy to handle in a pipelined implementation benefits while sacrificing average code size.

Summary

- The operations performed on data stored in registers are called microoperations.
- The symbolic notation used to described the microoperation transfers among registers is called register transfer language.
- Different types of microoperations are register transfer microoperation, arithmetic microoperation, logic microoperation and shift microoperation.
- Arithmetic microoperations perform arithmetic operations like addition, subtraction, increment, decrement on numeric data stored in registers.

- ➡ Logic microoperations perform bit manipulation operations like AND, OR, NOT, etc. on data stored in registers.
- ➡ Shift microoperations perform shift operations like shift-left, shift-right etc. on data stored in registers.
- ➡ Every register transfer statement implies a hardware implementation for transfer.
- ➡ A memory register or word is symbolized by the letter M.
- ➡ The transfer of information from a memory register to outside environment is called a read operation.
- ➡ The transfer of new information in a memory register is called a write operation.
- ➡ Data bus carries data between two units is bidirectional.
- ➡ Memory bus used to carry memory addresses is unidirectional.
- ➡ The process by which the requesting processor may be granted the access to the bus is known as bus arbitration.
- ➡ The two approaches to bus arbitration are serial arbitration and parallel arbitration.
- ➡ Serial arbitration process is also known as daisy-chain arbitration.
- ➡ The circuit that generates the arithmetic addition of two binary number of any length is called binary adder.
- ➡ The binary adder circuit is constructed using full-adder.
- ➡ An *n*-bit binary adder requires n full-adders.
- ➡ The subtraction of binary numbers can be done conveniently by means of complements.
- ➡ The subtraction $(A - B)$ can be done by taking the 2's complement of B and adding it to A, *i.e.* $A - B = A + \overline{B} + 1$.

Review Questions

1. Represent the given conditional control statement by two register transfer Statements with Control functions.
   If (P = 1), then R1 ← R2 else if (Q = 1) then R1 ← R3.
2. What do you understand by the term micro-operation? Explain Register and Arithmetic types of micro-operation. Show the hardware realisation of decrement microoperation *i.e.* $\alpha : A \leftarrow A - 1$.
3. Implement the following RTL code, using common bus and tri-state buffers.

   j : M ← A
   o : A ← Y
   h : R ← M
   n : Y ← R, M ← R

   Assume M, A, R and Y are to be one bit D flip-flop.
4. A register 'A' holds an 8-bit binary number 11011001. Determine the operand 'B' and the logic micro operation to be performed in order to change the value of 'A' to (*i*) 01101101 (*ii*) 11111101.
5. Draw the block diagram of Arithmetic Logic Unit (ALU) to perform the following operations. (use MUX) (Assume the length of the data).
   (*i*) AND (*ii*) OR
   (*iii*) NOR (*iv*) NAND
6. Explain the different shift operations with examples.
7. Design a hardware circuit by using common bus architecture to implement the following Register Transfer Languages.

   P : A1 ← A2
   Q : A2 ← A3
   R : A4 ← A1

$S : A3 \leftarrow A4, A1 \leftarrow A4$

Where A1, A2, A3, A4 are one bit register.

8. Give the hardware realization of 4-bit arithmetic circuit capable of doing addition, subtraction, increment, decrement etc. Give the function table and explain its operation.

9. Draw a neat block diagram of a 4-bit binary adder-subtractor. Then, explain its operation in detail.

10. What is a micro-operation? List and briefly explain the most commonly encountered arithmetic operations.

11. Design a 4 bit arithmetic circuit which is capable of performing the following micro-operations:

Addition (with and without carry)
Subtraction (with and without borrow)
Increment
Decrement
and Transfer

Substantiate the circuit diagram with the help of an example.

12. Give your comment on these following register transfer statements

(*i*) $GT : PC \leftarrow \overline{PC}, PC \leftarrow PC + AR$ (*ii*) $\overline{x}T : AC \leftarrow \overline{R_1} + 1 + AC$

(*iii*) $PT : \begin{cases} R_i \leftarrow 0, \text{ for } i = 0 \\ R_i \leftarrow R_{i-1} \text{ for } i \neq 0 \end{cases}$ (*iv*) $xT_1 + \overline{Y}T_2 : R_1 \leftarrow R_1 + R_2 + 1$

13. Draw the logic diagram of a logic circuit capable of performing X-OR, NOR, NAND, complement operation of two bits A and B. During complement operation, the circuit should be capable of complementing B.

14. Draw a diagram of bus system for four registers of 4-bits each. The bus is to be constructed with multiplexers.

15. A digital computer has a common bus system for 8 registers of 8 bits each. The bus is constructed with multiplexers.

(*i*) How many selection inputs are there in each multiplexer?
(*ii*) What size of multiplexers are needed?
(*iii*) How many multiplexers are there in the bus?

16. A digital computer has a common bus system for 16 registers of 32 bits each. The bus is constructed with multiplexers.

(*i*) How many selection inputs are there in each multiplexer?
(*ii*) How many multiplexers are there in the bus?

17. Convert the following :

(*i*) $(43.125)_{10} = (\ )_2$ (*ii*) $(6B.28)_{10} = (\ )_2$
(*iii*) $(76A)_{16} = (\ )_8$ (*iv*) $(11001010)_2 = (\ )_{GRAY}$

18. Describe process of error-detection and correction. Give an example of error detection code. Explain how it detects error.

19. Discuss various number systems.

20. Describe different numeric codes.

21. Differentiate between ASCII and EBCDIC codes.

22. Perform the following conversions :

(*i*) $(8620)_{10} = (\ )_2$ (*ii*) $(CDFE)_{16} = (\ )_2$
(*iii*) $(2489)_{10} = (\ )_8$ (*iv*) $(1101110.01)_{16} = (\ )_{10}$

23. Convert the octal number 235.4 to equivalent hexadecimal number.

24. Convert the following decimal numbers to binary, octal, hexdecimal, BCD : 251, 141.

25. Convert the following :

$$(5214)_{10} = (\ )_{16}$$
$$(1011)_{10} = (\ )_2$$
$$(5A8)_{16} = (\ )_8$$
$$(1011)_{Gray} = (\ )_{Binary}$$
$$(13)_{10} = (\ )_{BCD}$$

❑❑❑

3

Chapter

Basic Computer Organization and Design

INSIDE THIS CHAPTER

3.1. INTRODUCTION

The processor (really a short form for *microprocessor* and also often called the *CPU* or *central processing unit*) is the central component of the PC. It is the brain that runs inside the PC. All work that you do on your computer is performed directly or indirectly by the processor. Obviously, it is one of the most important components of the PC. It is also, scientifically, not only one of the most amazing parts of the PC, but one of the most amazing devices in the world of technology.

The first microprocessor (Intel 4004) was invented in 1971. It was a 4-bit calculation device with a speed of 108 kHz. Since then, microprocessor power has grown exponentially. So what exactly are these little pieces of silicone that run our computers ?

TIPS

Processor is the brain of a computer system that performs a bulk of data processing in order to convert it into useful information.

3.2. PROCESSOR

The processor plays a significant role in the following important aspects of a computer system :

Performance

The processor is probably the most important single determinant of system performance in the PC. While other components also play a key role in determining performance, the processor's capabilities dictate the maximum performance of a system. The other devices only allow the processor to reach its full potential.

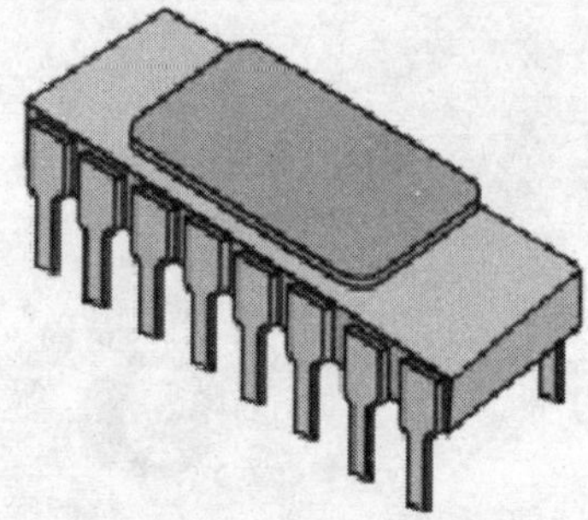

Fig. 3.1. Intel 4004 Processor.

Software Support

Newer, faster processors enable the use of the latest software. In addition, new processors such as the Pentium with MMX Technology, enable the use of specialized software not usable on earlier machines.

Reliability and Stability

The quality of the processor is one factor that determines how reliably system is running. While most processors are very dependable, some are not. This also depends to some extent on the age of the processor and how much energy it consumes.

Energy Consumption and Cooling

Originally processors consumed relatively little power compared to other system devices. Newer processors can consume a great deal of power. Power consumption has an impact on everything from cooling method selection to overall system reliability.

Motherboard Support

The processor you decide to use in your system will be a major determining factor in what sort of chipset you must use, and hence what motherboard you buy. The motherboard in turn dictates many facets of your system's capabilities and performance.

3.3. INSIDE THE PROCESSOR

3.3.1. Functional Units

The processor is made up of a group of interrelated units (or control units). Microprocessor architecture varies considerably from one design to another, but the main elements of a microprocessor are as follows :

(*i*) A **control unit** that links the incoming data, decodes it, and sends it to the execution unit. The control unit is made up of the following elements:

- **Sequencer** (or *monitor and logic unit*) that synchronizes instruction execution with the clock speed. It also sends control signals;
- **Program counter** that contains the address of the next instruction to be executed;
- **Instruction register** that contains the currently executing instruction.

(*ii*) An **execution unit** (or *processing unit*) that accomplishes tasks assigned to it by the control unit. The execution unit is made of the following elements :

- The **arithmetical and logic unit** (**ALU**). The ALU performs basic arithmetical calculations and logic functions (AND, OR, EXCLUSIVE OR, etc.);
- The **floating point unit** (**FPU**) that performs complex calculations which cannot be done by the arithmetical and logic unit.
- The **status register** that reflects the results of instruction.
- The **accumulator register** that is used to store operand, intermediate results and the final results.

A **bus management unit** (or *input-output unit*) that manages the flow of incoming and outgoing information and interfaces with system RAM.

The Fig. 3.2 shown below gives a simplified representation of the elements that make up the processor (the physical layout of the elements is different than their actual layout).

Do You Know

Cache memory is the fastest memory of a computer system which is used to store frequently needed data of a CPU.

3.3.2. Cache Memory

Cache memory (also called *buffer memory*) is local memory that reduces waiting times for information stored in the RAM. In effect, the computer's main memory is slower than that of the processor. There are, however,diffrent types of memory that are much faster, but which have a greatly increased cost. The solution is therefore to include this type of local memory close to the processor and to temporarily store the primary data to be processed in it. Recent model computers have many different levels of cache memory :

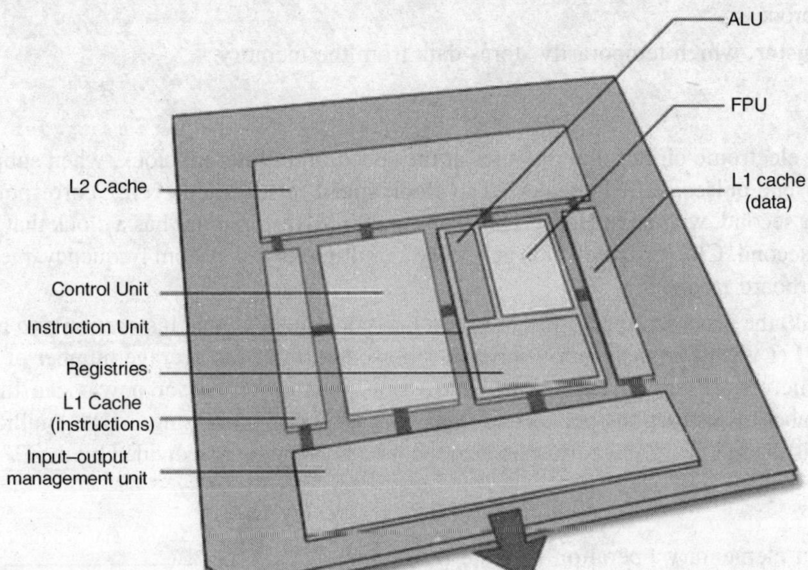

Fig. 3.2. Functional Units of a Processor

(*i*) **Level 1 cache memory** (called **L1 Cache**, or **Level 1 Cache**) is directly integrated into the processor. It is subdivided into two parts:

- The first part is the instruction cache, which contains instructions from the RAM that have been decoded as they came across the pipelines.
- The second part is the data cache, which contains data from the RAM and data recently used during processor operations.

Level 1 cache can be accessed very rapidly.

(*ii*) **Level 2 cache memory** (called **L2 Cache** or **Level 2 Cache**) is located in the case along with the processor (in the chip). The level 2 cache is an intermediate between the processor, with its internal cache, and the RAM. It can be accessed more rapidly than the RAM, but less rapidly than the level 1 cache.

(*iii*) Level 3 cache memory (called L3 Cache, or Level 3 Cache) is located on the motherboard.

All these levels of cache reduce the latency time of various memory types when processing or transferring information. While the processor works, the level 1 cache controller can interface with the level 2 controllers to transfer information without impeding the processor. As well, the level 2 cache interfaces with the RAM (level 3 cache) to allow transfers without impeding normal processor operation.

3.3.3. Registers

> **Do You Know**
> *Registers are used to temporarily store the data.*

When the processor executes instructions, data is temporarily stored in small, local memory locations of 8, 16, 32 or 64 bits called registers. Depending on the type of processor, the overall number of registers can vary from about ten to many hundreds. The main registers are :

- The **accumulator register** (*ACC*), which stores the results of arithmetic and logical operations;
- The **status register** (*PSW*, *Processor Status Word*), which holds system status indicators (carry digits, overflow, etc.);
- The **instruction register** (*RI*), which contains the current instruction being processed;
- The **ordinal counter** (*OC* or *PC* for *Program Counter*), which contains the address of the next instruction to process;
- The **buffer register**, which temporarily stores data from the memory.

3.3.4. Operation

The processor is an electronic circuit that operates at the speed of an internal clock, when subjected to an electrical current, send pulses, called "**peaks**". The **clock speed** (also called **cycle**), corresponds to the number of pulses per second, written in Hertz (Hz). Thus, a 200 MHz computer has a clock that sends 200,000,000 pulses per second. Clock frequency is generally a multiple of the system frequency, meaning a multiple of the motherboard frequency.

With each clock peak, the processor performs an action that corresponds to an instruction or a part of it. A measure called **CPI** (*Cycles Per Instruction*) gives a representation of the average number of clock cycles required for a microprocessor to execute an instruction. A microprocessor power can thus be characterized by the number of instructions per second that it is capable of processing. **MIPS** (millions of instructions per second) is the unit used and corresponds to the processor frequency divided by the *CPI*.

3.3.5. Instructions

> **Do You Know**
> *Instruction is a command given to computer in order to do an operation.*

An instruction is an elementary operation that the processor can accomplish. Instructions are stored in the main memory, waiting to be processed by the processor. An instruction has two fields:

- The **operation code**, which represents the action that the processor must execute;
- The **operand,** which defines the data on which operation is to be performed. The operand depends on the operation. It can be data or a memory address.

Operation Code	Operand Field

Fig. 3.3. Instruction Field

The number of bits in an instruction varies according to the type of data (between 1 and 4 bytes). Instructions can be grouped into different categories, according to the following factors :

Memory Access

Accessing the memory or transferring data between registers.

Arithmetic Operations

Operations such as addition, subtraction, division or multiplication.

Logic Operations

Operations such as AND, OR, NOT, EXCLUSIVE OR, etc.

Control

Sequence controls, conditional control, and iterative control etc.

3.3.6. Control Signals

Control signals are electronic signals that control the various processor units participating in the execution of an instruction. Control signals are sent using an element called a *sequencer*. For example, the *Read /Write* signal allows the memory to be told that the processor wants to read or write information in it.

3.4. BLOCK DIAGRAM OF PROCESSOR

As studied earlier, three main structural components :

1. **Central Processing Unit :** It controls the operation of the computer system and performs its data processing functions.
2. **Main memory :** It stores data.
3. **I/O :** It moves data between the computer and external environment and system interconnection (Some mechanism that provides for communication among CPU, main memory, and I/O).

The most interesting and the complex component is the CPU, and known as the processor. CPU is the brain of any computer system because it takes all major decisions, makes all sorts of calculations and directs different parts of the computer functions by activating and controlling the operations.

As one computer generation has evolved to the next, the size of the processor has become smaller and smaller, while its speed and capacity have increased tremendously. Figure 3.4 shows the functional units within a hypothetical processor.

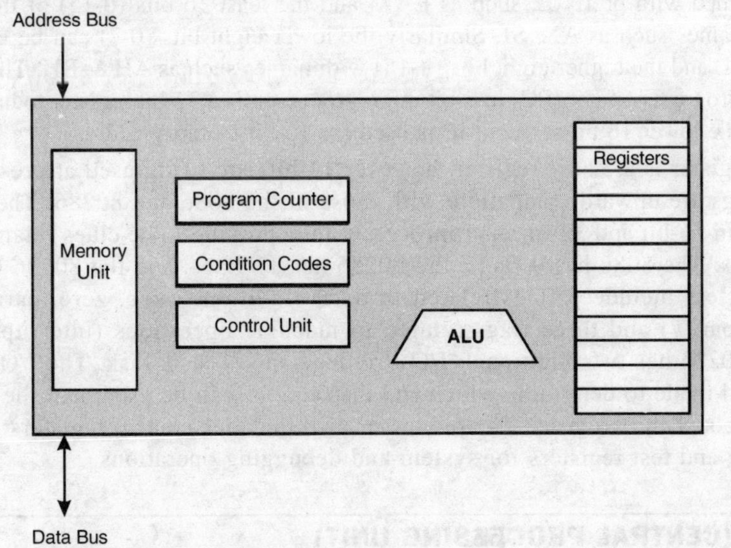

Fig. 3.4. Flow Diagram of Processors

The most common parts of modern microprocessors are : ALU, CU, Registers, Condition Code Register and program Counter. The parts of the CPU are usually connected by an electronic component referred to as an electronic highway between them.

Arithmetic Logical Unit : The actual processing of data and instruction are performed by Arithmetic Logical Unit. The major operations performed by the ALU are Addition, subtraction, multiplication, division, logic and comparison. Data is transferred to ALU from storage unit when required. After processing the output is returned back to storage unit for further processing or getting stored.

The ALU controls the speed of calculations and so receives a great deal of attention from computer designers to meet the needs of fast-growing world.

Control Unit : The control unit determines the sequence in which computer programs and instructions are executed. Things like processing of programs stored in the main memory, interpretation of the instructions and issuing of signals for other units of the computer to execute them. It coordinates the activities of computer's peripheral equipment as they perform the input and output.

Remind that the control unit does not execute instructions itself; rather it tells other parts of the computer system what to do.

Registers : While performing operations the ALU takes data from the temporary storage area inside the CPU named Registers. Registers are a group of the cells used for memory addressing, data manipulation and processing. Each cell is typically as wide as the processor's data bus (1 "word"). Some of the registers are general purpose and some are reserved for certain functions. It is a high-speed memory which holds only data for the next instruction, they are sent back to the main memory and registers are occupied by the new data used in the next instruction.

Condition Code Register : The condition code register contains a set of bits which indicates whether the result of the last command was "greater than zero", "equal to zero", or "Less than zero".

Program Counter : The program counter contains the address of the instruction being executed.

Register Set

The purpose of the Register is to hold temporary results, and control the execution of the program. General-purpose registers in Pentium are EAX, ECX, EDX, EBX, ESP, EBP, ESI, or EDI. The 32-bit registers are named with prefix E, such as EAX, and the least 16 bits (0-15) of these registers can be accessed with names such as AX, SI. Similarly the lower eight bits (0-7) can be accessed with names such as AL & BL and the higher eight bits (8-15) with names such as AH & BH. The instruction pointer EAP known as program counter (PC) in 8-bit microprocessor, is a 32-bit register to handle 32-bit memory addresses, and the lower 16 bit segment IP is used for 16-bit memory address.

The flag register is a 32-bit register, however 14-bits are being used at present for 13 different tasks; these flags are upward compatible with those of the 8086 and 80286. The comparison of the available flags in 16-bit and 32-bit microprocessor may provide some clues related to capabilities of these processors. The 8086 has 9 flags, the 80286 has 11 flags, and the 80286 has 13 flags. All of these flag registers include 6 flags related to data conditions (sign, zero, carry, auxiliary, carry, overflow, and parity) and three flags related to machine operations (interrupts, Single-step and Strings). The 80286 has two additional: I/O Privilege and Nested Task. The I/O Privilege uses two bits in protected mode to determine which I/O instructions can be used, and the nested task is used to show a link between two tasks. The processor also includes control registers and system address registers, debug and test registers for system and debugging operations.

3.5. CPU (CENTRAL PROCESSING UNIT)

CPU is made up of three major parts as shown in Fig. 3.5.

(*i*) Register Set

(*ii*) ALU (Arithmetic and Logic Unit)

(*iii*) CU (Control Unit)

Do You Know

The part of the computer that performs the bulk of data processing operations is known as Central Processing Unit or CPU.

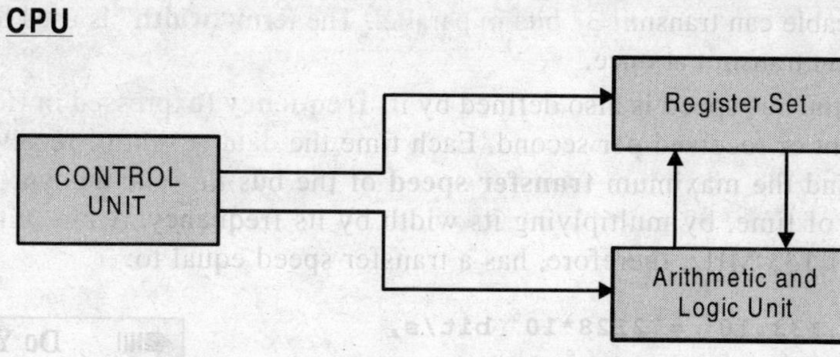

Fig. 3.5. Central Processing Unit

Register set are used to store intermediate results obtained during the execution of instructions. ALU performs the arithmetic and logic microoperations required for executing the instructions. Control unit instructs the ALU to perform the desired operation and supervises the transfer of information among different registers.

Generally, the computers have one or a combination of the following CPU organizations.

(*a*) Single accumulator organization.

(*b*) General register organization.

(*c*) Stack organization.

> **Do You Know**
>
> *Bus is a set of electronic signal that is used to transfer information and signals between the components inside or outside of a computer.*

3.5.1. Basic Components of CPU

Some of the general components upon which the processor is built are as follows :

Buses

A bus set is of electronic pathways that allows information and signals to travel between components inside or outside of a computer. In other words, a bus is a set of physical connections (cables, printed circuits, etc.) which can be shared by multiple hardware components in order to communicate with one another.

The purpose of buses is to reduce the number of "pathways" needed for communication between the components, by carrying out all communications over a single data channel as shown in Fig. 3.6. If only two hardware components communicate over the line, it is called a **hardware port** (such as a serial port or parallel port).

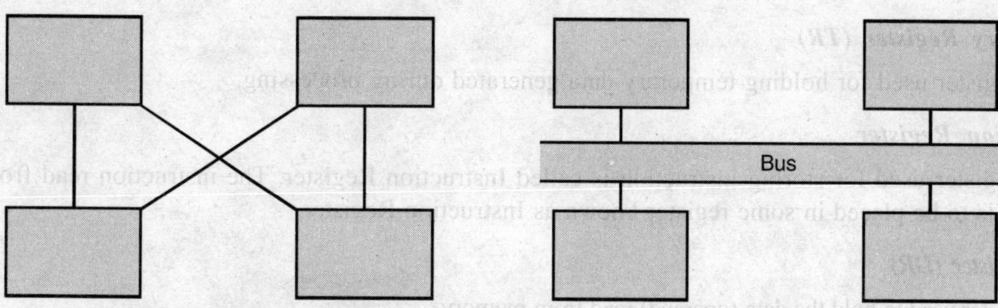

(a) Communication without Bus *(b) Communication using Bus*

Fig. 3.6.

Characteristics of a Bus

A bus is characterized by the amount of information that can be transmitted at once. This amount, expressed in bits, corresponds to the number of physical lines over which data is sent simultaneously.

A 32-wire ribbon cable can transmit 32 bits in parallel. The term "**width**" is used to refer to the number of bits that a bus can transmit at once.

Additionally, the bus speed is also defined by its **frequency** (expressed in Hertz), *i.e.* the number of data packets sent or received per second. Each time the data is sent or received is called a **cycle**. It is possible to find the maximum **transfer speed** of the bus *i.e.*, the amount of data which it can transport per unit of time, by multiplying its width by its frequency. A bus with a width of 16 bits and a frequency of 133 MHz, therefore, has a transfer speed equal to:

$$16 * 133.10^6 = 2128*10^6 \text{ bit/s,}$$
$$\text{or } 2128*10^6/8 = 266*10^6 \text{ bytes/s}$$
$$\text{or } 266*10^6 /1000 = 266*10^3 \text{ KB/s}$$
$$\text{or } 259.7*10^3 /1000 = 266 \text{ MB/s}$$

Do You Know

A processor register (AC) is required for doing operations on data.

The Primary Buses

There are generally two buses within a computer:

(*i*) The **internal bus** (sometimes called the *front-side bus* or *FSB* for short). The internal bus allows the processor to communicate with the system's central memory (*i.e.* RAM).

(*ii*) **The expansion bus** (sometimes called the *input/output bus*) allows various motherboard components (USB, serial, and parallel ports, cards inserted in PCI connectors, hard drives, CD-ROM and CD-RW drives, etc.) to communicate with one another. However, it is mainly used to add new devices using **expansion slots** connected to the input/output bus.

Registers

Different CPU Registers are :

Accumulator (AC)

The register holds data on which addition, subtraction, shift and logical operations are to be carried out. The result of an arithmetic and logical operation is automatically stored in the accumulator.

Thus, it is used for storing result and for doing operations (arithmetic as well as logical) on its content.

Program Counter (PC)

It deals with the order of the execution of instructions. It holds the address of the next instruction to be executed. Thus, it acts as a pointer which points to the memory location where the next instruction is stored.

Temporary Register (TR)

A register used for holding temporary data generated during processing.

Instruction Register

A register used for storing instruction is called Instruction Register. The instruction read from the memory is to be placed in some register known as Instruction Register.

Data Register (DR)

Register used to hold the data (operand) read from memory.

Address Register (AR)

Register used to hold the address of memory word.

INPR

Input register will hold/receives data from an input device.

OUTR

It holds data that need to be sent to output devices.

Table 3.1 describes the brief function and the number of bits that different registers contain.

TABLE 3.1. List of Registers.

Register Symbol	Number of bits	Register Name	Function of Register
DR	16	Data Register	Holds memory operand
AR	12	Address Register	Holds address for memory
AC	16	Accumulator	Processor Register
IR	16	Instruction Register	Holds instruction code
PC	12	Program Counter	Holds address of next instruction
TR	16	Temporary Register	Holds temporary data
INPR	8	Input Register	Holds input character
OUTR	8	Output Register	Holds output character.

Flags

There are number of indicators known as flags that shows the processor's status. Most of these flags represent the results of the last operations. For example, the addition of two numbers might produce a negative sign, an overflow, a carry, or a value of zero.

These flags are represented by a single bit such as if the result of an addition is negative, the sign flag would set to 1. If the result was not a negative number (zero or greater than zero), the sign flag would equal to 0.

These flags are organized *i.e.* grouped together into a single register called the flag resister or the processor status register. Since the values stored in these flags are in the form of bit which are the outcome of an arithmetic or logical operation, and hence the flag registers are connected to the mathematical unit of a processor.

Stacks

During the execution of operation, there are number of times when the processor needs to use a temporary memory to store different data values so that they can be used again when required. Every processor has a finite number of registers. But if an application needs more registers than available, the register values that are not needed immediately by processor can be stored in the temporary memory. Also, when a processor needs to jump to a subroutine or function, it needs to remember the instruction from where it jumped so that it can return back to the same place when the subroutine is completed. Hence, the return address must be stored and this return address are generally stored in this temporary memory. This temporary memory is known as Stack.

The Stack is a block of memory locations reserved to functions as temporary memory. When a processor puts a piece of data, on the top of stack, the data below it cannot be removed until the data above it is removed. This type of memory location is referred as "Last-In-First-Out" or LIFO.

The two main operation that the processor can perform on the stack are PUSH and POP. It can either store the value of a register to the top of the stack or can remove the top element from the stack. Storing the data to the stack is referred to as pushing and removing the top data from stack is referred as poping.

I/O Ports

Input/Output ports, referred as I/O ports, is any connections that exist between the processor and its external devices. For example, a USB printer can be connected to the computer system through an I/O (USB) port. Using this port the computer can issue commands and send to be printed.

Some I/O devices are connected directly to the memory bus. Sending data to the port is done by storing data to a memory address and retrieving data from port is done by reading from a memory address.

3.5.2. Single Accumulator Organization

Earlier when the hardware were very expensive, the processor was designed usually as a single address machine with accumulator. The accumulator used to store operand for the instructions. It is used for storing result and for doing operations on the content of the accumulator. Thus, accumulators used as source for the operands as well as the destination for storing the result of the operations performed on its content. Thus, it accumulate datas. The accumulator based organization system results in reduced size of instruction, since the instructions contains only one operand address and the other operand is in accumulator itself. Also, after performing the operations the result is stored in the accumulator. Thus, these accumulator based organization systems are also known as 1-address machine, since only one operand address is specified in the instruction. But there can be only one or a limited number of accumulators in the CPU that affects the performance of the systems when an arithmetic expression is to be evaluated containing many different terms.

3.5.3. General Register Organization

A bus organization for seven CPU registers is shown in Fig. 3.7. The output of each registers is connected to two multiplexers (MUX) to form the two buses *A* and *B*. The selection lines in

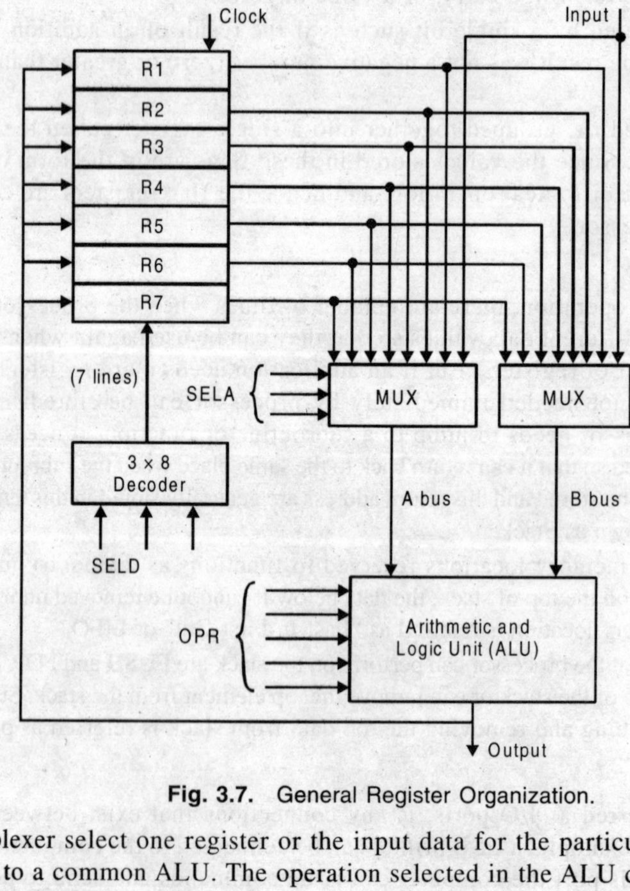

Fig. 3.7. General Register Organization.

each multiplexer select one register or the input data for the particular bus. The *A* and *B* buses form the inputs to a common ALU. The operation selected in the ALU determines the arithmetic or logic microoperations to be performed. The result of the microoperation is available for the output data and also goes into the inputs of seven registers. The register that receives the information from the output bus is selected by the decoder. The decoder activates one of the register load inputs and thus providing a transfer path between the output data bus and the inputs of the selected destination register.

Let the operation be $R1 \longleftarrow R2 + R3$

To perform this operation, the control must provide

(*i*) SELA : place the contents of *R*2 into bus *A*.

(*ii*) SELB : place the contents of *R*3 into bus *B*.

(*iii*) ALU operation selector OPR : provide the arithmetic addition *A* + *B*.

(*iv*) SELD : transfer the contents of the output bus into *R*1.

Do You Know

The group of binary bits assigned to perform a specified operation is known as control word.

At the beginning of a clock cycle, the four control selection variables generated *R*2 and *R*3 must be available in the control unit. Two source registers propagate through multiplexers and the ALU, to the output bus, and to the input of destination register, during the clock cycle interval. At the next clock transition, the information from output bus is transferred to the destination register *R*1.

Control Word

There are 14 binary selection inputs in the units, and their combined value specifies a control word. It consists of four field as shown in Fig. 3.8.

3	3	3	5
SELA	SELB	SELD	OPR

Fig. 3.8. Control Word.

Three fields contain three bits each, one field has five bits. The three bits of SELA select a source register for the *A* input of the ALU. The three bits of SELB select a register for the *B* input of the ALU. The three bits of SELD select a destination register using the decoder and its seven load outputs. The five bits of OPR select one of the operations in the ALU.

The 14-bit control word when applied to the selection inputs specify a particular microoperations. The encoding of register selections is specified in Table 3.2.

TABLE 3.2. Encoding of Register Selection Fields.

Binary code	SELA	SELB	SELD
000	input	input	none
001	*R*1	*R*1	*R*1
010	*R*2	*R*2	*R*2
011	*R*3	*R*3	*R*3
100	*R*4	*R*4	*R*4
101	*R*5	*R*5	*R*5
110	*R*6	*R*6	*R*6
111	*R*7	*R*7	*R*7

When the 3-bit binary code for SELA or SELB is 000, the respective multiplexer selects the external input data as shown in Fig. 3.8, and when the 3-bit binary code for SELD = 000, no destination register is selected and the content of output bus is for external output.

The OPR field has five bits. The encoding for five bit OPR field is specified in the Table 3.3.

Let the microoperation given by the statement is

$$R1 \longleftarrow R4 \wedge R5$$

TABLE 3.3. Encoding of ALU Operation.

OPR	Operation	Symbol
00000	Transfer *A*	TSFA
00001	Increment *A*	INCA
00010	Addition	ADD
00101	Subtract	SUB
00110	Decrement *A*	DECA
01000	AND *A* and *B*	AND
01010	OR *A* and *B*	OR
01100	XOR *A* and *B*	XOR
01110	Complement *A*	COMA
10000	Shift right *A*	SHRA
11000	Shift left *A*	SHLA

This statement specifies *R*4 for the *A* input of ALU, *R*5 for the *B* input of ALU, and *R*1 as the destination register. The microoperation to be performed is AND operation between *R*4 and *R*5. The control word for the above statement according to the Table 3.2 and Table 3.3 is as follows :

TABLE 3.4.

SELA	SELB	SELD	OPR
*R*1	*R*4	*R*5	AND
001	100	101	01000

Thus, the control word is 001 100 101 01000.

3.6. ADDITIONAL FEATURES OF CPU

The similarity between memory locations and individual registers within the processor, However, it should now be apparent that there is a major difference between the two in terms of the organisation and circuitry which is necessary to achieve access to locations. In more sophisticated processors, which may have several on-chip registers, there is the same requirement for a decoder and the difference in this case is merely one of scale. It is this difference in scale..in terms of both the number of gates within the decoder and the distance between the source and destination of data memory. Even with large memories, however, the response time is never so slow that it cannot be accommodated by the processor pausing for a few cycles during a memory access.

Apart form these relatively minor differences in organisation and speed of operation, of functions and implementation of memory and processor are so similar that there is no special arrangement required to allow the two devices to communicate with each other. The form of communication between the two is also particularly straightforward. The memory is totally dependent upon the processor for its operation, so that all signals and timing are supplied by the processor. The relationship between the two can be described as strictly master/slave.

In contrast to this, most of the other devices which we need to connect to the processor are electromechanical (an except being the VDU), so that the speeds of operation are far slower than for memory. In addition, there are difference in the way in which data is represented, both physically and logically, within the CPU and within devices. Finally, most devices are more complex than memory and require more control information form the processor to ensure correct operation. All these factors make for a larger disparity between the processor and devices than between the processor and memory. This requires a change in strategy used to communicate with the external devices-I/O devices or peripherals-which is apparent both in the hardware and software. Since the processor and peripheral are so different, it is usually not feasible to connect them directly as we did with memory. The peripheral will often be expecting data to be passed to it in a different format, and perhaps at a different speed, form the way in which it is represented and manipulated within the processor. A striking illustration of this is the representation of logic 0 and logic1. Within the processor, a typical representational scheme would be 0 = ~0.2 V, 1 = ~3.5 V; in many peripheral devices 0 = –12 V and 1 = +12 V is often used.

There is a clear requirement here for an intermediate device-called an interface, which can perform the conversion of the two electrical representations. By connecting the processor, peripheral and interface as shown in Fig. 3.9, it is possible for communication to take place between the processor and peripheral, despite their many differences in operating characteristics.

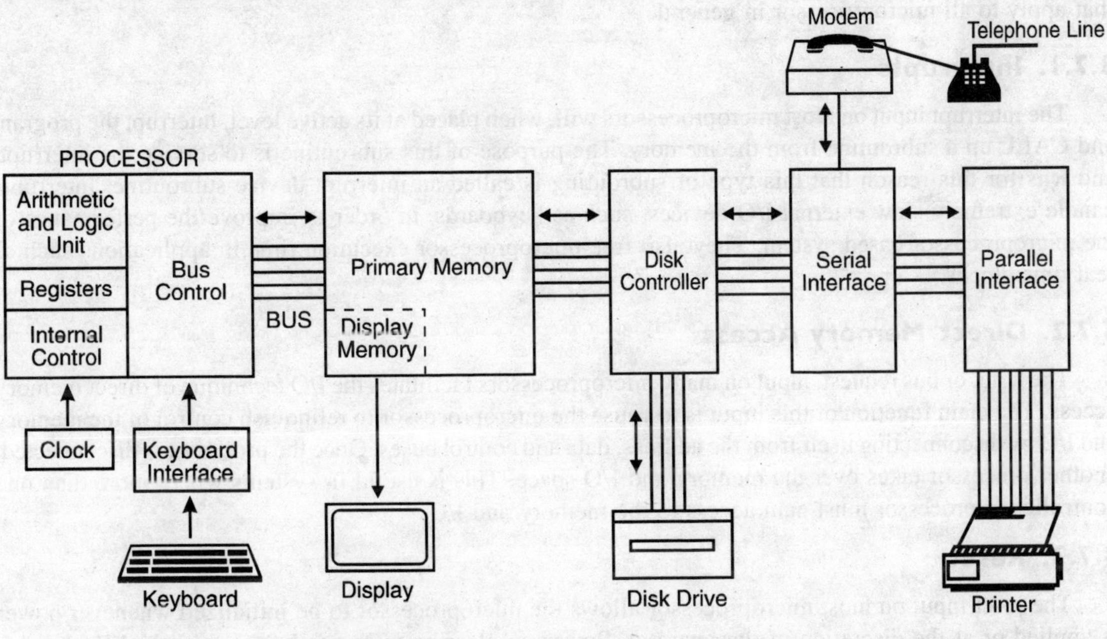

Fig. 3.9. The Connection of a Processor to Peripherals

We may expect there to be as many interfaces as there are peripherals. For specialised peripherals, this is usually so, and the interface is available as a separate chip, and this may meet the requirements of a wide variety of processor-peripheral connections. In this section, we will look at the general services which a peripheral interface would be expected to provide. We will then look at two specific examples of standard interfaces in the next section. You should then be in a position to deal with any specialised interface which you may encounter in the literature, although there is not the space to include them here.

3.7. I/O CONTROL

There are two different techniques for handling the control of I/O devices, which examining the control signals of different microprocessor will illustrate. One of these techniques is the memory-mapped I/O, in which I/O devices are treated as memory. In fact, with regard to software and hardware control signals, the I/O, is indeed memory.

The other technique is isolated I/O or I/O-mapped I/O, in which the I/O device is a separate I/O device. That is, the I/O device has a unique I/O address, not a memory address. Processors that use this type of I/O have special instructions, IN and OUT, to transfer data to and from this separate I/O space. Both techniques have advantages and disadvantages.

In the memory-mapped I/O scheme, there are two basic control signals: read and write In the isolated I/O scheme, there are four basic control signals: memory read, memory write, I/O read, and I/O write. The main advantage of the memory-mapped I/O system is that it requires only two control signals to function. The isolated I/O scheme requires four, but is does not require any of the memory space to function. Which one of the schemes is better? It depends upon the application and the microprocessor chosen.

In addition to these basic memory and I/O control signals, there are special inputs and outputs, such as interrupt control, direct memory access control, ready, reset, and various other processor-dependent control signals. This chapter will not discuss in detail processor-dependent control signals, only those that apply to all microprocessor in general.

3.7.1. Interrupts

The interrupt input on most microprocessors will, when placed at its active level, interrupt the program and CALL up a subroutine from the memory. The purpose of this subroutine is to service the interrupt, and it is for this reason that this type of subrouting is called an interrupt device subroutine. Interrupts handle extremely slow external I/O devices, such as keyboards, in order to improve the performance of the microprocessor based system. They also free microprocessor execution time in applications such as real time clocks.

3.7.2. Direct Memory Access

The hold, or bus request, input on many microprocessors facilitates the I/O technique of direct memory access. The main function of this input is to cause the microprocessor to relinquish control of the memory and I/O by disconnecting itself from the address, data and control buses. Once the processor is disconnected, another processor takes over the memory and I/O space. This is useful in systems where more than one controller or processor must gain access to the memory and I/O.

3.7.3. Reset

The reset input on most microprocessor allows the microprocessor to be initialized whenever power is applied or at the discretion of the operator. Power on clearing is normally accomplished by using a simple RC circuit. Whenever Vcc, or 5 V,is applied, the capacitor, which is initially discharged, begins to charge towards 5 V. this takes some time because the RC time constant of the circuit allows the microprocessor ample time to feel a logic zero on it while Vcc is applied to reset itself. The push button switch allows the operator to reset the microprocessor whenever it might be required. Upon resetting, many microprocessors are internally initialized, so that they fucnction at a predefined memory location fetching an instruction. This reset location, in some microprocessors a vector, is very important because it is at this location that the system program must begin. In the case of a vector, the vector points to the begining of the system software.

3.7.4. Wait States

The READY or MRDY pin provided on many microprocessors indicates to the microprossor that the memory and/or the I/O circuitry is ready to function. If a slow memory device.. a device with a longer-than-normal access time..is to be connected to the microprocessor, the ready input causes the microprocessor to enter into waiting states. Wait states are multiples of the system clock pulses and cause the system to allow more time for the memory to access data. Because of the wide range of timing among microprocessors, the best source of information about the operation and timing of this pin is the literature provided by the manufacturer, including, in some cases, the circuit recommended to cause wait states.

3.7.5. Clock Period

The clocking period will be discussed at this point, since it does not depend on understanding memory and I/O interfacing. There are normally two discrete limits to the clocking period that are determined by the internal structure of the microprocessor and must be observed for proper operation. The shortest clocking period depends on the type of logic family that comprises the microprocessor and is caused by internal propagation delay times. The longest allowable clocking period is determined by the fact that most microprocessors used dynamic storage in their register arrays. Dynamic register requires periodic refreshing to maintain the integrity of the data stored in them. If the clocking period is longer than that recommended by the manufacturer, the integrity of the data stored internally is lost.

The clock waveform or waveforms are normally generated by a crystal controlled oscillator that is incorporated into the microprocessor or available as an ancillary component from the manufacturer. The system hardware generally governs the operating period within the constraints allowed by the manufacturer.

Summary

- Processor is the brain of a computer that directly or indirectly does all the processing of computer.
- Program counter contains the address of next instruction to be executed by the CPU.
- Cache memory is the fastest memory of a computer system which is used to store frequently needed data of a CPU.
- Level 1 cache is directly integrated into the processor which is used to store both instruction and data.
- Level 2 cache memory is located in the case along with the processor which act as intermediate between internal cache and processor.
- Level 3 cache memory is located on the motherboard.
- Registers are used to temporarily store the data. There are registers of 4, 8, 16, 32 or 64 bit.
- Processor packaging means to package the processor in a material that protects them, allows them to dissipate heat properly, and has connectors of a standard size and shape.
- Processor socket or CPU slot is an electrical component that attaches to a printed circuit board (PCB) and is designed to house a microprocessor.

Review Questions

1. Explain the role of processor in a personal computer?
2. Write down the various processor generations and processor families?
3. Explain the components of a typical processor with its block diagram?
4. What is the need for standardized sockets and slots?
5. What do you mean by processor packaging? Explain any three processor packing?
6. Define overdrive processors. How do overdrive processors differ from regular processors?
7. What are the main criteria for selecting the right CPU?

□□□

4

Chapter

Design of Control Unit and Central Processing Unit

INSIDE THIS CHAPTER

4.1. CENTRAL PROCESSING UNIT

It is made up of three major parts as shown in Fig. 4.1.

(*i*) Register Set

(*ii*) ALU (Arithmetic and Logic Unit)

(*iii*) CU (Control Unit)

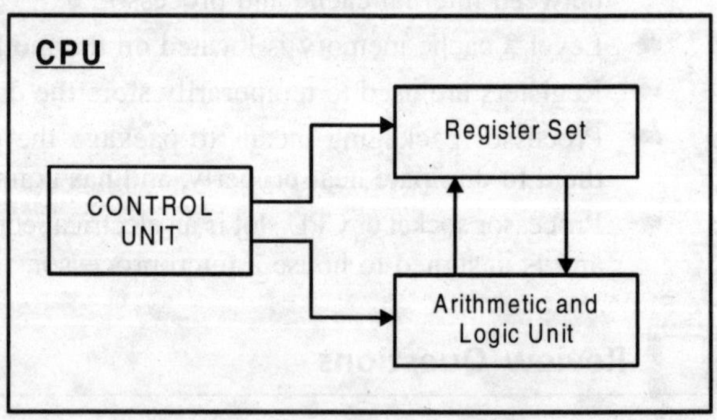

Fig. 4.1. Central Processing Unit.

Register set are used to store intermediate results obtained during the execution of instructions. ALU performs the arithmetic and logic microoperations required for executing the instructions. Control unit instructs the ALU to perform the desired operation and supervises the transfer of information among different registers.

Definition The part of the computer that performs the bulk of data processing operations is known as Central Processing Unit or CPU.

Generally, the computers have one or a combination of the following CPU organizations.

(*a*) Single accumulator organization.

(*b*) General register organization.

(*c*) Stack organization.

4.2. COMPONENTS OF CPU

Some of the general components upon which the processor is built are as follows :

4.2.1. Buses

In order to communicate with memory, a processor needs three types of connections data, address and control. The data lines are used to send or to receive data from memory. There is an individual connection or wire for each bit of data.

The address lines are controlled by the processor and are used to specify which memory location the processor wants to communicate with. The address is an unsigned binary integer that identifies a unique location where data elements are to be stored or retrieved.

The control lines consist of the signals that manages the transfer of data.

By using a bus, the processor can communicate with exactly one device at a time even though it is physically connected to many devices. If only one device on the bus is enable at a time, the processor can perform a successful data transfer. If two devices tried to drive the data lines simultaneously, the result will be data lost called bus contention.

4.2.2. Registers

Different CPU Registers are :

(*i*) **Accumulator (AC) :** A processor register (AC) is required for doing operations on data. This register holds data on which addition, subtraction, shift and logical operations are to be carried out. The result of an arithmetic and logical operation is automatically stored in the accumulator.

Thus, it is used for storing result and for doing operations (arithmetic as well as logical) on its content.

(*ii*) **Program Counter (PC) :** It deals with the order of the execution of instructions. It holds the address of the next instruction to be executed. Thus, it acts as a pointer which points to the memory location where the next instruction is stored.

(*iii*) **Temporary Register (TR) :** A register used for holding temporary data generated during processing.

(*iv*) **Instruction Register :** A register used for storing instruction is called Instruction Register. The instruction read from the memory is to be placed in some register known as Instruction Register.

(*v*) **Data Register (DR) :** Register used to hold the data (operand) read from memory.

(*vi*) **Address Register (AR) :** Register used to hold the address of memory word.

(*vii*) **INPR :** Input register will hold/receives data from an input device.

(*viii*) **OUTR :** It holds data that need to be sent to output devices.

Table 4.1 describes the brief function and the number of bits that different registers contain.

TABLE 4.1. List of Registers.

Register Symbol	*Number of bits*	*Register Name*	*Function of Register*
DR	16	Data Register	Holds memory operand
AR	12	Address Register	Holds address for memory
AC	16	Accumulator	Processor Register
IR	16	Instruction Register	Holds instruction code
PC	12	Program Counter	Holds address of next instruction
TR	16	Temporary Register	Holds temporary data
INPR	8	Input Register	Holds input character
OUTR	8	Output Register	Holds output character.

4.2.3. Flags

There are number of indicators known as flags that shows the processor's status. Most of these flags represent the results of the last operations. For example, the addition of two numbers might produce a negative sign, an overflow, a carry, or a value of zero.

These flags are represented by a single bit such as if the result of an addition is negative, the sign flag would set to 1. If the result was not a negative number (zero or greater than zero), the sign flag would equal to 0.

These flags are organized *i.e.* grouped together into a single register called the flag resister or the processor status register. Since the values stored in these flags are in the form of bit which are the outcome of an arithmetic or logical operation, and hence the flag registers are connected to the mathematical unit of a processor.

4.2.4. Stacks

During the execution of operation, there are number of times when the processor needs to use a temporary memory to store different data values so that they can be used again when required. Every processor has a finite number of registers. But if an application needs more registers than available, the register values that are not needed immediately by processor can be stored in the temporary memory. Also, when a processor needs to jump to a subroutine or function, it needs to remember the instruction from where it jumped so that it can return back to the same place when the subroutine is completed. Hence, the return address must be stored and this return address are generally stored in this temporary memory. This temporary memory is known as Stack.

The Stack is a block of memory locations reserved to functions as temporary memory. When a processor puts a piece of data, on the top of stack, the data below it cannot be removed until the data above it is removed. This type of memory location is referred as "Last-In-First-Out" or LIFO.

The two main operation that the processor can perform on the stack are PUSH and POP. It can either store the value of a register to the top of the stack or can remove the top element from the stack. Storing the data to the stack is referred to as pushing and removing the top data from stack is referred as poping.

4.2.5. I/O Ports

Input/Output ports, referred as I/O ports, is any connections that exist between the processor and its external devices. For example, a USB printer can be connected to the computer system through an I/O (USB) port. Using this port the computer can issue commands and send to be printed.

Some I/O devices are connected directly to the memory bus. Sending data to the port is done by storing data to a memory address and retrieving data from port is done by reading from a memory address.

4.3. SINGLE ACCUMULATOR ORGANIZATION

Earlier when the hardware were very expensive, the processor was designed usually as a single address machine with accumulator. The accumulator used to store operand for the instructions. It is used for storing result and for doing operations on the content of the accumulator. Thus, accumulators used as source for the operands as well as the destination for storing the result of the operations performed on its content. Thus, it accumulate datas. The accumulator based organization system results in reduced size of instruction, since the instructions contains only one operand address and the other operand is in accumulator itself. Also, after performing the operations the result is stored in the accumulator. Thus, these accumulator based organization systems are also known as 1-address machine, since only one operand address is specified in the instruction. But there can be only one or a limited number of accumulators in the CPU that affects the performance of the systems when an arithmetic expression is to be evaluated containing many different terms.

4.4. GENERAL REGISTER ORGANIZATION

A bus organization for seven CPU registers is shown in Fig. 4.2. The output of each registers is connected to two multiplexers (MUX) to form the two buses *A* and *B*. The selection lines in each multiplexer select one register or the input data for the particular bus. The *A* and *B* buses form the inputs to a common ALU. The operation selected in the ALU determines the arithmetic or logic microoperations to be performed. The result of the microoperation is available for the output data and also goes into the inputs of seven registers. The register that receives the information from the output bus is selected by the decoder. The decoder activates one of the register load inputs and thus providing a transfer path between the output data bus and the inputs of the selected destination register.

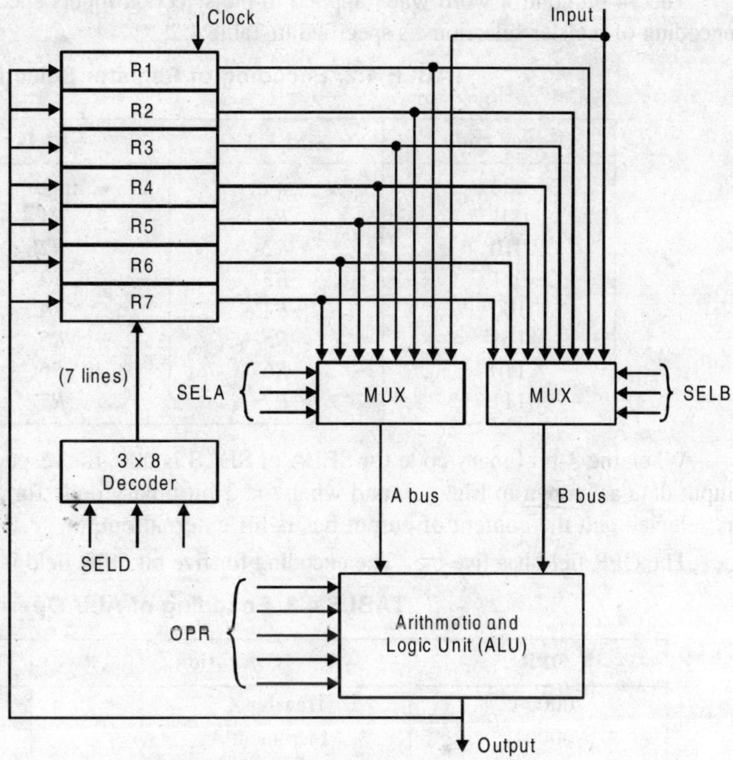

Fig. 4.2. General Register Organization.

Let the operation be $R1 \longleftarrow R2 + R3$

To perform this operation, the control must provide

(*i*) SELA : place the contents of $R2$ into bus A.

(*ii*) SELB : place the contents of $R3$ into bus B.

(*iii*) ALU operation selector OPR : provide the arithmetic addition $A + B$.

(*iv*) SELD : transfer the contents of the output bus into $R1$.

At the beginning of a clock cycle, the four control selection variables generated $R2$ and $R3$ must be available in the control unit. Two source registers propagate through multiplexers and the ALU, to the output bus, and to the input of destination register, during the clock cycle interval. At the next clock transition, the information from output bus is transferred to the destination register $R1$.

Control Word

Definition The group of binary bits assigned to perform a specified operation is known as control word.

There are 14 binary selection inputs in the units, and their combined value specifies a control word. It consists of four field as shown in Fig. 4.3.

3	3	3	5
SELA	SELB	SELD	OPR

Fig. 4.3. Control Word.

Three fields contain three bits each, one field has five bits. The three bits of SELA select a source register for the *A* input of the ALU. The three bits of SELB select a register for the *B* input of the ALU. The three bits of SELD select a destination register using the decoder and its seven load outputs. The five bits of OPR select one of the operations in the ALU.

The 14-bit control word when applied to the selection inputs specify a particular microoperations. The encoding of register selections is specified in Table 4.2.

TABLE 4.2. Encoding of Register Selection Fields.

Binary code	SELA	SELB	SELD
000	input	input	none
001	*R*1	*R*1	*R*1
010	*R*2	*R*2	*R*2
011	*R*3	*R*3	*R*3
100	*R*4	*R*4	*R*4
101	*R*5	*R*5	*R*5
110	*R*6	*R*6	*R*6
111	*R*7	*R*7	*R*7

When the 3-bit binary code for SELA or SELB is 000, the respective multiplexer selects the external input data as shown in Fig. 4.3, and when the 3-bit binary code for SELD = 000, no destination register is selected and the content of output bus is for external output.

The OPR field has five bits. The encoding for five bit OPR field is specified in the Table 4.3.

TABLE 4.3. Encoding of ALU Operation.

OPR	Operation	Symbol
00000	Transfer *A*	TSFA
00001	Increment *A*	INCA
00010	Addition	ADD
00101	Subtract	SUB
00110	Decrement *A*	DECA
01000	AND *A* and *B*	AND
01010	OR *A* and *B*	OR
01100	XOR *A* and *B*	XOR
01110	Complement *A*	COMA
10000	Shift right *A*	SHRA
11000	Shift left *A*	SHLA

Let the microoperation given by the statement is

$$R1 \longleftarrow R4 \wedge R5$$

This statement specifies *R*4 for the *A* input of ALU, *R*5 for the *B* input of ALU, and *R*1 as the destination register. The microoperation to be performed is AND operation between *R*4 and *R*5. The control word for the above statement according to the Table 4.2 and Table 4.3 is as follows :

SELA	SELB	SELD	OPR
*R*1 001	*R*4 100	*R*5 101	AND 01000

Thus, the control word is 001 100 101 01000.

4.5. STACK ORGANIZATION

Definition A stack is an ordered collection of items which permits the insertion or deletion of an item to occur only at one end.

The stack is also known as last-in first-out list. The stack can be consider as a storage method in which the item that stored last is the first item to be removed. The most common example of the stack phenomenon, is a pile of trays in a cafeteria. A tray which is placed last on the top of pile is the first to be taken off.

The stack in a digital computer is a part of memory unit. Also, with the stack an address register is associated that holds the address of the last element stored in the stack. This address register is known as *Stack Pointer* (SP). Thus, the stack pointer always points to the top most element of the stack.

Definition The insertion operation is known as push and the deletion operation is known as pop.

4.5.1. Push and Pop Operation

Insertion and deletion of items are the operations related with the stack. The process of inserting an item into the stack is known as *push operation*. The process of deleting an item from the stack is known a *pop operation*. These operation are done by incrementing or decrementing the Stack Pointer (SP).

4.5.2. Register Stack

A stack can be organized by a finite number of registers or a stack can be a finite number of memory words. The stack pointer contains the address of the word that is currently on the top of the stack. A 32-word register stack is shown in Fig. 4.4. The stack pointer contains a binary value. Currently, there are four items X1, X2, X3 and X4 are placed in the stack with X4 at the top of stack so the content of stack pointer is 4. The items are removed from the stack by using POP instruction. When we remove the top item X4 from the stack, X3 is now on top of stack and the content of SP is decrement so that the SP holds the address 3. To insert a new item, first the SP will incremented and then the item is inserted so that SP points to the top of the stack.

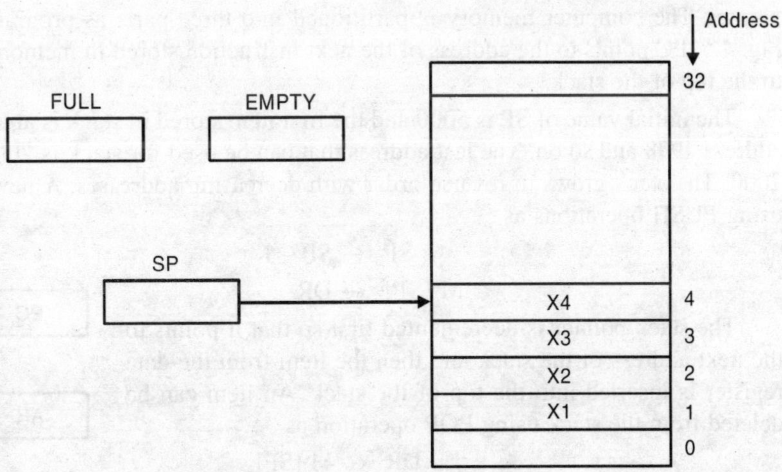

Fig. 4.4. Register Stack.

In a 32-word register stack, the address of each location will be of five bits since $2^5 = 32$. Thus, stack pointer will be of five bits and cannot exceed the value 11111. Thus, when the SP pointer content is 11111, the one-bit register FULL is set to 1, indicating that the stack is full and there is no location empty for any more item. Similarly when the content of SP = 00000 another one-bit register EMPTY is set to 1 indicating that the stack is empty and there is no element in the stack that can be deleted from the stack. The data register DR holds the items that is to written into the stack or read out of the stack.

Initially, the SP is cleared to 0 so the stack pointer points to the word at address 0. Also, the one-bit register FULL is cleared to 0, indicating that the stack is not full and the register EMPTY is set to 1. A new item is inserted into the stack by push operation. The PUSH operation will be the set of following microoperations :

$SP \leftarrow SP + 1$ Increment stack pointer

$M[SP] \leftarrow DR$ Add item on the top of stack

If $(SP = 0)$ then $(FULL \leftarrow 1)$ Check if stack is full

$EMPTY \leftarrow 0$ Mark the stack not empty.

If the stack is not empty, an item can be deleted from the stack using the POP operation. The POP operation is implemented by the following set of microoperations.

$DR \leftarrow M[SP]$ Read item from the top of stack

$SP \leftarrow SP - 1$ Decrement stack pointer

If $(SP = 0)$ then $(EMPTY \leftarrow 1)$ Check if stack is empty

$FULL \leftarrow 0$ Mark the stack not full.

The top item is read from the stack into DR, then the SP is decremented by 1 so that it points to top of stack. The SP is checked whether it is zero on not. If zero, EMPTY sets to 1 indicating that the stack is empty.

4.5.3. Memory Stack

A stack can also be implemented using random-access memory attached to the CPU. This can be implemented by assigning a portion of memory for stack operation using the processor register as a stack pointer. The computer memory is partitioned into three parts as program, data, and stack as shown in Fig. 4.5. PC points to the address of the next instruction stored in memory. The stack pointer (SP) points to the top of the stack.

The initial value of SP is 5000 and the first item stored in stack is at address 4999, the second item at address 4998 and so on. The last address that can be used for stack is 2000 *i.e.* the final value of stack is 2000. The stack grows in reverse order with decreasing addresses. A new item into the stack is inserted using PUSH operations as :

$$SP \leftarrow SP - 1$$
$$M[SP] \leftarrow DR$$

The stack pointer is decremented first so that it points to the next address of the stack and then the item from the data register is inserted into the top of the stack. An item can be deleted from the stack using POP operation as

$$DR \leftarrow M[SP]$$
$$SP \leftarrow SP + 1$$

The top item of the stack is read into the data register DR and then the stack pointer is incremented by 1 so that it points to the top item of the stack.

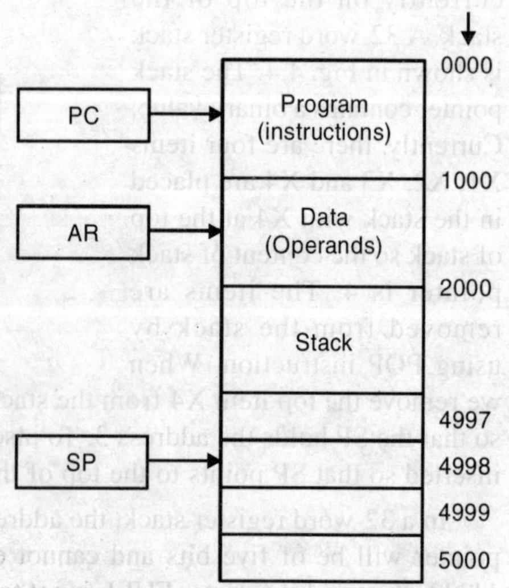

Fig. 4.5. Computer Memory Divided into Segments Program, Data and Stack.

Most of the computer does not provide any method to check the stack overflow or underflow to check whether the stack is full or empty. One possible method is to used two processor registers holding the addresses 2000 (upper limit) and 5000 (lower limit) respectively. Then, stack pointer is compared every time the push operation takes place with the upper-limit register and its with the lower-limit register, after the pop operation takes place.

4.5.4. Reverse Polish Notation

Let us consider an expression $x + y$. The plus operator is placed in between the two operands x and y. Such a notation is known as *infix notation*. If the operator is placed before the two operands as $+xy$, the notation is said to be *prefix notation*, also known as polish notation. If the operator is placed after the two operands as $xy+$, the notation is said to be *postfix notation*, also known as **Reverse polish notation**. Thus, the three notation are

$x + y$ Infix notation

$+ xy$ Prefix or polish notation

$xy +$ Postfix or reverse polish notation.

For Stack manipulation the reverse polish notation is best suited. The reverse polish notation for the expression $A * B + C * D$ is $AB * CD * +$

Conversion to Reverse Polish Notation

The conversion of an expression from infix notation to the reverse polish notation must be done according to the operational hierarchy that follows for infix notation. First perform all arithmetic operations inside the inner parentheses, then inside outer parantheses, then do multiplication and division operations and lastly the addition and subtraction operations.

Example 4.1. Convert the infix expression $(x + y) * [z * (w + v) + s]$ into reverse polish notation.

Solution. The two sub-expression $(x + y)$ and $(w + v)$ will solved first. Thus the postfix expression of these subexpression will be $xy+$ and $wv+$ respectively.

Now, in the square bracket z will be multiplied by $(w + v)$. Thus, the postfix of this multiplication is $zwv + *$.

This multiplication result is then added to s will result in $zwv + *s +$

Finally, $xy+$ and $zwv + *s +$ will be multiplied together to get

$$xy + zwv + *s + *.$$

The procedure is shown again as

$$\begin{aligned}(x + y) * [z * (w + v) + s] &= xy + * [z * wv ++ s]\\ &= xy + * [zwv + * + s]\\ &= xy + * zwv + *s +\\ &= xy + zwv + *s + *\end{aligned}$$

Example 4.2. Convert the infix notation $A * B + A * (B * D + C * E)$ into RPN.

Solution. $A * B + A * (B * D + C * E)$

$$\begin{aligned}&= AB* + A * (BD* + CE*)\\ &= AB* + A * BD * CE * +\\ &= AB * + ABD * CE * + *\\ &= AB * ABD * CE * + * +\end{aligned}$$

4.5.5. Evaluation of Arithmetic Expression

Consider an expression $A * B + C * D$ in infix notation. Its reverse polish notation is $AB * CD * +$. This postfix expression will be evaluated as follows : scan the expression from left to right. Whenever an operator is found, perform the operation with the two operands on the left side of operator. Remove the operator and the two operands and replaced them by the result obtained by performing that operation. Continue in the same manner and repeat the procedure for every operator found until there are no more operators.

Thus, for the reverse polish notation $AB * CD* +$ first we find the operator $*$ and the two operands to the left of $*$ are A and B. Thus, we perform $A * B$ and replace A, B and $*$ by the product we get

$$(A * B)\ CD* +$$

The next operator is $*$ and the two operands to the left of $*$ are C and D. Thus, we perform $C * D$ and replace C, D and $*$ by the product, we get

$$(A * B)\ (C * D) +$$

The next operator is + and the two operands to the left + are the two products $(A * B)$ and $(C * D)$, hence the result obtained is

$$A * B + C * D.$$

Any arithmetic expression can be evaluated using stack as follows :

(*i*) Convert the given infix expression into its equivalent reverse polish notation.

(*ii*) Scan the expression from left to right.

(*iii*) While scanning when operands are found, pushed them into the stack as they appears.

(*iv*) When operators are found, pop two top most operands from the stack, perform the operation involving the operator and then pushed back the result into the stack.

(*v*) Continue, scanning the expression until there are no more operators.

(*vi*) Finally, the result of the expression will remain on the top of the stack.

To illustrate this, consider the expression $(A + B) * (C + D)$. In reverse polish notation this expression is $AB + CD + *$. The stack operation is shown in Fig. 4.6. The arrow (→) points to the top of the stack.

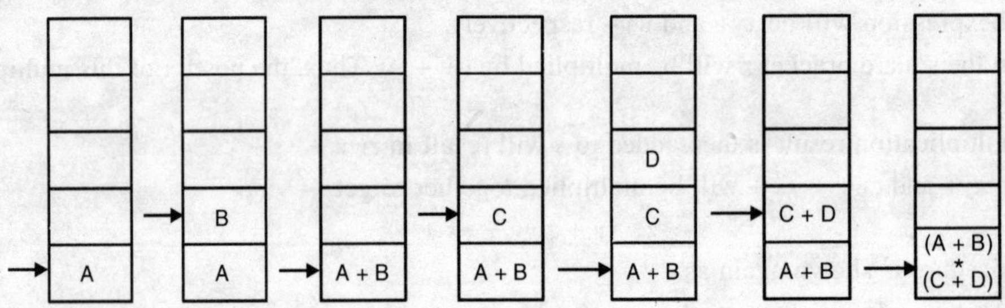

Fig. 4.6. Stack Operation to Evaluate (A + B) and (C + D).

4.5.6. Uses of a Stack

(*a*) Temporary storage.

(*b*) Reversing strings.

(*c*) Evaluation of Reverse Polish Notation (RPN) expressions.

(*d*) Conversion of Infix to Reverse Polish Notation.

(*e*) Subroutine calls.

(*f*) Zero address instructions.

4.6. INSTRUCTION FORMATS

Definition An instruction is a command given to a computer to perform a specified operation on some given data and the format in which the instruction is specified is known as Instruction Format.

The most common fields found in the instruction are :

(*i*) An operation code field that specifies the operation to be performed. It is known as opcode field.

(*ii*) An address field that designates the register address and/or a memory address.

(*iii*) A mode field that specifies the way the operands or the effective address is determined.

For example,

ADD *R*1, *R*0. ADD is the opcode and *R*1, *R*0 are the address field.

Operations specified by computer instructions are executed on some data stored in memory or some registers. Operands residing on memory are specified by memory address and operands residing on processor registers are specified by register address. A register address is a binary number of K-bits that defines one of 2^K registers in the CPU. Thus, if a CPU has processor registers R0 to R15, then address of each registers will be of four bits. For example, the binary information 0101, is the address of register *R*5.

The instruction may be of several different lengths containing different number of addresses. The number of address fields in the instruction format of a computer system depends on the internal architecture/organization of registers. The different types of CPU organization are :

(1) Single Accumulator Organization

(2) General Register Organization

(3) Stack Organization

4.6.1. Accumulator-Type Organization

All operations are performed with an implied accumulator register. The instruction format uses one address field, *i.e.* only one operand address is specified in the instruction. The other operand is in the accumulator. The result is placed in the accumulator.

For example,

ADD *X*, $\quad AC \longleftarrow AC + M[X]$

The ADD *X* instruction means add the content at memory location *X*, symbolizes as *M*[*X*], with the content of accumulator. Thus, the previous value of accumulator will be lost and the accumulator contain the result of above instruction.

4.6.2. General-Register Organization

The instruction format in this type of computer needs two or three addresses. The number of addresses in the instruction can be reduced to two from three if the destination register is same as one of the source registers.

In two address instructions both operand address are specified. The result is placed in one of the specified addresses. In three-address instruction two addresses are specified for the two operands and one address of the result. Thus, general-register-type computers employ two or three address fields in the instruction format. Each address field may specify a processor register or a memory location.

Examples :

ADD *R*1, *R*2, *R*3 $\quad R1 \longleftarrow R2 + R3$

The above instruction contains three register addresses. The operation performed is the add operation between the content of processor register *R*2 and *R*3 and result is to be placed into the destination register *R*1.

ADD *R*1, *R*2 $\quad R1 \longleftarrow R1 + R2$

The above instruction consists of only two register addresses. *R*1 and *R*2 are source registers where *R*1 also serves the purpose of destination register. The instruction specifies the add operation between the contents of *R*1 and *R*2 and result to be stored into *R*1.

MOV *R*1, *R*2 $\quad R1 \longleftarrow R2$

Mnemonic MOV is used for transfer instruction. The instruction contains only two register address *R*1 and *R*2 where *R*2 is the source register and *R*1 is the destination. Thus, in transfer-type instruction only two addresses are required. The instruction specifies move the content of *R*2 into register *R*1.

Add $R1$, X $R1 \longleftarrow R1 + M[X]$

This instruction has two address field, $R1$ the register address and X a memory address.

4.6.3. Stack Organization

Stack-oriented machines do-not contain any accumulator or general-purpose registers. Computers with stack organization have PUSH and POP instructions which requires an address field. Thus the instruction

PUSH X TOP $\longleftarrow M[X]$

will push the word/data at address X to the top of the stack. The SP is automatically updated. The operation instruction does not contain any address field because the operation is performed on two top most operands of the stack. For example,

ADD

The instruction ADD consist of only operation code with no address field. This instruction pops the top two operands from the stack, add the numbers and then PUSH the result into the stack.

Example 4.3. To show how the number of address affects a computer program, we will evaluate the arithmetic statement

$$X = (A + B) * (C + D)$$

using three, two, one or zero address instruction.

Solution. ADD, SUB, DIV and MUL are used for arithmetic operations, MOV for the transfer-operation. LOAD and STORE for transfers to and from memory and AC register. We will assume that the operands are in memory addresses A, B, C and D and the result must be stored in memory address X. $R1$, $R2$ are the register and T is the address of temporary memory location used to store intermediate result.

Three-Address Instruction

ADD R1, A, B	R1 ⟵ M[A] + M[B]
ADD R2, C, D	R2 ⟵ M[C] + M[D]
MUL X, R1, R2	X ⟵ R1 ∗ R2

The symbol M[A] denotes the operand at memory address symbolized by A.

The advantage of the three-address format is that it results in short programs when evaluating arithmetic expressions.

Two-Address Instruction

MOV R1, A	R1 ⟵ M[A]
ADD R1, B	R1 ⟵ R1 + M[B]
MOV R2, C	R2 ⟵ M[C]
ADD R2, D	R2 ⟵ R2 + M[D]
MUL R1, R2	R1 ⟵ R1 ∗ R2
MOV X, R1	M[X] ⟵ R1

One-Address Instruction

One address instruction use an accumulator (AC) register for all data manipulation.

LOAD A	AC ⟵ M[A]
ADD B	AC ⟵ AC + M[B]
STORE T	M[T] ⟵ AC
LOAD C	AC ⟵ M[C]

ADD D	AC ⟵ AC + M[D]
MUL T	AC ⟵ AC ∗ M[T]
STORE X	M[X] ⟵ AC

All operations are done between the AC register and a memory operand. T is the address of a temporary memory location required for storing intermediate result.

Zero-Address Instruction

To evaluate arithmetic expression for zero-address machine, the expression must be in reverse-polish notation. Also, the instructions like ADD, MUL does not requires an operand field. It simply pop-up the two top most operands from the stack, perform the operation and places the result on the top of the stack. However, PUSH and POP instructions requires an address field to specify the operand that communicates with the stack. TOS stands for top of stack.

The reverse polish notation of expression.

X = (A + B) ∗ (C + D) is evaluated as

= (AB+) ∗ (CD+)

= AB + CD + ∗

PUSH A	TOS ⟵ A
PUSH B	TOS ⟵ B
ADD	TOS ⟵ A + B
PUSH C	TOS ⟵ C
PUSH D	TOS ⟵ D
ADD	TOS ⟵ C + D
MUL	TOS ⟵ (A + B) ∗ (C + D)
POP X	M[X] ⟵ TOS

Example 4.4. Evaluate the following arithmetic expressions using 3-address, 2-address, 1-address and 0-address instructions to show the effect on computer program.

X = (A + B) * (C – D)

Solution. In the above arithmetic expression, variables A, B, C and D are given. We will assume that the operands are stored at memory addresses, A, B, C and D. The result will be stored at memory address X. In the program, ADD SUB and MUL are used for arithmetic operations. MOV is used for the transfer operations. LOAD and STORE are used for transfers to and from memory and AC register. R1, R2 are general purpose registers and T is the address of tempory memory location used to store intermediate result.

Three-Address Instructions

ADD R1, A, B	R1 ⟵ M[A] + M[B]
SUB R2, C, D	R2 ⟵ M[C] – M[D]
MUL X, R1, R2	M[X] ⟵ R1 * R2

Two-Address Instruction

MOV R1, A	R1 ⟵ M[A]
ADD R1, B	R1 ⟵ R1 + M[B]
MOV R2, C	R2 ⟵ M[C]
SUB R2, D	R2 ⟵ R2 – M[D]
MUL R1, R2	R1 ⟵ R1 * R2
MOV X, R1	M[X] ⟵ R1

One-Address Instruction

LOAD A	AC ⟵ M[A]
ADD B	AC ⟵ AC + M[B]
STORE T	M[T] ⟵ AC
LOAD C	AC ⟵ M[C]
SUB D	AC ⟵ AC – M[D]
MUL T	AC ⟵ AC * M[T]
STORE X	M[X] ⟵ AC

Zero-Address Instruction

Given expression is

$$X = (A + B) * (C - D)$$

Reverse polish notation of the given expression is evaluated as

$$X = AB + * CD -$$
$$= AB + CD - *$$

PUSH A	TOS ⟵ A
PUSH B	TOS ⟵ B
ADD	TOS ⟵ (A + B)
PUSH C	TOS ⟵ C
PUSH D	TOS ⟵ D
SUB	TOS ⟵ (C – D)
MUL	TOS ⟵ (A + B) * (C – D)
POP X	M[X] ⟵ TOS

4.7. ADDRESSING MODES

Definition The techniques for specifying the address of the operands are known as addressing modes. The address of an operand is known as effective address.

Each instruction needs data on which it has to perform the specified operation. The operand (data) may be in accumulator, general purpose register or at some specified memory location. Thus, there are various ways of specifying the address of the data, known as *addressing modes.*

The different addressing modes are:

(1) Implied addressing
(2) Immediate addressing
(3) Direct (absolute) addressing
(4) Indirect addressing
(5) Register addressing
(6) Register indirect addressing
(7) Relative addressing
(8) Index addressing
(9) Base register addressing

Instruction Cycle

(1) Fetch the instruction from the memory.
(2) Decode the instruction
(3) Execute the instruction.

Program Counter (PC) counter keeps track of the instructions in the program stored in the memory. PC holds the address of the next instruction to be executed and is incremented each time an instruction is fetched from the memory. The decoding determines the operation to be performed, the addressing mode of the instruction and the location of the operands. The computer then executes the instruction and then returns to the step 1 to fetch the next instruction in sequence.

OPCODE	MODE	ADDRESS

There are two modes that need no address field at all. They are implied and immediate. Zero-Address instructions in a stack-organized computer are implied mode instructions since the operands are implied to be on top of the stack.

4.7.1. Implied Addressing Mode

Implied addressing mode is also known as implicit or inherent addressing mode. The operands are specified implicitly in the definition of the instruction itself.

For example,

"Complement Accumulator" is an implied mode instruction because the operand in the accumulator register is implied in the definition of the instruction.

CMA : Take complement of the content of AC.

RLC : Rotate the content of the Accumulator.

All reference instruction that use an accumulator are implied-mode instructions.

4.7.2. Immediate Addressing Mode

In this mode the operands is specified in the instruction itself, *i.e.* in immediate addressing mode, instruction has an operand field rather than address field. The operand field contains the actual operand. This mode is useful for initializing registers to a constant value.

INSTRUCTION

	OPERAND

For example,

MVI 06	Move 06 to the accumulator
ADD 05	Add 05 to the content of the accumulator

Merits : The operand is available in the instruction as soon as the instruction fetch is over. Hence, the instruction executes fast.

Demerits : The value of the operand is limited by the length of the operand field in the instruction.

4.7.3. Register Addressing Mode

In Register addressing mode the operands are in registers that reside within the CPU, *i.e.* the contents of the register is the operand itself. The instruction contains the register number that has the operand. This addressing mode is very useful to a long program in storing the intermediate results in the registers rather than in memory.

OPCODE	REGISTER NUMBER (R = OPERAND)

For example,

MOV R1, R2	R1 ← R2, Transfer the content of register R2 to that of register R1
ADD R1	AC ← AC + R1, Add the content of register R1 to that of accumulator
LD R1	AC ← R1 LOAD, the content of register R1 to the accumulator

Register Addressing Mode

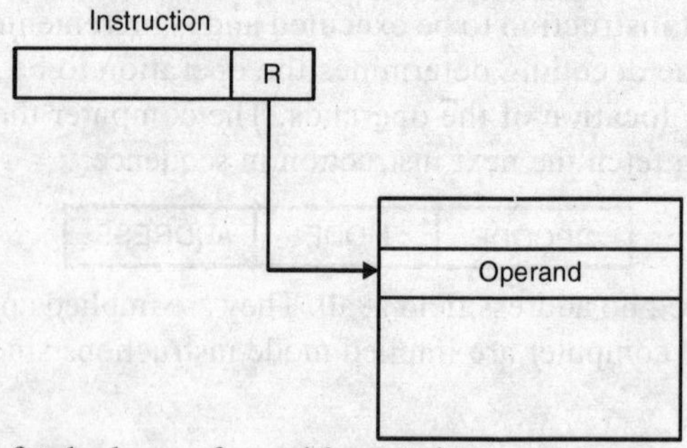

Merits : Operands are fetched very fast without using memory.

Demerits : Since the number of registers are very limited, hence the utilization of those register is done very carefully.

4.7.4. Register Indirect Addressing Mode

The instruction specifies a register in the CPU whose contents give the address of the memory location where the operand is stored, *i.e.* the selected register contains the memory address of the operand rather the operand itself. In this mode the register acts as the memory address register.

For example,

LD(R1) AC ← M[R1]

Register Indirect Addressing Mode

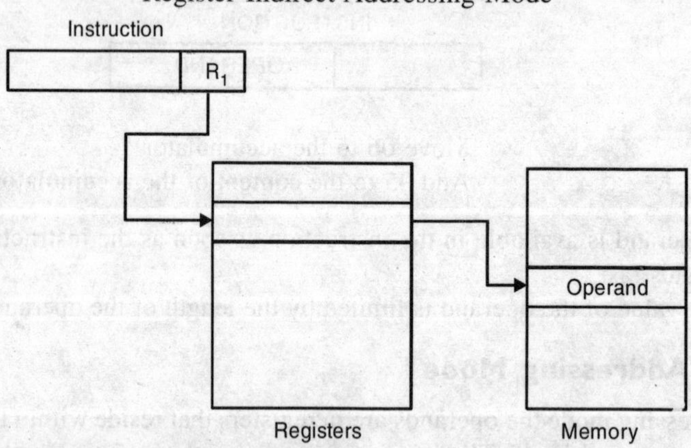

Merits : Since the register number is specified with bits, the length of the instruction must be decided very carefully.

4.7.5. Direct Addressing Mode

Direct addressing mode is also known as *absolute addressing mode*. In this mode the address of data *i.e.* operand is specified in the instruction itself, *i.e.* the operand resides in the memory and its address are given directly by the address field of the instruction.

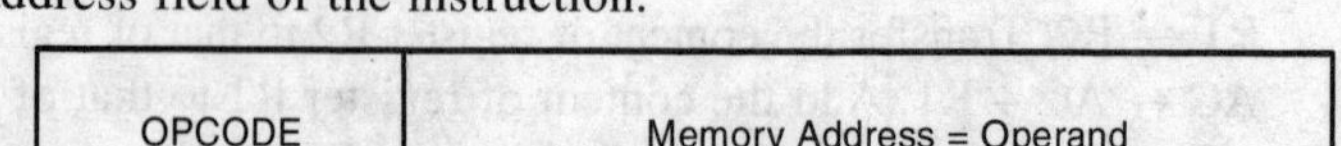

Direct Addressing Mode

For example,

LD ADR   AC ⟵ M[ADR]

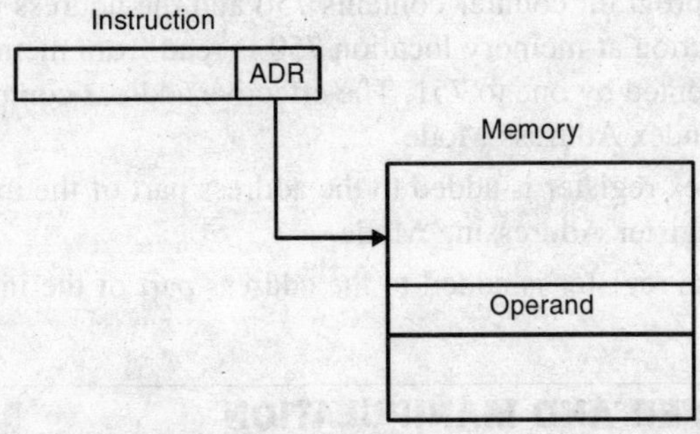

Merits : The instruction itself contains the operand address, there is no need to find the effective address of the operand. Hence, instruction executes fast.

Demerits : For specifying operand address the number of bits is limited.

4.7.6. Indirect Addressing Mode

In this mode the address field of the instruction gives the memory address where the operand is stored in the memory.

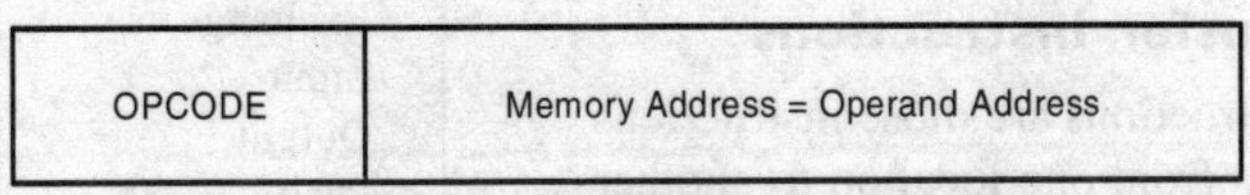

Indirect Addressing Mode

For Example,

LD @ ADR   AC ⟵ M[M[ADR]]

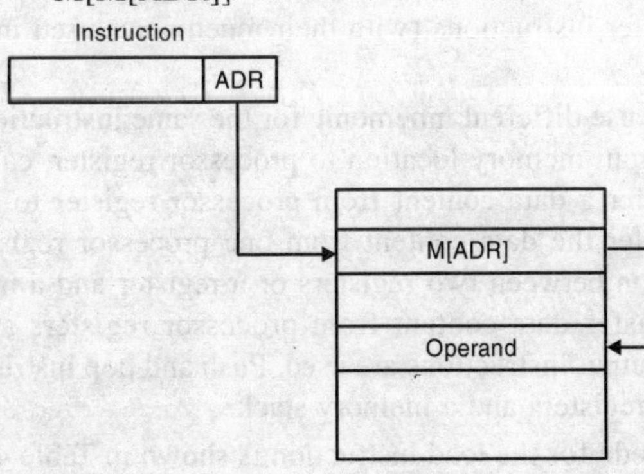

4.7.7. Displacement Addressing Mode

Displacement addressing modes are of three types:

(*i*) Relative Addressing Mode.

(*ii*) Indexed Addressing Mode

(*iii*) Base Register Addressing Mode

These addressing modes require that the address field of the instruction be added to the content of a specific register in the CPU to get the effective address. Thus, effective address = address part of instruction + content of the CPU register

The CPU register used in the computation of effective address may be program counter, an index register or a base register. Relative Addressing Mode.

The content of the program counter is added to the address part of the instruction in order to obtain the effective address.

For example, Let the program counter contains 750 and the address part of the instruction contains the number 35. The instruction at memory location 750 is read from memory during fetch phase and the program counter is incremented by one to 751. The effective address computation for the relative address mode is 751 + 35 = 786. Index Address Mode.

The content of the index register is added to the address part of the instruction in order to obtain the effective address. Base Register Addressing Mode.

The content of the base register is added to the address part of the instruction in order to obtain the effective address.

4.8. DATA TRANSFER AND MANIPULATION

Computer instructions are broadly classified into three different categories

(*i*) Data transfer instructions

(*ii*) Data manipulation instructions

(*iii*) Program control instructions.

4.8.1. Data Transfer Instructions

Data transfer instructions are those instructions that transfers the data from one location to another without changing the data content. These transfers can be between the two processor registers or between the memory location and processor registers or between the processor registers and input or output. Different data transfer instructions (with their mnemonic) used in many computers are listed in Table 4.4.

TABLE 4.4. Data Transfer Instructions.

Name	Mnemonic
Load	LD
Store	ST
Move	MOV
Exchange	XCH
Input	IN
Output	OUT
Push	PUSH
Pop	POP

Different computer can use different mnemonic for the same instruction. The load instruction is used to transfer a data content from memory location to processor register, called an accumulator. The store instruction is used to transfer a data content from processor register to some memory location. Move instruction is used to transfer the data content from one processor register to another. Whenever it is required to swap information between two registers or a register and a memory location, the exchange instruction is used. To transfer data content from processor registers and input or output terminals, respectively the input and output instructions are used. Push and pop instructions are used to transfer data content between processor registers and a memory stack.

Different addressing mode for the load instruction is shown in Table 4.5, where ADR is the memory address, NBR is the number or operand, X is index register, $R1$ is the processor register and AC is accumulator.

TABLE 4.5. Addressing Mode for Load Instruction.

Addressing Mode	Instruction	Register Transfer
Direct	LD ADR	AC ← M[ADR]
Indirect	LD @ ADR	AC ← M[M[ADR]]
Register	LD R1	AC ← R1
Register Indirect	LD (R1)	AC ← M[R1]
Relative	LD$ ADR	AC ← M[PC + ADR]
Index	LD ADR (X)	AC ← M[ADR + XR]
Immediate	LD # NBR	AC ← NBR

The character @ before memory address indicates indirect address. In case of register indirect mode, the register that holds the memory address is enclosed in parentheses. The character $ before memory address makes the address relative to the program counter PC. The character # before the operand indicates immediate mode instruction.

4.8.2. Data Manipulation Instructions

Data manipulation instructions are those instructions that perform arithmetic, shift or logic operations to manipulate the data. Thus, data manipulation instructions are broadly divided into three basic categories :

(*i*) Arithmetic instructions
(*ii*) Shift instructions
(*iii*) Logic instructions

4.8.2.1. Arithmetic Instruction

Addition, subtraction, multiplication and division are the four basic arithmetic operations. Most of the computers provide the instructions to perform these operations. Increment (or decrement) instructions adds 1 (or subtracts 1) to the value stored in a register or some memory word. A list of standard arithmetic instructions is shown in Table 4.6.

TABLE 4.6. Arithmetic Instructions.

Name of Instruction	Mnemonic
Add	ADD
Subtract	SUB
Multiply	MUL
Divide	DIV
Increment	INC
Decrement	DEC
Negate (2's complement)	NEG
ADD with carry	ADD *C*
Subtract with Borrow	SUB *B*

4.8.2.2. Shift Instruction

Shift operations can be a circular or arithmetic shift or can be simple logical shift in which the bits of a word moved to the left or to the right. For both the cases, logical shift-left or logical shift right, 0 is inserted at the end bit position. Rotate instructions produces circular shift. It circulates the bits around

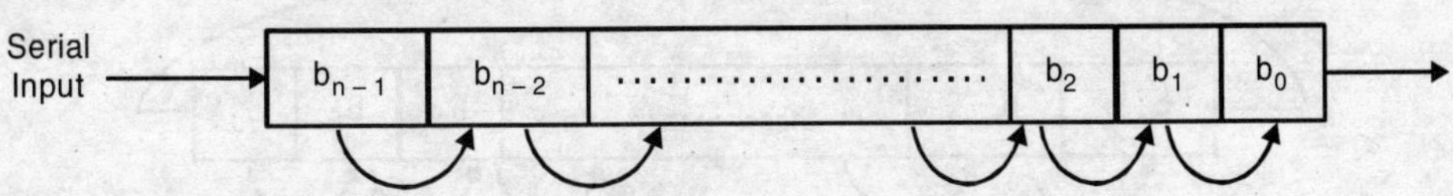

(*a*) Shift right operation.

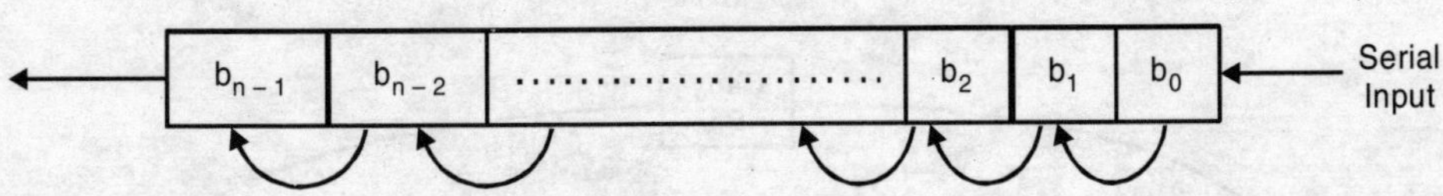

(*b*) Shift left operation.

Fig. 4.7.

the two end without loss of information. The instructions like rotate through carry treats a carry bit as an extention of the register whose word is being rotated. Thus, rotate-left through carry instructions transfers the carry bit into the rightmost bit position and transfers the leftmost bit into the carry and at the same time shifts the entire register to left. The arithmetic shift-left instruction inserts 0 at the rightmost bit position. The arithmetic shift right instructions leaves the sign bit unchanged and shifts the bit (including the sign bit) to the right, and the rightmost bit is lost. Basic shift instructions are listed in Table 4.7.

The shift instructions are used to shift the contents of the operand one bit to the right or left. During a shift-right operation, the serial input transfers a bit into the leftmost bit position and during a shift-left operation, the serial input transfers a bit into the rightmost position as shown in Fig. 4.7.

Circular Shift

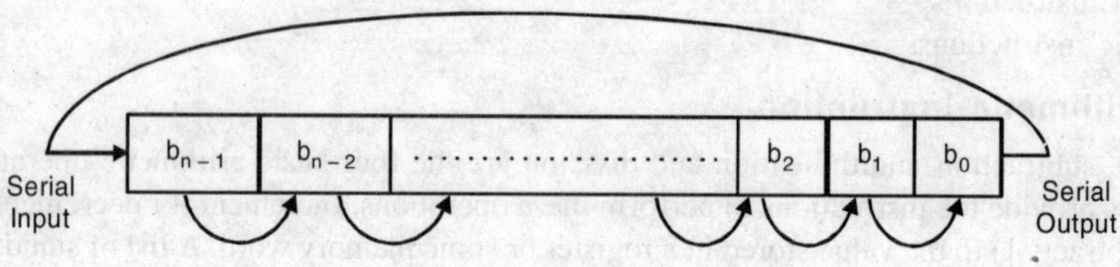

(a)

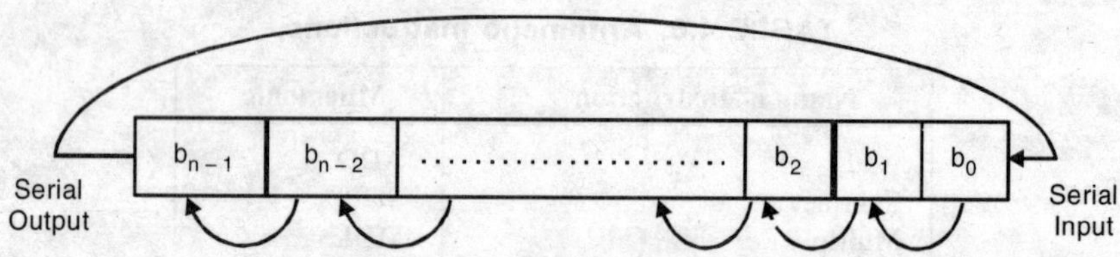

(b)

Fig. 4.8. (a) Rotate right, (b) Rotate left.

It is also known as rotate operation. In this operation, the bits are circulated in the register around the two ends without the loss of any information. This circular shift operation is achieved by connecting the serial input of the shift register to the serial output as shown in Fig. 4.8.

Rotate Through Carry

This operation treats the carry bit as a part of operand in the register. The rotate left through carry

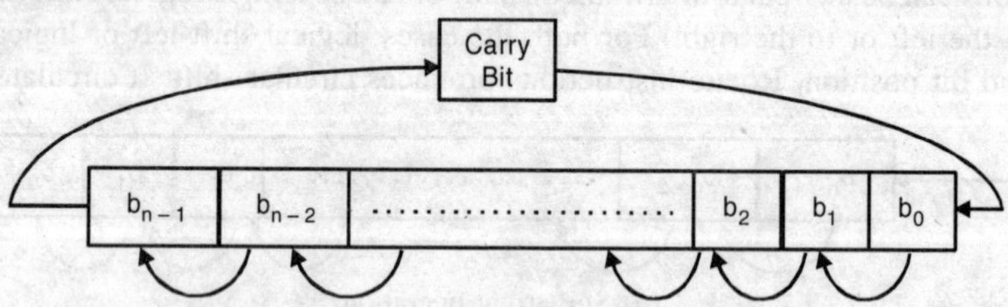

(a)

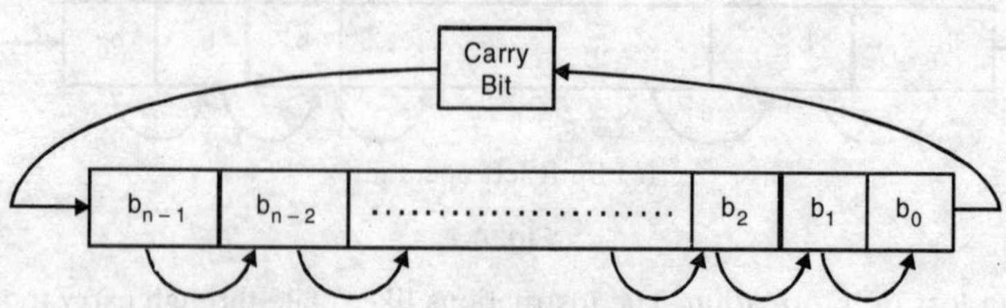

(b)

Fig. 4.8. (a) Rotate left through carry, (b) Rotate right through carry.

transfers the carry bit into the rightmost bit position (*i.e.* least significant bit position) and transfers and leftmost bit (*i.e.* bit at the most significant bit position) into the carry and at the same time all other bits are shifted to the left. Similarly, the rotate right through carry transfers the carry bit into the leftmost bit position (*i.e.* most significant bit position) and transfers the rightmost bit (*i.e.* bit at the least significant bit position) into the carry and at the same time all other bits are shifted to the right. The two operations are shown in Fig. 4.9.

Arithmetic Shift

The arithmetic shift operation shifts a signed binary number to the left or right. An arithmetic shift right operation divides the binary number by two and an arithmetic shift left operation multiplies the binary number by two. The two arithmetic shift operations leave the sign-bit unchanged. Multiplying or dividing the number by two does not change the sign of the original number. Hence, the sign of the number must be same. The two operation are shown in Fig. 4.10.

The most significant bit, b_{n-1}, holds the sign bit and the remaining bits represents the binary number. The arithmetic shift right leaves the sign bit unchanged and shifts the bit, including the sign bit, to the right. The rightmost bit is lost. The arithmetic shift left inserts 0 into b_0 and all other bits are shifted towards left. The bit, b_{n-1}, is shifted into a flag register and is replaced by bit b_{n-2}. If the bit b_{n-1} and b_{n-2}, differ in their values a sign reversal occurs. Thus, b_{n-1} changes in value after the arithmetic shift left operation. This happens if the multiplication by 2 results in an overflow. The overflow occurs if before shift operation, bit $b_{n-1} \neq b_{n-2}$. This overflow stituation can be found using

$$V_S = b_{n-1} \oplus b_{n-2}.$$

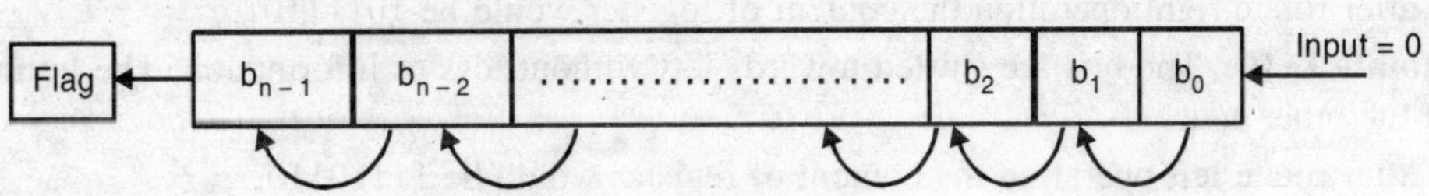

(*a*) Arithmetic shift left.

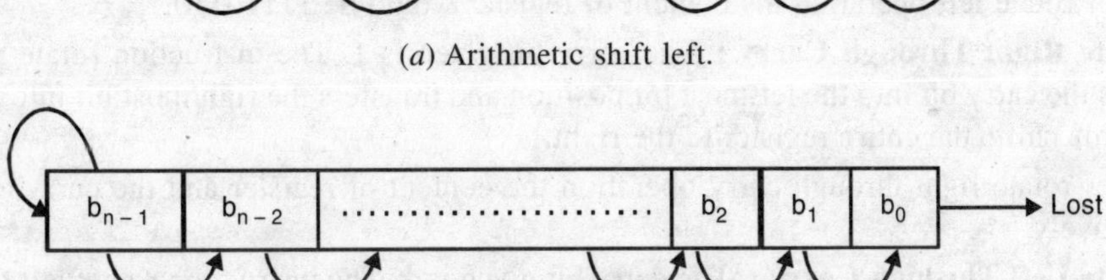

(*b*) Arithmetic shift left.

Fig. 4.10.

If $V_S = 0$, it means $b_{n-1} = b_{n-2}$ in values and there will be no overflow. But if $V_S = 1$, it means $b_{n-1} \neq b_{n-2}$ and there will be an overflow and hence a sign reversal is required after the shift operation.

TABLE 4.7. Shift Instructions.

Name of Instruction	*Mnemonic*
Logical shift left	SHL
Logical shift right	SHR
Rotate left	ROL
Rotate right	ROR
Arithmetic shift left	SHLA
Arithmetic shift right	SHR A
Rotate left through carry	ROL C
Rotate right through carry	ROR C

Example 4.5. An 8-bit register contains the value 01111011 and the carry bit is equal to 1. Perform the following shift operations given by instructions. Each time, start from the initial value given above.

(*a*) Logical shift right (SHR)

(*b*) Logical shift left (SHL)

(*c*) Rotate right (ROR)

(*d*) Rotate left (ROL)

(*e*) Rotate right through carry (RORC)

(*f*) Rotate left through carry (ROLC)

(*g*) Arithmetic shift right (SHRA)

(*h*) Arithmetic shift left (SHLA)

Solution. Given, the content of 8-bit register 01111011.

(*a*) **Logical Shift Right :** In logical shift right operation, the bits are shifted to the right and the rightmost bit is lost and 0 is inserted at the other end.

Thus, after logical shift right operation the content of the 8-bit register would be 00111101.

(*b*) **Logical Shift Left :** In logical shift left operation, the bits are shifted to the left and the leftmost bit is lost and 0 is inserted at the other end.

Thus, after logical shift left operation the content of the 8-bit register would be 11110110.

(*c*) **Rotate Right :** The bits are shifted towards right without loss of information. The rightmost bit is inserted at the other end.

Thus, after rotate right operation the content of register would be 10111101.

(*d*) **Rotate Left :** The bits are shifted towards left without loss of information. The leftmost bit is inserted at the other end.

Thus, after rotate left operation the content of register would be 11110110.

(*e*) **Rotate Right Through Carry :** The carry bit given is 1. The instruction rotate right through carry transfers the carry bit into the leftmost bit position and transfers the rightmost bit into the carry and at the same time shifts the entire register to the right.

Thus, after rotate right through carry operation the content of register and the carry are 10111101 and 1, respectively.

(*f*) **Rotate Left Through Carry :** The carry bit given is 1. The instruction rotate left through carry transfers the carry bit into the rightmost bit position and transfers the leftmost bit into the carry and at the same time shifts the entire register to the left.

Thus, after rotate left through carry operation the content of register and the carry are 11110111 and 0, respectively.

(*g*) **Arithmetic Shift Right :** The arithmetic shift right instruction leaves the sign bit unchanged and shifts the bit (including the sign bit) to the right, and the rightmost bit is lost.

Thus, after arithmetic shift right operation the content of register would be 00111101.

(*h*) **Arithmetic Shift Left :** In arithmetic shift left the bits are shifted left and the sign bit is transfer into flag. For the given register content, the two rightmost bits are not same, hence overflow occurs.

After operation the content of register is 11110110 (overflow condition).

Example 4.6. Given the 16-bit value 1001101011001101. What operation must be performed in order to :

(*a*) Clear to 0 the first eight bits ?

(*b*) Set to 1 the last eight bits ?

(*c*) Complement the middle eight bits ?

Solution. Given, the 16-bit value 1001101011001101.

(*a*) The first eight bits of the given 16-bit value will cleared to 0 if AND operation is performed with the value 0000000011111111.

```
        1001101011001101
AND     0000000011111111
        ----------------
        0000000011001101
```

(*b*) The last eight bits of the given 16-bit value will set to 1 if OR operation is performed with the value 0000000011111111

```
        1001101011001101
OR      0000000011111111
        ----------------
        1001101011111111
```

(*c*) The middle eight bits of the given 16-bit value complements if XOR operation is performed with the value 0000111111110000.

```
        1001101011001101
XOR     0000111111110000
        ----------------
        1001010100111101
```

Example 4.7. Perform the logic AND, OR and XOR with the two binary strings 10011100 and 10101010.

Solution. AND operation with 10011100 and 10101010

```
        10011100
AND     10101010
        --------
        10001000
```

ORoperation with 10011100 and 10101010

```
        10011100
OR      10101010
        --------
        10111110
```

XOR operation with 10011100 and 10101010

```
        10011100
XOR     10101010
        --------
        00110110
```

4.8.2.3. Logical Instruction

Different logical instructions are listed in the Table 4.8. AND, OR and XOR instructions provide the corresponding logical operations. The complement instruction produces the 1's complement of the operand. Clear instruction replaces all bits of the operand by 0's. Clear carry, set carry and complement carry are instructions that are performed on the individual bits. If the instruction is clear carry, the carry bit is cleared to 0. If it is set carry, the carry bit sets to 1. Similarly, if the instruction is complement carry the carry, bit complements and the bit changes from 0 to 1 or from 1 to 0.

4.8.3. Program Control Instructions

TABLE 4.8. Logical Instructions.

Name of Instruction	*Mnemonic*
AND	AND
OR	OR
Exclusive-OR	XOR
Complement	COM
Clear	CLR
Clear Carry	CLR C
Set Carry	SET C
Complement carry	COM C
Enable Interrupt	EI
Disable Interrupt	DI

Program control instructions are those instructions that can alter the flow of control. Generally, the programs executed in a straight line, with one instruction sequentially following the another. Most programs consist of a number of loops in which a series of instructions repeats until a specific requirement achieved. Also, a program can consist of various test to determine which of several actions to take. Thus, a transfer of control to the address of an instruction that does not immediately follow the one currently executing is required. This transfer of control may be forward, to execute a new series of instructions or backward, to re-execute the same instructions. The address of the next instruction to be executed is contained in the program counter and is automatically incremented each time after the instruction is fetched from the memory, so that it contains the address of the next instruction in sequence. Hence, it is the program control type of instruction that can change address of program counter and causes the flow of control of program to be altered.

Different program control instructions are listed in the Table 4.9. Branch and Jump instructions can be conditional or unconditional. The conditional branch instructions are those instructions that contains some condition and the program counter is loaded with the branch address when the specified condition in the instruction is fulfilled, and the next instruction is fetch from that address. If the condition is false, the program counter is not changed and the next instructions is fetched from the next location in sequence. Both Branch and Jump instructions can used interchangeably and is usually a one-address instruction, as BR ADR where ADR is a symbolic name given to an address. When the branch instruction is executed, it causes the value ADR to transfer into the program counter and the next instructions will fetched from this location ADR.

Skip instruction is a zero-address instruction and hence address field is not required. It can be conditional or unconditional. If the condition is met the skip instruction will skip the next instruction. If the condition is not met, control executes the next instruction in the sequence.

The compare and test do not directly change the program sequence. But they are used in setting conditions for conditional branch instructions. The compare instruction compare two operands by subtracting the two operands, and sets certain status bit like Carry (C), Sign (S), Zero (Z), Overflow (V) etc, but the result of subtraction operation is not retained.

TABLE 4.9. Program Control Instructions.

Name of Instruction	Mnemonic
Branch	BR
Jump	JMP
Skip	SKP
Call	CALL
Return	RET
Compare	CMP
Test	TST

Similarly, the test instruction performs the logical AND operation between two operands. In this case also certain status bits are updated without retaining the result of logical AND operation. These status bits are called condition-code bits or *flag bits*. The status bits Carry (C), Sign (S), Zero (Z), and Overflow (V) sets or cleared depending on the operation performed in the ALU.

Carry flag (C) : The carry flag bit C is set to 1, if the operation performed generates a carry and it is cleared to 0 if no carry is generated.

Sign flag (S) : The sign flag bit S is set to 1 if the most significant bit i.e. the highest order bit is 1 else it is cleared to 0.

Zero flag (Z) : The zero flag bit (Z) is set to 1 if the output of the operation performed by ALU contains all 0's. It is cleared to 0 otherwise.

Overflow flag (V) : The overflow flag bit (V) is set to 1 if there is an overflow after the operation. Otherwise it is cleared to 0.

Conditional Branch Instructions

Different conditional instructions includes conditional branches, conditional jumps and subroutine calls. If the condition with in the branch instruction evaluates to true, the control is transferred to the address specified in the instruction else the next instruction in sequence will execute. These conditional instruction are used in conjunction with the jump, skip, call and return instruction. Some of the common conditional branch instructions are listed in Table 4.10.

TABLE 4.10. Conditional Branch Instructions

Symbol Used	Branch Condition	Condition
BZ	Branch if zero	Z = 1
BNZ	Branch if not zero	Z = 0
BC	Branch if carry	C = 1
BNC	Brnach if no carry	C = 0
BP	Branch if plus	S = 0
BM	Branch if minus	S = 1
BV	Branch if overflow	V = 1
BNV	Branch if no overflow	V = 0

Subroutine Call and Return

A subroutine is a self-contained sequence of instructions that performs a given computational task. During the program execution, a subroutine can be called many times to perform its operations. Each time a subroutine is called, a branch is executed to start executing its set of instruction and a branch is made back to the main program when the subroutine has been executed. The instruction that causes the transfer of program control to a subroutine is known by different names. Some common names are call subroutine, jump to subroutine or branch and save address.

The call subroutine instructions consists of an operation code and the address that specifies the beginning of subroutine. The instruction is executed by performing two operations (*i*) the address of the next instruction *i.e.* the return address in the program counter is stored at some temporary location, (*ii*) the control is transferred at the beginning of the subroutine. The last instruction of every subroutine, called as return from subroutine, transfers the return address stored in temporary location into the program counter. This return address helps in a transfer of program control to the instruction whose address was stored in temporary location. This temporary location can be the first memory location of the subroutine, or some fixed memory location or a processor register or can be a memory stack. But most efficient way is to store the return address in a memory stack.

Program Interrupt (Unconditional Transfers of Control)

Program interrupt refers to the transfer of control from the program in execution to another service as of result of some internal requests. Once the service program completes its execution the program control again returns to the original program in execution. This is unconditional transfer of program control because of some interrupts. There are three type of interrupts.

(*a*) **External Interrupts :** It includes interrupts from some external devices such as I/O devices when data need to be transferred. It also occurs when some time out has occured because of power failure.

(*b*) **Internal Interrupts :** Internal interrupts can occurs by some exceptional condition resulting out of incorrect program syntax or logic. It can occurs both at compile time or at runtime. For example, use of invalid opcode, divide by zero, stack overflow, register overflow etc.

(*c*) **Software Interrupts :** Software interrupts are initiated by executing an instruction. It occurs under such conditions when a user program requests the operating system for handling some special, privileged operations such as input-output operations. It is a special call instruction that behaves as an interrupt and thus can be used by programmer to initiate an interrupt routine at any desired point in the program. For example, int *n* where '*n*' is the starting address of the procedure.

An interrupt is same as of a subtroutine call except for the following :

(*a*) The interrupt is basically because of an external or internal signal rather than from the execution of an instruction except for the software interrupts.

(*b*) The hardware determines the address of the interrupt service routine rather than from the address field of an instruction.

(*c*) The interrupt procedure stores all information necessary to define the state of the CPU rather than just storing the return address.

4.9. RISC-REDUCED INSTRUCTION SET COMPUTER

Instruction sets decides the way how machine language programs can be constructed for a particular computer. In early computers, instructions sets are simple and small. As the time progress, the computer instructions tends to increase both in number and complexity. Instruction sets of many computer contains more than 100 instructions. These computers also contains a large number of addressing modes and employ various types of data types. These computers were classified as complex instruction set computer or CISC.

RISC, which stands for reduced instruction set computer, gives a new generation of faster and cheaper machines. The basic idea behind RISC is to use a small set of instructions with simple constructs so that they can executes much faster within the CPU.

The major characteristics of a RISC processors are :

(*i*) Relatively few and simple instruction types.

(*ii*) Fixed and easily decoded instruction formats.

(*iii*) Memory access limited mainly to load and store instructions.

(*iv*) Relatively few addressing modes

(*v*) A large number of register sets.

(*vi*) Fast single cycle instruction execution.

(*vii*) Hardwired rather than microprogrammed control

(*viii*) Use of instruction pipeline and cache memory.

(*ix*) Use of compilers to optimize object code performance.

Simple and smaller instruction set leads to simple hardware and faster execution. The small instruction set mostly consists of register-to-register operations with simple load and store operations for memory access. All operation are done within the CPU registers. Thus, with the help of load instruction, first the operands are brought into processor registers and with the help of store instructions the results are transferred into memory. Also, the RISC processors uses only few addressing modes, and thus all instructions have only simple register addressing. The use of hardwired control rather than microprogrammed helps in faster execution of operations. Another important aspects of RISC instruction format is that the instruction length are fixed and are easy to decode. RISC processors has the ability to execute one instruction pipelining by overlapping the fetch, decode and execute phases of two or more instructions. A large number of registers in the RISC processor is useful for storing intermediate results and in minimizing the register memory operations by keeping the most frequent accessed operands in register.

4.10. CISC ARCHITECTURE

Processor executes with scalar data. The simplest scalar processor executes integer instructions using fixed-point operands. More capable scalar processors execute both integer and floating-point operations. A modern scalar processor may possess both an integer unit and a floating-point unit. Based on a complex instruction set, a CISC scalar processor can be built either with a single chip or with multiple chips mounted on a processor board.

Three representative CISC scalar processors are listed.

- The VAX 8600 processor is built on a PC board.
- M68040 are single-chip microprocessors.
- The i486

4.10.1 The Digital Equipment VAX 8600 Processor Architecture

The VAX 8600 was introduced by Digital Equipment Corporation in 1985. This machine implements a typical CISC architecture with micro programmed control. The instruction set contains about 300 instructions with 20 different addressing modes. As shown in the Figure 4.12, the VAX 8600 executes the same instruction set, runs the same VMS operating system, and interfaces with the same I/O buses (such as SBI and Unibus) as the VAX 11/780.

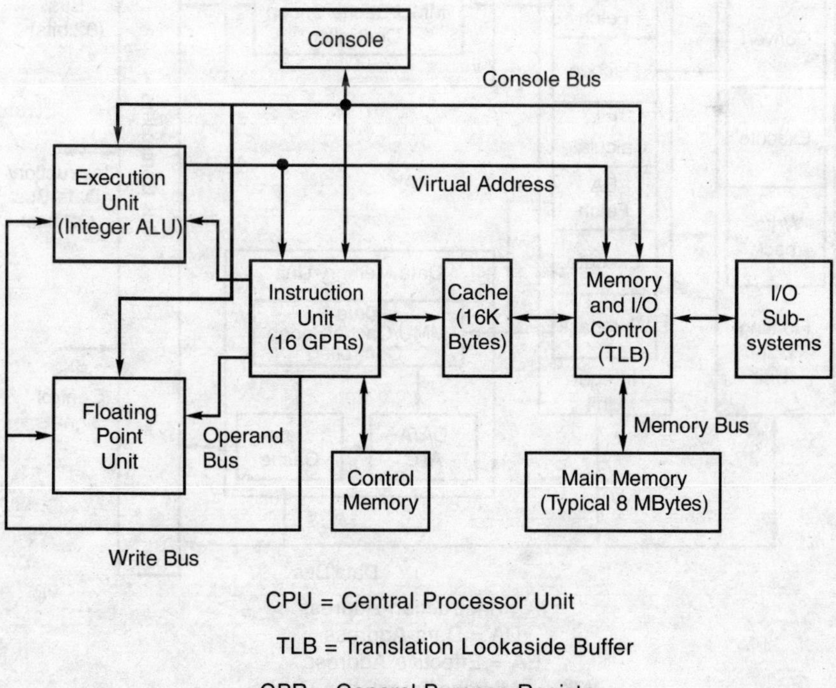

Fig. 4.12. The VAX 8600 CPU

The CPU in the VAX 8600 consists of two functional units for concurrent execution of integer and floating-point instructions. The unified cache is used for holding both instructions and data. There are 16 GPRs in the instruction unit. Instruction pipelining has been built with six stages in the VAX 8600, as in most CISC machines. The instruction unit prefetches and decodes instructions, handles branching operations, and supplies operands to the two functional units in a pipelined fashion.

A *translation look aside buffer* (TLB) is used in the memory control unit for fast generation of a physical address from a virtual address. Both integer and floating-point units are pipelined. The CPI of a VAX 8600 instruction varies within a wide range from 2 cycles to as high as 20 cycles. For example, both multiply and divide may tie up the execution unit for a large number of cycles. This is caused by the use of long sequences of microinstructions to control hardware operations.

4.10.2 CISC Microprocessor Families

In 1971, the Intel 4004 appeared as the first microprocessor based on a 4-bit ALU. Since then, Intel has produced the 8-bit 8008, 8080, and 8085. Intel's 16-bit processors appeared in 1978 as the 8086, 8088, 80186 and 80286. In 1985, the 80386 appeared as a 32-bit machine. The 80486 are 80586 are the latest 32-bit microprocessors in the Intel 80x86 family.

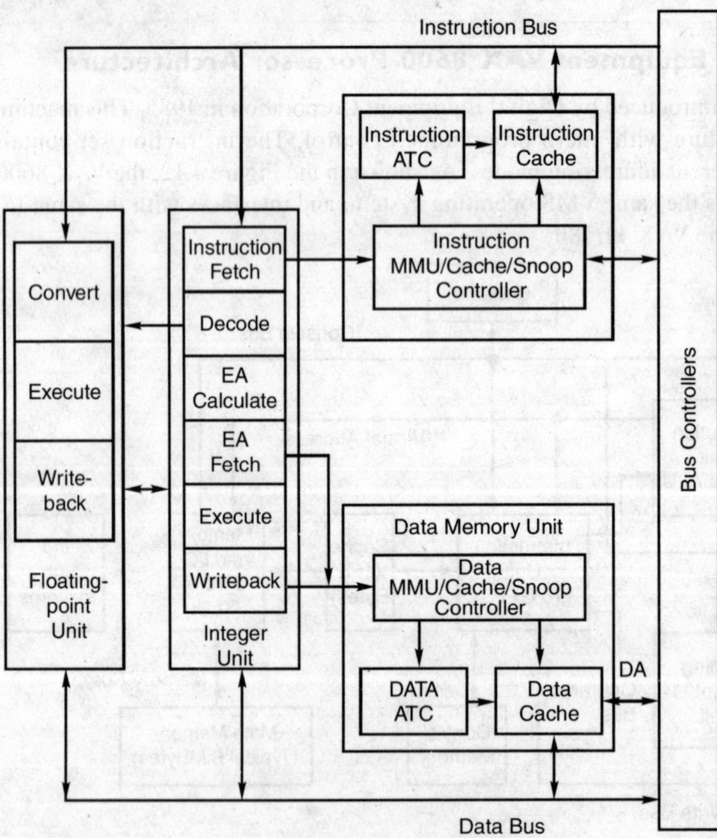

IA = Instruction Address
DA = Data Address
EA = Effective Address
ATC = Address Translation Cache
MMU = Memory Management Unit

Fig. 4.13. Architecture of the MC68040 processor

Motorola produced its first 8-bit microprocessor, the MC6800 in 1974, then moved to the 16 bit 68000 in 1979, and then to the 32 bit 68020 in 1984. The latest are the MC68030 and MC68040 in the Motorola MC680x0 family. National semiconductor's latest 32 bit microprocessor is the NS 325 introduced in 1988. These CISC microprocessor families are widely used in the *personal computer* (PC) industry.

In recent years, the parallel computer industry has begin to build experimental systems with a large number of open-architecture microprocessors. Both CISC and RISC microprocessors have been employed in these efforts.

The Motorola MC68040 Microprocessor Architecture

The MC68040 is a 0.8 mm HCMOS microprocessor containing more than 1.2 million transistors, comparable to the i80486. Figure 4.13 shows MC68040 architecture. The processor implements over 100 instructions using 16 general-purpose registers, a 4-Kbyte data cache, and a 4-Kbyte instruction cache, with separate *memory management units* (MMUs) supported by an *address translation cache* (ATC), which is equivalent to the TBB used in other systems. The data formats range from 8 to 80 bits, based on the IEEE floating-point standard.

Eighteen addressing modes are supported, including register direct and indirect, indexing, memory indirect, program counter indirect, absolute, and immediate modes. The instruction set includes data movement, integer, BCD, and floating-point arithmetic, logical, shifting, bit-field manipulation, cache maintenance, and multiprocessor communications, in addition to program and system control and memory management instructions.

4.11. HARDWIRED CONTROL UNIT

DEFINITION▸

When the control signals are generated by hardware, it is called hardwired control unit.

When the control signals are generated by hardware using conventional logic design techniques, the control unit is said to be *hardwired*. Logic gates, flip-flops, decoders and other digital circuits are used to implement hardwired control organization. As the name suggests, if the design has to be modified or changed, a hardwired control requires changes in the wiring among the various components. The block diagram of the hardwired control unit is shown in Fig. 4.14.

It consist of two decoders, a sequence counter and a number of control logic gates. Instruction read from memory is placed in the instruction register (IR), which is divided into three parts : the I bit, the operation code and bits 0 through 11. Bit 15 of the instruction is transferred to a flip-flop designated by I. The operation code bits 12 through 14 are decoded with a 3 × 8 decoder. The eight outputs of the decoder are designated by the symbols D_0 through D_7. Bits 0 through 11 are applied to the control logic gates. The 4-bit sequence counter can count in binary from 0 through 15, which are decoded into 16 timing signals T_0 through T_{15}. This sequence counter are generally incremented so that it provide the sequence of the timing signals. Very rarely, the sequence counter is cleared to 0 and thus, the next active timing signal will be T_0.

Consider a case, when sequence counter is incremented to provide timing signals of T_0, T_1, T_2, T_3 and T_4 in sequence and at time T_4, sequence counter SC is cleared to 0 if decoder output D_3 is active. This timing sequence as shown in Fig. 4.15 can be symbolically expressed as

$$D_3 T_4 : SC \leftarrow 0$$

Initially the clear CLR input of the sequence counter is active and SC responds to positive transition of the clock. With first positive transition of clock, SC clears to 0 and timing signal T_0 activates and remains active during one clock cycle. SC is incremented with every positive transition unless its clear input becomes active, which produces the sequence as T_0, T_1, T_2, T_3, T_4 and so on. This timing signals will continue upto T_{15} and back to T_0, if sequence counter in not cleared. The timing diagram for D_3T_4 : SC ← 0 is shown in Fig. 4.15. The figure shows how SC is cleared when $D_3T_4 = 1$. At the end of timing signal T_2 the output D_3 from decoder becomes active. When timing signal T_4 becomes active, the sequence counter cleared to 0 which causes the timing signal T_0 to become active instead of T_5 which would be active signal if sequence counter is incremented instead of cleared.

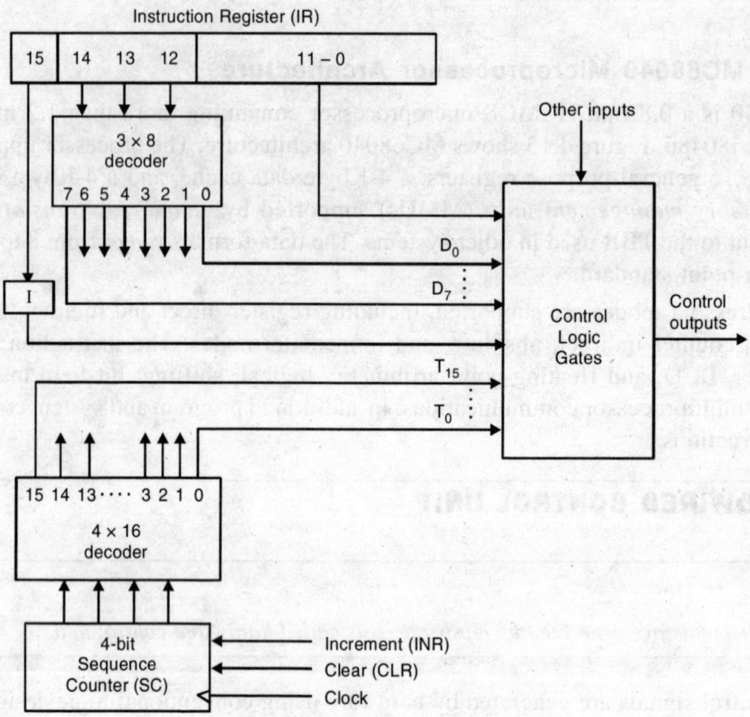

Fig. 4.14. Hardwired Control Unit.

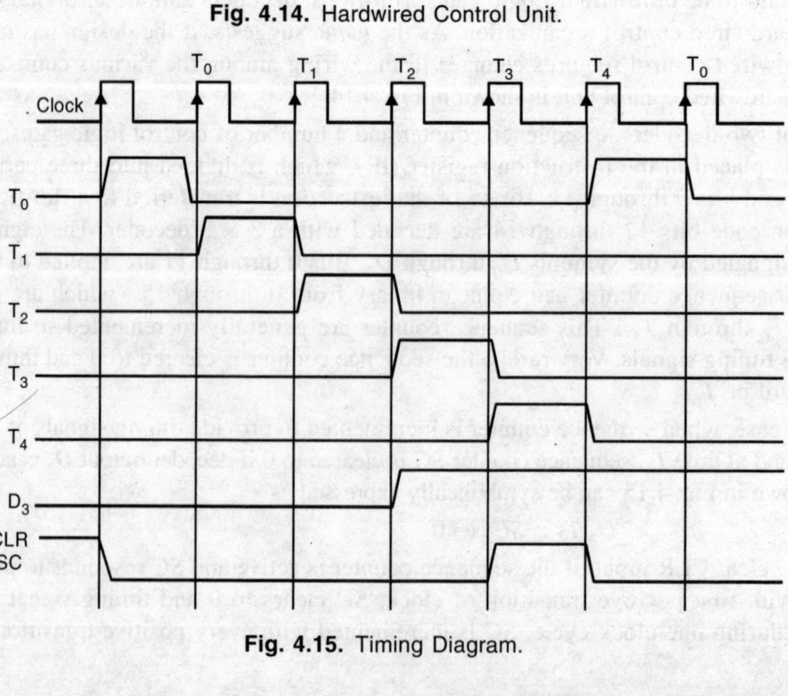

Fig. 4.15. Timing Diagram.

4.12. MICROPROGRAMMED CONTROL UNIT

DEFINITION ▸

The control unit which generates control signals according to microprogram rather than using hardware is called a microprogrammed control unit.

Microprogramming is a second alternative for designing the control unit of a digital computer. The control function that specifies a microoperation is a binary variable and when the binary variable is in one binary state, the corresponding microoperation is execution.

A computer employing the microprogrammed control consist of two separate memories, a control memory and a main memory. The main memory is for user for storing the programs. Its content may alter whenever the data are manipulated and every time the program is changed. But the control memory holds a fixed microprogram that cannot be changed by general user. These microprogram consists of a microinstruction that specifies various control signals that helps in the execution of register microoperations.

The block diagram of a microprogrammed control unit is shown in the Fig. 4.16.

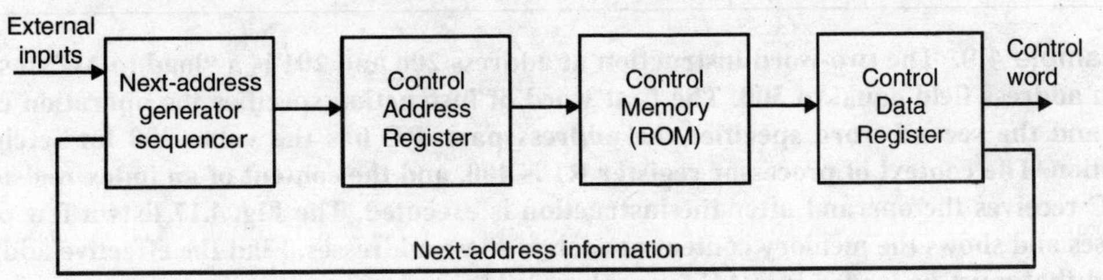

Fig. 4.16. Microprogrammed Control Organization.

It is the control memory, within which all control information are permanently stored. This control memory is also assumed to be ROM. The address of the microinstruction is specified in the control memory address register and the microinstruction read from the memory is in control data register *i.e.* the control data register holds the microinstruction which is read from memory location specified by control address register. The microinstruction contains a control word that specifies one or more microoperation for the data processor. The control data register is also known as pipeline register since it allows the execution of the microoperations specified by control word simultaneously with the generation of next microinstruction. Once these microoperations are executed, the control must also determine the address of the next microinstruction which may be the one next is sequence or it may be located at some other location within the control memory. It is the function of the next address generator to determine the address sequence that is read from control memory. Sometime the next address generator is also referred to as microprogram sequencer. It depends on the input of the sequencer that decides the address of the next microinstructions. Basic function of microprogram sequencer are:

(*i*) incrementing the control address register by one,
(*ii*) loading an address from control memory into the control address register,
(*iii*) transferring an external address,
(*iv*) loading an initial address to start the control operations.

Solved Examples

Example 4.8. Specify the control word that must be applied to the processor of Fig. 4.2 to implement the following microoperations.

(*a*) **R1 ← R2 + R3**

(*b*) **R4 ← $\overline{\textbf{R4}}$**

(*c*) **R5 ← R5 – 1**

(*d*) **R6 ← shl R1**

(*e*) **R7 ← input.**

Solution. The control word of the microoperations given in the example are listed in Table 4.11.

Table 4.11.

Microoperation	*SEL A*	*SEL B*	*SEL D*	*OPR*	*Control Word*			
(a) R1 ← R2 + R3	R2	R3	R1	ADD	010	011	001	00010
(b) R4 ← R4	R4	–	R4	COMA	100	×××	100	01110
(c) R5 ← R5 – 1	R5	–	R5	DECA	101	×××	101	00110
(d) R6 ← shl R1	R1	–	R6	SHLA	001	×××	110	11000
(e) R7 ← input	Input	–	R7	TSFA	000	×××	111	00000

Example 4.9. The two-word instruction at address 200 and 201 is a "load to AC" instruction with an address field equal to 500. The first word of instruction specifies the operation code and mode, and the second word specifies the address part. PC has the value 200 for fetching this instruction. The context of processor register R1 is 400, and the content of an index register XR is 100. AC receives the operand after the instruction is executed. The Fig. 4.17 lists a few pertinent addresses and shows the memory content at each of these addresses. Find the effective address and operand that must be loaded into AC for each possible mode.

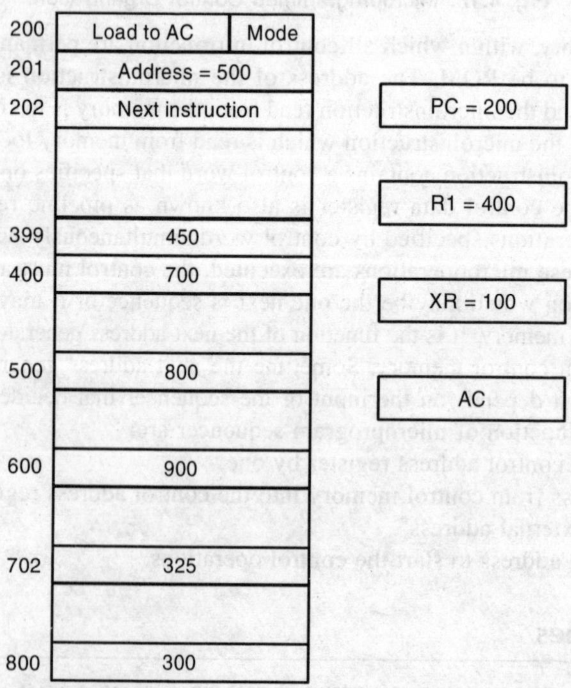

Fig. 4.17. Numerical Example for Addressing Modes.

Solution.

(*i*) **Immediate.** In immediate mode the second word of the instruction is operand rather than address, so 500 is loaded into AC. Effective address in this case is 201.

(*ii*) **Direct.** In this mode the effective address is the address part of the instruction 500 and the operand loaded into AC is 800.

(*iii*) **Indirect.** In this mode, the effective address is stored in memory at address 500. So the effective address is 800 and the operand loaded into AC is 300.

(*iv*) **Register.** In this mode, the operand is in R1 and hence 400 is loaded into AC.

(*v*) **Register Indirect.** In this mode, the effective address is in processor register R1 *i.e.* 400 is effective address and hence the operand 700 is loaded into AC.

(*vi*) **Index.** In this mode, the effective address is address part of instruction + content of index register *i.e.* 500 + 100 = 600, so 900 is loaded into AC.

(*vii*) **Relative.** In this mode, the effective address is 500 + 202 = 702, and hence 325 is loaded into AC.

Example 4.10. A two word instruction is stored in memory at an address designated by symbol *W*. The address field of the instruction (stored at *W* + 1) is designated by the symbol *Y*. The operand used during the execution of the instruction is stored at an address symbolized by *Z*. An index register contains the value *X*. State how *Z* is calculated from the other addresses if the addressing mode of the instruction is

(*a*) Direct **(*b*) Indirect**

(*c*) Relative **(*d*) Indexed**

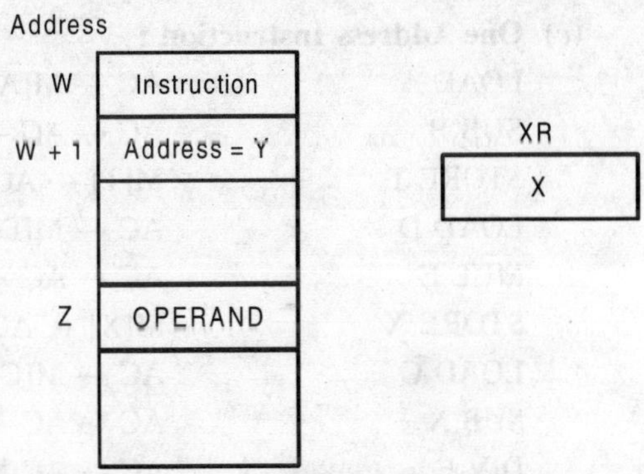

Fig. 4.18. Memory Content.

Solution. According to question, the memory figure is shown in Fig. 4.18.

(*a*) In direct addressing mode the address part of the instruction is the address of the operand.

Thus, $Z = Y.$

(*b*) In indirect addressing mode the address part of the 0 instruction gives the memory address, what is the effective address of the operand Π.

Thus, $Z = M[Y]$

(*c*) In case of relative addressing mode, the effective address is the address part of the instruction (*Y*) plus the content of program counter. The instruction is stored at *W* with address part at *W* + 1. Thus, when instruction is fetched from memory the PC will increment and points to *W* + 2.

Thus, $Z = Y + W + 2.$

(*d*) In indexed addressing mode, the effective address is the address part of instruction plus the content of index register.

Thus, $Z = Y + X$

Example 4.11. Write a program to evaluate arithmetic expression

$$X = (A - B) * (((C - D * E)/F)/G)$$

(*a*) Using a general register computer with three address instructions.

(*b*) Using a general register computer with two address instructions.

(*c*) Using an accumulator type computer with one address instructions.

(*d*) Using stack organized computer with zero address instructions.

Solution. $X = (A - B) * (((C - D * E)/F)/G)$

(*a*) **Three Address Instructions :**

SUB R1, A, B	R1 ← M[A] – M[B]
MUL R2, D, E	R2 ← M[D] – M[E]
SUB R2, C, R2	R2 ← M[C] – R2
DIV R2, R2, F	R2 ← R2/M[F]
DIV R2, R2, G	R2 ← R2/M[G]
MUL X, R1, R2	M[X] ← R1 * R2

(*b*) **Two address Instructions :**

MOV R1, A	R1 ← M[A]
SUB R1, B	R1 ← R1 – M[B]
MOV R2, D	R2 ← M[D]
MUL R2, E	R2 ← R2 * M[E]
MOV X, R2	M[X] ← R2
MOV R2, C	R2 ← M[C]
SUB R2, X	R2 ← R2 – M[X]
DIV R2, F	R2 ← R2/M[F]
DIV R2, G	R2 ← R2/M[G]
MUL R1, R2	R1 ← R1 * R2
MOV X, R1	M[X] ← R1

(*c*) **One Address Instruction :**

LOAD A	AC ← M[A]
SUB B	AC ← AC – M[B]
STORE T	M[T] ← AC
LOAD D	AC ← M[D]
MUL E	AC ← AC * E
STORE X	M[X] ← AC
LOAD C	AC ← M[C]
SUB X	AC ← AC – M[X]
DIV F	AC ← AC/M[P]
DIV G	AC ← AC/M[G]
MUL T	AC ← AC * M[T]
STORE	M[X] ← AC

(*d*) **For stack organized computer, the expression must be in reverse polish notation :**

$$
\begin{aligned}
X &= (A - B) * (((C - D * E)/F)/G) \\
&= AB - * (((C - DE*)/F/G) \\
&= AB - *(((CDE*-)/F)/G) \\
&= AB - *(CDE* - F/)/G) \\
&= AB - *CDE* -F/G/ \\
&= AB - CDE* - F/G/*
\end{aligned}
$$

Thus, the RPN of given arithmetic expression is $AB - * \ CDE* -F/G/*$

Zero Address Instruction :

PUSH A	TOS ← A
PUSH B	TOS ← B
SUB	TOS ← A – B

PUSH C	TOS ← C
PUSH D	TOS ← D
PUSH E	TOS ← E
MUL	TOS ← D ∗ E
SUB	TOS ← C – D ∗ E
PUSH F	TOS ← F
DIV	TOS ← (C – D ∗ E)/F
PUSH G	TOS ← G
DIV	TOS ← ((C – D ∗ E)/F)/G
MUL	TOS ← (A – B) ∗ (((C – D ∗ E)/F)/G)
POP X	M[X] ← TOS

Example 4.12. The memory unit of a computer has 256K words of 32 bits each. The computer has an instruction format with four fields: an operation code field, a mode field to specify one of the seven addressing modes, a register address field to specify one of 60 processor registers, and a memory address. Specify the instruction format and number of bits in each field if the instruction is in one word memory.

Solution. Memory unit = 256 K words

$= 2^{18}$ words

Hence, 18 bits are required for memory address.

There are 60 processor registers and we know that $2^5 < 60 < 2^6$.

Hence, 6 bits are required for register address. Also, there are 7 addressing modes and $2^2 < 7 < 2^3$

Thus, we require 3 bits to specify 7 addressing modes.

Total number of bits in instruction format is 32. Hence,

Number of bits in opcode field $= 32 - (18 + 6 + 3)$

$= 32 - 27$

$= 5$ bits.

Thus, the instruction format is

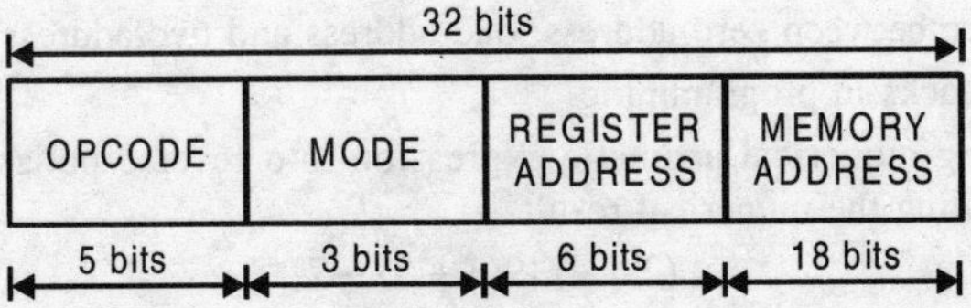

Fig. 4.19.

Summary

- Central Processing Unit (CPU) is the brain of computer. It contains ALU, control units and registers.
- ALU performs arithmetic and logical operations.
- Control unit control the entire operation of the computer.
- The data stored in the primary memory are transferred to the main memory before processing takes place.
- The group of binary bits assigned to perform a specific operation is known as control word.
- Register is a storage unit capable of holding binary information.

- PUSH and POP are two operations of STACK.
- Stack is a storage device which stores information in such a manner that the item stored last is first item to be retrieved.
- Stack can be organized as a collection of finite number of memory words or registers.
- An arithmetic expression can be expressed in infix notation, prefix notation or postfix notation.
- Postfix notation is also known as Reverse Polish notation.
- Instruction format defines the layout of bits of instruction of different fields of the instruction.
- Instructions can be zero-address, one-address, two-address or three-address instruction.
- Addressing modes is a technique for specifying the address of the operands.
- The address of an operand is known as effective address.
- Relative, indexed, and base-register addressing modes are known as displacement addressing modes.
- Data transfer instructions cause transfer of data from one location to another without changing the binary information content.
- Data manipulation instructions performs arithmetic, logic or shift operations.
- Program control instructions gives decision-making capabilities and change the path of program control.
- The full form of RISC is Reduced Instruction Set Computer.
- The full form of CISC is Complex Instruction Set Computer.

Review Questions

1. What do you mean by addressing modes ? Discuss different types of addressing modes with their merits and demerits.
2. Discuss various displacement addressing modes in details.
3. Instructions use operands and machine registers mainly in four different ways. Discuss them in brief.
4. Discuss all the characteristics of a good instruction format. Discuss structure of a typical instructio format.
5. What do you mean by Instruction format ? Also discuss the execution cycle of instruction.
6. Discuss in brief (with their advantage and disadvantage) :
   - RISC (reduced instruction set computer)
   - CISC (complex instruction set computer)
7. Explain register organization with the help of block diagram.
8. What is the difference between zero-address, one-address and two address instructions ?
9. Explain the role of stacks in programming.
10. Convert the following numerical artimetic expression into reverse polish notation and show the stack operations for evaluating the numerical result

$$(2 + 4)\ [9(3 + 5) + 7]$$

11. Write a program to evaluate the following arithmetic expressions for all four machines :
    (*a*) $X = (A + B * C) * (C - D * E)/F$
    (*b*) $X = (A/B + C * D)/(D*E - F + C/A)$
    (*c*) $X = A * (B + C * (D + E))/(F*(G + H))$
    (*d*) $X = (A + B * C)/(D - E * F + G * H)$
    (*e*) $X = A * B + A * (B * D + C * E)$
12. An instruction is stored at location 300 with its address field at location 301. The address field has the value 400. A processor register R1 contains the number 200. Valuate the effective address if the addressing mode of the instruction is
    (*a*) Direct
    (*b*) Immediate
    (*c*) Relative

(*d*) Index with R1 as index register

(*e*) Register indirect

13. A digital computer has a memory unit with 24 bits per word. The instruction set consists of 190 different operations. Each instruction is stored in one word of memory and consists of an op-code part and address part.

(*a*) How many bits are needed for operation code ?

(*b*) How many bits are left or address part of the instruction ?

(*c*) How many words can be accommodated in the memory unit ?

14. How many bits are needed to represent an instruction in a computer that has 53 different instructions, all of which are two addresses, when the computer has 65000-word memory ?

15. Convert the following arithmetic expressions from infix to reverse Polish notation.

(*a*) $A * B + C * D + E * F$

(*b*) $A + B * [C * D + E * (F + G)]$

(*c*) $\dfrac{A * [B + C * (D + E)]}{F * (G + H)}$

16. A computer has 32-bit instructions and 12-bit addresses. If these are 250 two-address instructions, how many one-address instructions can be formulated ?

17. The concent of the top of a memory stack is 5320. The content of the stack pointer SP is 3560. A two-word call subroutine instruction is located in memory at address 1120 followed by the address field of 6720 at location 1121. What are the content of PC, SP, and the top of the stack :

(*a*) Before the call instruction is fetched from memory ?

(*b*) After the call instruction is executed ?

(*c*) After the return from subroutine ?

18. A computer uses a memory unit with 256K words of 32 bits each. A binary instruction code is stored in one word of memory. The instruction has four parts: an indirect bit, an operation code, a register code part to specify one of 64 registers and an address part.

(*i*) How many bits are there in the operation code, the register code part and the address part?

(*ii*) Draw the instruction word format and indicate the number of bits in each part.

(*iii*) How many bits are there in the data and address inputs of the memory?

19. What is the difference between a direct and an indirect address instruction? How many references to memory are needed for each type of instruction to bring an operand in to a processor register?

20. Write a Program to evaluate the arithmetic statement

$$P = \frac{(a - b + c) * (d * e - f)}{g + h * i}$$

by using

(*i*) Two address instructions

(*ii*) One address instructions

(*iii*) Zero address instructions.

21. Convert the following arithmetic expression from infix notation to RPN.

A*B+B*(B*D+C*E)

22. How stack is implemented in a general microprocessor system

23. What do you mean by program control instructions? With a neat diagram, explain how the status register containing overflow, zero, sign and carry flags works with the status of the accumulator content obtained from ALU.

24. What are interrupts? Explain different types of interrupts.

25. What is an instruction? What are the different parts of an instruction? Explain the significance of each part of an instruction with an example?

26. If a Computer has 128 operation codes and 512 k addresses, how many bits would be required for (*i*) Single address instruction (*ii*) Two address instruction.

27. Explain with an example, how effective address is calculated in different types of addressing modes.

28. How does a basic computer handle an interrupt? Explain what happens during the interrupt with the help of an example. Also, give the register transfer statements.

29. Convert the following arithmetic expression from infix notation to Reverse Polish Notation (RPN).

(*i*) A+B*A+(B*E+A*D)

(*ii*) A*[C+(B*D)+(A*E)]

(*iii*) A*B/C+D

30. Justify the statement "Stack computer consists of an operation code only with no address field".

31. Write a program to evaluate the arithmetic statement

$$Y = \frac{A - B + C}{G + H}$$

(*i*) Using an accumulator type computer with one address instruction.

(*ii*) Using a stack organized computer with zero-address instructions.

32. Write down the three types of CPU organization with the help of examples.

33. What is an instruction? With examples explain three, two, one, zero address instructions.

34. Give stack operations to evaluate 3*4+5*6.

Objective Type Questions

1. A stack organized computer uses instruction of

(*a*) Indirect addressing (*b*) Two addressing
(*c*) Zero addressing (*d*) Index addressing

2. In a program using subroutine call instruction, it is necessary to

(*a*) Initialize program counter (*b*) Clear the accumulator
(*c*) Reset the microprocessor (*d*) Clear the instruction register

3. When CPU is executing a program that is a part of operating system, it is said to be in

(*a*) Interrupt mode (*b*) System mode
(*c*) Half mode (*d*) Simplex mode

4. ________ register keeps track of the instructions stored in program stored in memory.

(*a*) Address register (AR) (*b*) Index register (XR)
(*c*) Program counter (PC) (*d*) Accumulator (AC)

5. The addressing mode used in an instruction of the form ADD X, Y is

(*a*) Absolute (*b*) Indirect
(*c*) Index (*d*) None of these

6. 'Aging registers' are

(*a*) Counters which indicate how long ago their associated pages have been referenced.
(*b*) Registers which keeps track of when the program was last accessed.
(*c*) Counters to keep track of last accessed instruction.
(*d*) Counters to keep track of the latest data structures referred.

7. The BSA instruction is

(*a*) Branch and store instruction
(*b*) Branch and save instruction
(*c*) Branch and shift instruction
(*d*) Branch and second instruction

8. A group of bits that tell the computer to perform a specific operation is known as

(*a*) Instruction code (*b*) Micro-operation
(*c*) Accumulator (*d*) Register

9. The load instruction is mostly used to designate a transfer from memory to a processor register known as
   (*a*) Accumulator (*b*) Instruction register
   (*c*) Program counter (*d*) Memory address register
10. Content of the program counter is added to the address part of the instruction in order of obtain the effective address is called
   (*a*) Relative addressing mode (*b*) Index addressing mode
   (*c*) Register addressing mode (*d*) Implied addressing mode
11. Status bit is also known as
   (*a*) Binary bit (*b*) Flag bit
   (*c*) Signed bit (*d*) Unsigned bit
12. Which of the following is not a characteristics of RISC architecture
   (*a*) Large instruction set (*b*) One instruction per cycle
   (*c*) Simple addressing modes (*d*) Register to register operation
13. Stack overflow causes
   (*a*) Hardware interrupt (*b*) External interrupt
   (*c*) Internal interrupt (*d*) Software interrupt
14. Zero address instruction format is used for
   (*a*) RISC architecture (*b*) CISC architecture
   (*c*) Von-Neumann architecture (*d*) Stack-organized architecture
15. After fetching the instruction from the memory, the binary code of the instruction goes to
   (*a*) Instruction register (*b*) Program counter
   (*c*) Accumulator (*d*) Instruction pointer
16. In reverse polish notation, A + B * C becomes
   (*a*) AB * C + (*b*) ABC*+
   (*c*) ABC+* (*d*) None of these
17. In _________ addressing mode, the address part of the instruction designate an address of a memory word in which the address of the operand is found.
   (*a*) Indirect (*b*) Direct
   (*c*) Random (*d*) All of the above.
18. Zero address instructions are applied in a special memory organization called
   (*a*) Program counter (*b*) Instruction decoder
   (*c*) Stack (*d*) None of the above
19. While performing operations the ALU takes data from the temporary storage area inside the CPU named
   (*a*) Floppy disk (*b*) RAM
   (*c*) Registers (*d*) CD
20. In reverse polish notation, expression A * B + C * D is written as
   (*a*) AB*CD*+ (*b*) AB*C+D*
   (*c*) AB*CD+* (*d*) *AB*CD+
21. Computers use addressing mode techniques for _______
   (*a*) Giving programming versatility to the user by providing facilities as pointers to memory counters for loop control
   (*b*) To reduce number of bits in the field of instructions
   (*c*) Specifying rules for modifying or interpreting address field of the instruction
   (*d*) All of the above

ANSWERS

1. (*c*)	**2.** (*d*)	**3.** (*b*)	**4.** (*c*)	**5.** (*c*)
6. (*a*)	**7.** (*b*)	**8.** (*a*)	**9.** (*a*)	**10.** (*a*)
11. (*b*)	**12.** (*a*)	**13.** (*c*)	**14.** (*c*)	**15.** (*a*)
16. (*b*)	**17.** (*a*)	**18.** (*c*)	**19.** (*c*)	**20.** (*a*)
21. (*d*)				

❑❑❑

5 Chapter

Input-Output Organization

INSIDE THIS CHAPTER

5.1. PERIPHERAL DEVICES

The input and output units of a computer system are known as Input/Output units. They provide an efficient way of communication between the computer system and the outside world. A computer must have a system to receive information and must be able to communicate results to the external world. The most familiar input devices connected to the computer is keyboard. Devices that are under the control of the computer are said to be connected online. Input/output devices that are attached to the computer are also known as peripherals. Most common peripherals devices are keyboards, mouses, display units, printers etc.

Definition The devices which are connected to a computer and controlled by processor are called peripheral devices.

5.1.1. Keyboards

Keyboard is the most familiar input device for entering information into a computer. The layout of the keyboard is similar to that of the traditional typewriter with some extra command and functions keys. When a key on the keyboard is pressed, a binary coded character is send to the computer and the corresponding character is displaced on the screen. If a wrong key is pressed, it can be erased and the correct character can be retyped. The fastest possible speed of entering the information using keyboard depends on the person's typing speed. A keyboard used in a computer system is shown in Fig. 5.1.

Fig. 5.1. Keyboard.

5.1.2. Mouse

Fig. 5.2. Mouse.

Mouse is a pointing device. It can be moved on a smooth surface so that it points to the desired location on the screen. It can be optical or mechanical. User can move the mouse, stop it at a desired location and with the help buttons can makes selection of choices. The mouse used in a computer system is shown in Fig. 5.2.

5.1.3. Display Devices

Fig. 5.3. Display Monitor.

One of the most important peripheral of the computer system is the display moniter. There are different types of display moniter but the most popular is cathode ray tube (CRT). Others are liquid crystal display (LCD) and projection displays. The CRT contains a beam of electrons (cathode rays) emitted by the electron gun, passes through focusing and deflection systems that direct the beam toward specified position on the phosphor-coated screen. The phosphor then emits a small spot of light at each position contacted by the electron beam. The monitor used in a computer system is shown in Fig. 5.3.

5.1.4. Printers

Fig. 5.4. Printer.

Printers are used for getting the output on paper. Printers can be classified according to the print quality and the speed of the printing. The three basic types of character printers are daisywheel, dot matrix and laser printers. The daisywheel printers contains a wheel and the characters are placed along the circumference. When the printer has to print a character, the wheel rotates to the proper position and an energized magnet then presses the letter against the ribbon. Dot matrix printers uses impact technology and a print head containing banks of wires moving at high speeds against inked ribbon and paper. In laser printer a laser beam creates a charge distribution on a rotating drum coated with a photoelectric material, which is used to imprint the character images. The pattern is then transferred onto the paper in the same manner as a copying machine. The printer used in a computer system is shown in Fig. 5.4.

5.2. INPUT-OUTPUT INTERFACE

Definition A hardware unit that plays an important role between the CPU and peripheral devices to supervise the input and output transfers is known as interface.

Input-output interface provides a method for transferring information between internal storage and external I/O devices. Peripherals connected to a computer need special communication links for interfacing them with the central processing unit. The purpose of the communication link is to resolve the differences that exist between the central computer and each peripheral.

The major differences are :

(1) Peripherals are electromechanical and electromagnetic devices and their manner of operations are different from the operation of the CPU and memory, which are electronic devices. Therefore, the conversion of signals are required.

(2) The data transfer rate of peripherals are usually slower than the transfer rate of the CPU and hence, a synchronization mechanism are needed.

(3) Data codes and formats in peripherals are different from the word format in the CPU and memory.

(4) The operating modes of peripherals are different from each other and each must be controlled so as not to disturb the operation of other peripherals connected to the CPU.

To resolve these differences, computer systems include special hardware known as "interface units" between the CPU and peripherals to supervise and synchronize all input and output transfers.

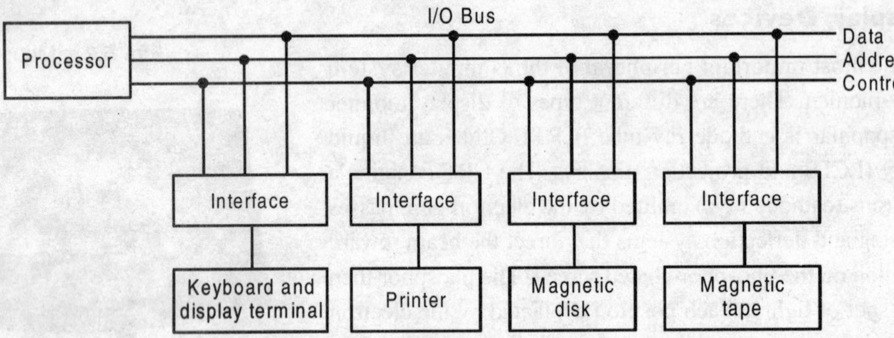

Fig. 5.5.

A typical communication link between the processor and several peripherals are shown in Fig. 5.5. The I/O bus consists of data lines, address lines and control lines.

Each peripheral devices has its associated interface unit. Each interface unit decodes the address and control received from I/O bus, interprets them for the peripheral and provides signals for the peripheral controller. It also synchronizes the data flow and supervises the transfer between peripherals and processor. Each peripheral has its own controller that operates the particular electromechanical device. For example, the printer controller controls the paper motion, the printing timing and the selection of printing characters.

The I/O bus from the processor is attached to all peripherals interfaces. To communicate with a particular device, the processor places a device address on the address lines. Each interface attached to the I/O bus contains an address decoder that monitors the address lines. When the interface detects its own address, it activates the path between the bus lines and the devices that it controls. All peripherals whose address does not correspond to the address in bus are disabled by their interfaces.

At the same time that the address is made available in the address lines, the processor provides a function code in the control lines. The interface selected responds to the function code and proceeds to execute it. These function codes are referred to as I/O commands. There are four types of commands that the interface may receive. They are :

(*i*) **Control Command :** A control command is issued to activate the peripheral and to inform it what to do. The particular control command issued depends on the peripheral.

For example, a magnetic tape unit may be instructed to back space the tape by one record, to rewind the tape or to start the tape moving in the forward direction.

(*ii*) **Status :** A status command is used to test various status conditions in the interface and the peripherals.

(*iii*) **Output Data :** A data output command causes the interface to respond by transferring data from the bus into one of its registers.

(*iv*) **Input Data :** The data input command is the opposite of the data output. In this case the interface receives an item of data from the peripheral and places it in its buffer register. The processor checks if the data are available by means of a status command and then issues a data input command. The interface places the data on the data lines, where they are accepted by the processor.

5.2.1. I/O Versus Memory Bus

The processor must communicate to the memory unit along with the I/O unit. The memory bus also contains the data, address and readwrite control lines as in I/O bus. The computer buses can communicate to memory and I/O in three different ways :

(*i*) Use two separate buses, one of memory and one for I/O.

(*ii*) Use one common bus for both memory and I/O bus having separate control lines for each.

(*iii*) Use one common bus and common control lines for both memory and I/O.

5.2.2. Isolated Versus Memory-Mapped I/O

The transfer of information between the CPU and the memory or I/O are done by means of a common bus. By enabling the specific read or write lines, the CPU specifies whether the address on the address bus is for a memory word or for an interface registers. During a memory transfer, the memory read or memory write control lines are enabled and during I/O transfer, the I/O read or I/O write control lines are enabled. In this way the I/O interface addresses are isolated from the memory addresses and is termed as isolated I/O method for assigning addresses in a common bus.

In isolated I/O configuration, the input and output instructions for both interface register and memory are different. When the CPU fetches and decodes the instructions for I/O interfaces, it enables the I/O read or I/O write control line and informs the other components attached to the common bus that the address in the address bus is for interface register and not for memory word. Similarly, when the CPU fetches and decodes the instruction from memory, it places the memory address on the address bus and enables the memory read or memory write control line, which inform the external components that the address is for memory word and not for I/O interface. Hence, the memory and I/O addresses are isolated in this method.

Alternative, method is known as memory mapped I/O. In this method the computer use the same address space for memory and I/O, both. In memory-mapped I/O there is no specific input or output instructions. The same instruction can be used for manipulating I/O data residing in interfaces and can also be used to manipulate memory words. Computer with memory-mapped I/O can use memory-type instructions to access I/O data and hence, allows the computer to use the same instructions for memory transfers or input-output transfers.

5.3. ASYNCHRONOUS DATA TRANSFER

The data transfer between two independent unit is said to be synchronous data transfer if they share a common clock. But if two independent unit uses its own clock for the data transfer between them the data transfer is said to be asynchronous data trasnfers. Most communications between the computer and the I/O devices are asynchronous. This asynchrouns way of data transfer is achieved by two methods :

(*i*) One way is by means of a strobe pulse supplied by one of the units to indicate to the other unit when the transfer has to occur. This method is known as *strobe control*.

(*ii*) Another method commonly used is to accompany each data item being transferred with a control signal that indicates the presence of data in the bus. The unit receving the data item responds with another control signal to acknowledge receipt of the data. This method of transferring data between two independent units is known as *handshaking*.

5.3.1. Strobe Control

The strobe control method of asynchronous data transfer employs a single control line. The strobe may be activated by either the source or the destination unit. The data bus carries the binary information from source unit to the destination unit. The bus has multiple lines to transfer an entire byte or word. The strobe is a single line that informs the destination unit when a valid data word is available in the bus.

Source-Initiated Strobe for Data Transfer

The strobe initiated by the source unit is shown in Fig. 5.6. The source unit first places the data on the data bus. After a brief delay to ensure that the data settle to a steady value, the source activates the strobe pulse. The information on the data bus and strobe control signal remain in the active state for a sufficient period of time to allow the destination unit to receive the data. Actually, the destination unit uses the falling edge of the strobe control to transfer the contents of data bus to one of its internal registers. The source removes the data from the data bus after it disables its strobe pulse. New valid data will be available only after the strobe is enabled again.

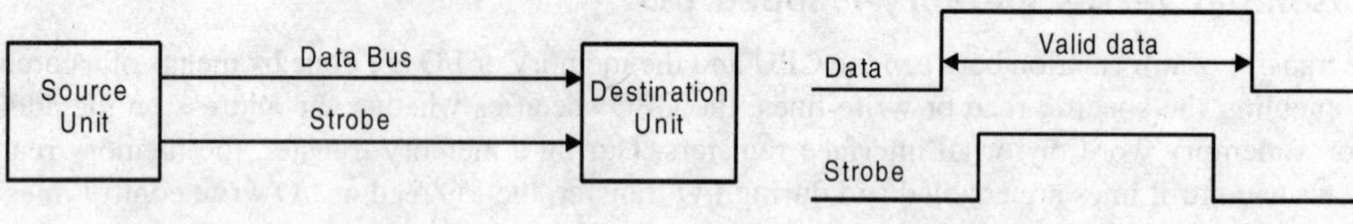

(*a*) Block diagram. (*b*) Timing diagram.

Fig. 5.6. Source-Initiated Strobe.

Destination-Initiated Strobe for Data Transfer

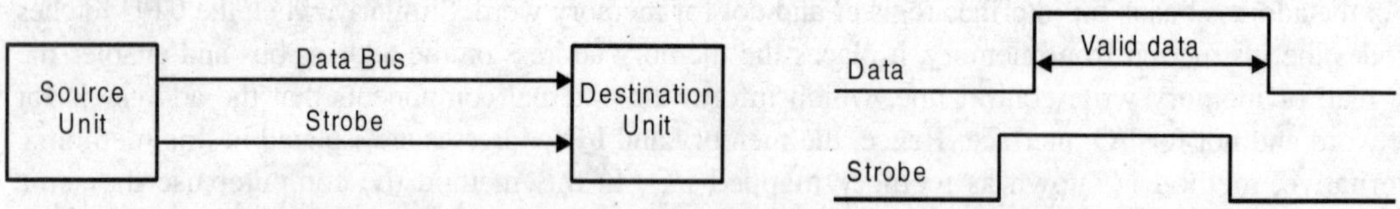

(*a*) Block diagram. (*b*) Timing diagram.

Fig. 5.7. Destination Initiated Strobe.

The strobe initiated by the destination unit is shown in Fig. 5.7. In this case the destination unit activates the strobe pulse, informing the source to provide the data. The source unit responds by placing the requested binary information on the data bus. The falling edge of the strobe pulse is used to transfer the content of the data bus to one of its internal registers. The destination unit then disables the strobe. The source then removes the data from the data bus after a predetermined period of time.

In the first case, the strobe may be a memory-write control signal from the CPU to a memory unit. The source, CPU places the word on the data bus and informs the memory unit, which is the destination, that this is a write operation.

In the second case, the strobe may be a memory-read control signal from the CPU to a memory unit. The destination, the CPU, initiates the read operation to inform the memory, which is a source unit, to place a selected word into the data bus.

5.3.2. Handshaking

The disadvantage of the strobe method is that the source unit that initiates the transfer has no way of knowing whether the destination unit has actually received the data that was placed in the bus. Similarly, a destination unit that initiates the transfer has no way of knowing whether the source unit has actually placed the data on the bus.

The handshaking problem solves this by introducing a second control signal line that provides a reply to the unit that initiates the transfer.

One control line is in the same direction as the data flow in the bus from the source to destination. It is used by the source unit to inform the destination unit whether there are valid data in the bus. The other control line is in the other direction from the destination to the source.

Source-Initiated Handshaking

The handshaking initiated by the source unit is shown in Fig. 5.8. The two handshaking lines are data valid, which is generated by the source unit and data accepted, generated by the desination unit.

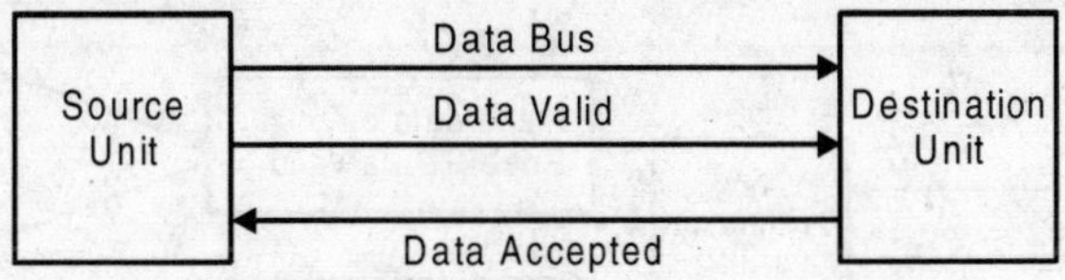

(*a*) Block Diagram.

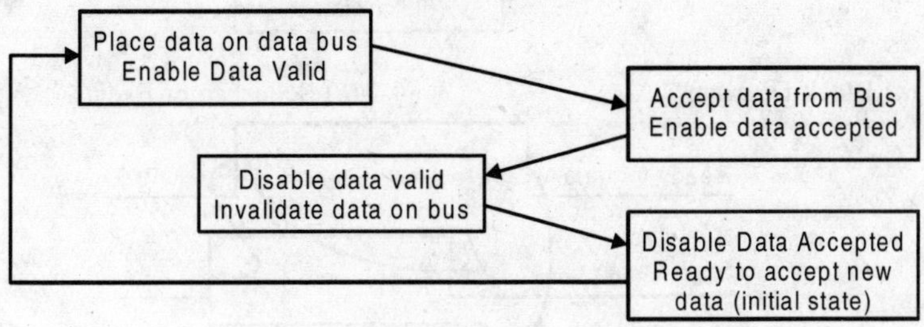

(*b*) Sequence of Events.

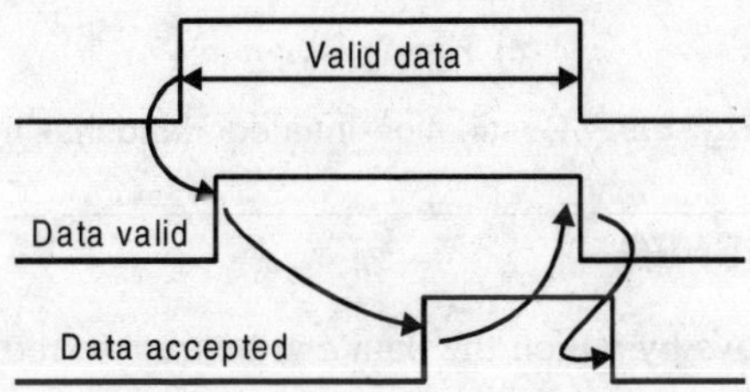

(*c*) Timing Diagram.

Fig. 5.8. Source-Initiated Handshaking.

The source unit initiates the transfer by placing the data on the bus and enabling its data valid signal. The data accepted signal is then activated by the destination unit after it accepts the data from the bus. The source unit then disable its data valid signal, which invalidates the data on the bus. After this the destination unit disables its data accepted signal and the system goes into initial state. The source does not send the next data item until after the destination unit shows its readiness to accept new data by disabling its data accepted signal. The sequence of event is described in Fig. 5.8(*b*) and the timing diagram is shown in Fig. 5.8(*c*) which shows the exchange of signals between two units.

Destination-Initiated Handshaking

The handshaking initiated by the destination unit is shown in Fig. 5.9. The two handshaking lines are data valid, generated by the source unit and ready for data, generated by the destination unit. The source unit does not place the data on the data bus until it receives the ready for data signal from the destination unit. From there on the handshaking process is same as that of source initiated. The sequence of event is described in Fig. 5.9(*b*) and the timing diagram is shown in Fig. 5.9(*c*).

The handshaking process provides a high degree of flexibility and reliability becaue the successful completion of a data transfer relies on active participation by both units. If one unit is faulty, the data transfer will be not completed. Such an error is detected by means of a timeout mechanisms, which produces an alarm if the data transfer is not completed within a predetermined time period. The time out is implemented by means of an internal clock that starts counting time when the unit enables one of its handshaking control signals. If return handshake signal does not respond within a given time period, the unit assumes that an error has been occurred. The time out signal can be used to interrupt the processor and hence execute a service routine that takes appropriate error recovery action.

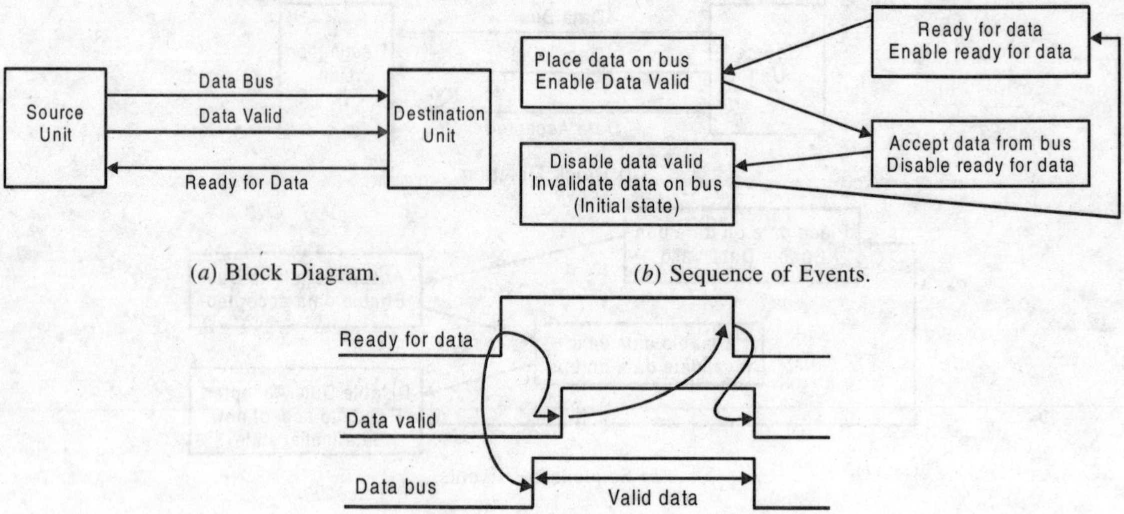

(*a*) Block Diagram. (*b*) Sequence of Events.

(*c*) Timing Diagram.

Fig. 5.9. Destination-Intiated Handshaking.

5.4. MODES OF TRANSFER

There are three different ways by which the data can be transferred from input/output devices to the memory. They are

(*i*) Programmed I/O

(*ii*) Interrupt Initiated I/O

(*iii*) Direct Memory Access (DMA).

5.4.1. Programmed I/O

In this mode of data transfer the input/output instructions are written in the form of computer program. The instruction written in the program basically initiates the data transfer to and from a CPU register and Input/Output peripheral devices. Programmed input/output requires that all input/output operations be executed under the direct control of the CPU *i.e.* every data transfer operation involving an input/output devices requires the execution of an instruction by the CPU. Thus, in programmed input/output mode of transfer, the CPU stays in program loop until the input/output indicates that it is ready for data transfer. This is time-consuming process and keeps the processor busy. This mode of transfer is useful in small, low speed systems where hardware costs must be minimized.

5.4.2. Interrupt Initiated I/O

In interrupt-initiated I/O mode of transfer the CPU concentrate on some other program. This method use interrupt signal to inform the CPU that the data are available from the peripheral device and the input-output flag is set to 1. When the flag is set the CPU deactivates its task to take care of the input-output transfer. After the transfer has been completed, the CPU returns to continue the previous program.

When the interrupt signal is generated, the CPU responds to it by storing the return address from program counter into the memory stack and the control branches to a service routine that processes the required the input/output trasnfer. There are two method to choose the branch address of the service routine. One is called vectored interrupt and the other is called non-vectored interrupt. In vectored interrupt the source that interrupts, gives the branch information to the computer and this information is known as interrupt vector. In non-vectored interrupt, the branch address is assigned to a fixed location in memory.

5.4.3. Direct Memory Access (DMA)

In DMA mode of transfer, the interface transfers data into and out of the memory unit through memory bus. The CPU initiates the transfer. It supply the starting address and number of words to be transferred to the interface unit and then busy with other tasks. When the transfer is made, the DMA disables the bus request line. The transfer of data can be made in several ways. A block sequence consisting of a number of memory words can be transferred into a continuous burst. Alternative way allows to transfer one data word at a time, after which it must return control of the buses to the CPU. Direct memory access is discussed in details in section 5.6.

5.5. INTERRUPT

Definition An interrupt signal is a signal sent by an I/O interface to CPU when it is ready to send information to the memory or receive information from the memory.

The word interrupt is used for any exceptional event that causes the CPU to temporarily transfer the control from its current program to another program, an interrupt handler which will service the interrupt. This program is known as *Interrupt Service Routine (ISR).* Interrupts are the primary means by which input/output devices obtain the services of the CPU. Various sources, internal or external to the CPU can generate interrupts. I/O interrupts are external requests to the CPU to initiate or terminate an I/O operation, such as data transfer with a hard disk. Interrupts are also produced by hardware or software error-detection circuits that invoke error-handling routines within the operating system. An attempt by an instruction to divide-by-zero is an example of software generated interrupts. A power-supply failure can generate an interrupt that requests interrupt handler to save critical data about the system's state.

In general, interrupts can be classified in the following three ways :

(*i*) Hardware and Software interrupt

(*ii*) Vectored and Non-vectored interrupts

(*iii*) Maskable and Non-maskable interrupts.

Hardware interrupts is a type of interrupt generated either externally by the hardware devices such as I/O ports, keyboard, and disk drive etc. or internally by the microprocessor. External hardware interrupts are used by devices to request attention from CPU. Internal hardware interrupts are generated by the CPU to control events.

The *software interrupts* are program instructions. These instructions are inserted at desired location in a program. A program generated interrupt, also called trap, which stops current processing in order to request a service provided by the CPU. While running a program, if software interrupt instruction is encountered the CPU initiates an interrupt. For example, a program might generate a software interrupt to read input from keyboard.

When interrupt signal is generated, the CPU responds to the interrupt signal by storing the return address from program counter into memory stack and then control is transfer or branches to the service routine that processes the interrupt. The processor chooses the branch address of the service routine in two different ways. One is called *vectored interrupt* and the other is called as *non-vectored interrupt.* In *non-vectored interrupt*, the branch address is assigned to a fixed location in memory. In *vectored interrupt*, the source that initiated the interrupt supplies the branch information. This information is called the interrupt vector.

In certain situations it may be desired that some of the several interrupts should not occur while CPU is busy in performing some important task. This can be done by masking. The interrupt that can be masked off is called *maskable interrupt.* When interrupts are masked, they are not withdrawn. They remain pending. Once the masking is removed, interrupt takes place.

Certain interrupts have to be serviced without delay, else something serious damage may be caused to program or data or results. Thus, the CPU must have some means to distinguish between urgent and non-urgent interrupts. Such interrupts are known as *non-maskable interrupts* and CPU does not mask (ignore) them. For example, memory failure, that must be serviced immediately.

5.5.1. Priority Interrupt

Definition A priority interrupt is a system that determines the priority among various requests when two or more device request arrive simultaneously.

When two or more requests arrive simultaneously, there must be a system that decides which condition is to be serviced first. This system is priority interrupt. It also decides which condition are permitted to interrupt the computer when another interrupt is being already serviced. When two devices interrupt the computer at same time, the devices with highest priority is serviced first. Devices with high-speed transfers such as magnetic disks are given high priority and devices with slow speed transfers such as keyboards are given low priority.

A hardware priority-interrupt unit function plays an important role in an interrupt system environment. It receives interrupt request from many devices and determines the highest priority request. This hardware priority unit function can be established by serial or parallel connection of interrupt lines. The serial connection is also known as daisy-chaining method.

5.5.2. Daisy-Chaining Priority

The daisy-chaining priority consists of a serial connection of all devices that requests an interrupt. The devices with the highest priority is place first, then lower-priority devices upto the lowest priority device which is placed last. Fig. 5.10 shows the connection of four devices with the CPU. Interrupt

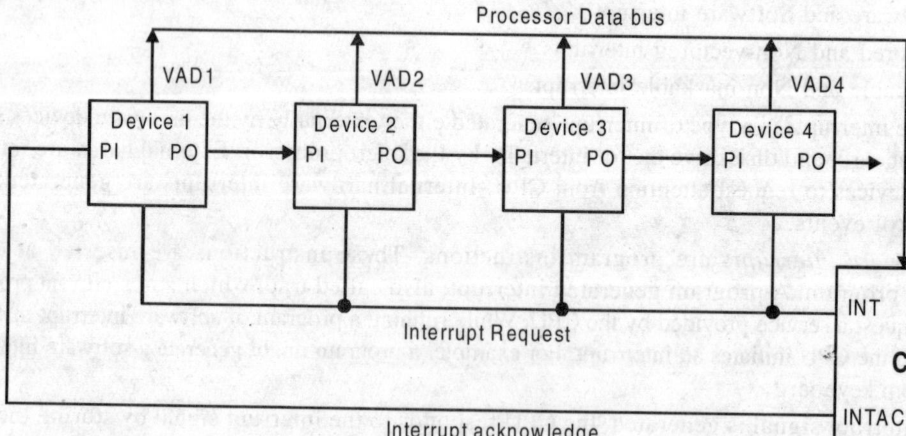

Fig. 5.10. Daisy-chain Priority Interrupt.

request line is common to all devices. When any device has its interrupt signal in the low level state, the interrupt line goes to the low level state and enables the interrupt input in the CPU. When there is no interrupts pending, the interrupt line stays in high level state and CPU recognized no interrupts.

Whenever there is an interrupt the CPU responds to the request by enabling the interrupt acknowledge line. This signal is received by device 1 at its priority in (PI) input. If device 1 is not requesting an interrupt the acknowledge signal is passed to next device through its priority out (PO) output. If device 1 has a pending interrupt, it places 0 in the PO output and blocks the acknowledge signal from the next device. Then, device 1 inserts its own interrupt vector address (VAD) into the processor data bus.

To inform the lower-priority devices that acknowledge signal has been blocked a device with 0 in its PI input places 0 in its PO output. If the device does not have any pending interrupts, it simply transmits the acknowledge signal by placing 1 in its PO output to the next device. A device having 1 in its PI input and is requesting interrupts will intercepts the acknowledge signal and places 0 in its PO output. Thus, the device with PI = 1 and PO = 0 is the one that is requesting the interrupt and places its VAD on the data bus.

5.6. DIRECT MEMORY ACCESS (DMA)

Direct memory access are system pathways that are used by many I/O devices for transferring information directly to and from memory. Today DMAs are used most commonly by floppy drive, tape drives and sound cards.

> **Do You Know**
>
> *Direct memory access is a technique for moving data directly between the main memory and the peripheral devices without the intervention of CPU.*

Thus, removing the CPU from the path and letting the peripherals device manage the memory buses directly would improve the speed of transfer. This transfer technique is called *Direct Memory Access.*

During DMA transfer, the CPU is idle and has no control of the memory buses. A DMA controller takes over the buses to manage the transfer directly between the I/O device and memory.

Fig. 5.11. Bus Signals for DMA Transfer.

The Fig. 5.11 shows two control signals in the CPU that facilitate the DMA transfer. The bus request (BR) input is used by the DMA controller to request the CPU to relinguish the control of the buses. When this input is active, the CPU terminates the execution of the current instruction and places the address bus, the data bus and the read and write lines into a high impedance state. Then the CPU activates the bus grant (BG) output to inform the external DMA that the buses are in the high impedance state. The DMA that originated the bus request, can now take control of the buses to conduct memory transfers without processor intervention. When the DMA terminates the transfer, it disables the bus request line. The CPU then disables the bus grant line and takes control of the buses and returns to the normal operation.

When DMA takes control of the bus system, it communicates directly with the memory. The transfer can be made in several ways.

(*i*) **Burst Transfer :** In DMA burst transfer, a block sequence consisting of a number of memory words is transferred into a continuous burst while the DMA controller is master of the memory buses.

(*ii*) **Cycle Stealing :** In this method, the DMA controller transfers one data word at a time, after which it must return control of the buses to the CPU. The CPU merely delays its operation for one memory cycle to allow the direct memory I/O transfer to "Steal" one memory cycle.

5.6.1. DMA Controller

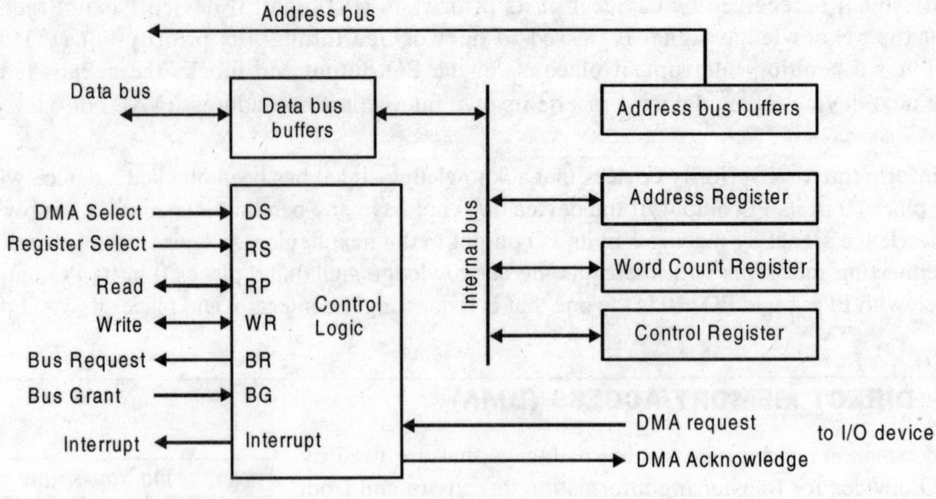

Fig. 5.12. Block Diagram of DMA Controller.

The DMA controller needs the circuits of an interface to communicate with the CPU and I/O device. It also needs an address register, a word count register, a set of address lines. The address register and address lines are used for direct communication with the memory. The word count register specifies the number of words that must be transferred.

The Fig. 5.12 shows the block diagram of DMA controller. The unit communicates with the CPU via the data bus and control lines. The registers in the DMA are selected by the CPU through the address bus by enabling the DS (DMA select) and RS (register select) inputs. When the BG (bus grant) input is 0, the CPU can communicate with the DMA registers through the data bus to read from or write to the DMA registers. When BG = 1, the CPU relinguished the buses and the DMA can communicate directly with the memory by specifying an address in the address bus and activating the RD or WR control.

The DMA controller has three registers. The *address register* contains an address to specify the desired location in memory and the address bits goes into the address bus through address bus buffers. The address register is incremented after each word that is transferred to memory. The *word count register* holds the number of words to be transferred to memory. This register is decremented by one after each word is transferred into memory and internally tested for zero. The *control register* specifies the mode of transfer. All registers in DMA appears to the CPU as I/O interface registers and hence the CPU can read from or write into the DMA registers.

5.6.2. DMA Initialization

The CPU initializes the DMA by sending the following information through the data bus :

(*i*) The starting address of the memory block where data are available (for read) or where data are to be stored (for write).

(*ii*) The word count, which is the number of words in the memory block.

(*iii*) Control to specify the mode of transfer such as read and write.

(*iv*) A control to start the DMA transfer.

The starting address is stored in the address register. The word count is stored in the word count register and the control information is stored in the control register. Once the DMA is initialized, the CPU stops communicating with the DMA unless it receives an interrupt signal or if it wants to check how many words have been transferred.

5.6.3. DMA Transfer

Figure 5.13 shows the DMA transfer in a computer system. With the help of address and data bus, CPU communicates with the DMA. DMA has its own address, which activates DS and RS lines. The CPU initializes the DMA through data bus. The DMA can start data transfer between the peripheral device and the memory once it receives the start control command.

When the peripheral device sends a DMA request, the DMA controller activates the BR line, requesting the CPU to give control over the buses. Then CPU activates BG line and informs the DMA controller that its buses are now disabled. Then the DMA sends a DMA acknowledge to the device by putting the current value of its address register into address bus and initiates the RD or WR signals. After receiving the acknowledge signal, the device puts a word in the data bus (for write) or receives a word from data bus (for read). The DMA increments its address register and decrements the word count register after each word that is transfered. Internally, the word count register is tested for zero. If the word count register reaches zero, the DMA stops any further transfer and disables its bus request signals and informs the CPU through an interrupt. When CPU receives the interrupt signals, it reads the word-count register. The zero value in word-count register signifies that all the words are transferred successfully.

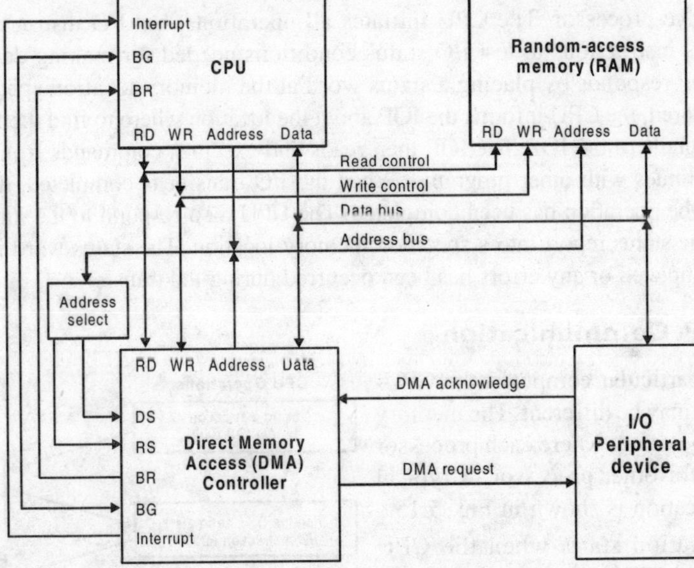

Fig. 5.13. DMA Transfer in a Computer System.

DMA transfer is very useful in many applications. It is used for fast transfer of information between magnetic disk and memory.

5.7. INPUT-OUTPUT PROCESSOR

Definition An I/O processor is a processor that is dedicated to handle only input and output operations.

An input-output processor (IOP) is a processor having capability of direct memory access that is dedicated to handle the input and output operations. It is designed to handle the details of I/O processing. Thus, in computers having IOP, the IOP takes care of input and output tasks and the CPU is relived from the tasks involving I/O transfers. The input-output processor can also perform other processing such as arithmetic, logic branching and code translation. The block diagram of the computer having CPU and IOP both is shown in Fig. 5.14.

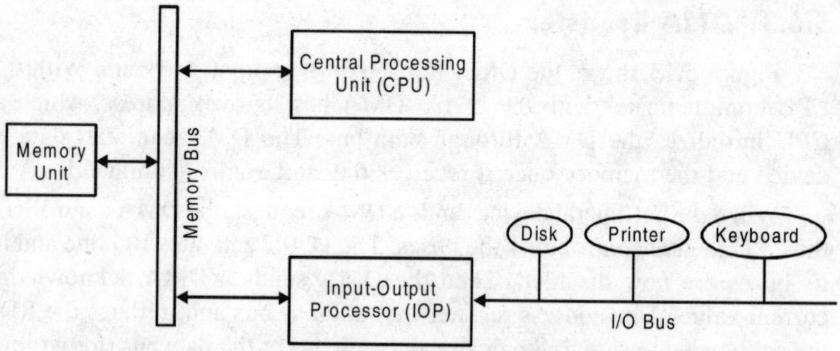

Fig. 5.14. Block Diagram of a Computer with I/O Processor.

With the help of memory bus, the memory can communicate with both processor. CPU is responsible for processing data when ever there is any computational task. Input-output processor communicates with the peripherals devices through I/O bus. IOP provides a path for transfer of data between various peripheral devices and the memory. CPU only initiates the I/O program and then IOP operates independent of the CPU and transfers data from external devices to the memory. But the problem is the data formats of the peripheral devices, of memory and of CPU are different from each other. Thus, it is the responsibility of the input-output processor to structure the data words from many different sources.

In most computers, CPU is superior than IOP and hence the CPU plays the role of master processor and the IOP is a slave processor. The CPU initiates all operations, but I/O instructions are executed in IOP. The CPU sends instructions to test I/O status conditions needed for making decisions on various I/O activities. The IOP responds by placing a status word at the memory location specified by CPU. When a I/O operation is desired, the CPU informs the IOP about the location where to find the I/O program and then leaves the transfer details to the IOP. The IOP then reads and executes commands from the memory. At the same time CPU continues with other programs. When the I/O transfer is completed, the IOP interrupts the CPU to inform that the operation has been completed. The CPU then respond to the interrupts by issuing an instruction to read the status report into a specified memory location. The status word indicates whether the transfer has been completed or any errors has been occurred during the transfer.

5.7.1. CPU-IOP Communication

Depending on particular computer, the CPU-IOP communication may be different. The memory unit acts as a message center where each processor leaves information for other processor. A typical CPU-IOP communication is shown in Fig. 5.15.

The communication starts when the CPU sends an instruction to IOP to test IOP path. Then, the IOP send a status-word to memory for the CPU to check the status. This status-word contains the condition of IOP and I/O devices, such as device is busy, or device is ready for I/O transfer etc. The CPU then decides and if everything is in correct way, the CPU sends an instruction to IOP to start I/O transfer. This instruction also contains an memory address to tell the IOP where to find its program. The CPU also continues with another program while the IOP is busy with I/O transfers. When IOP terminates the execution, it sends an interrupt signal to CPU and the CPU responds to the

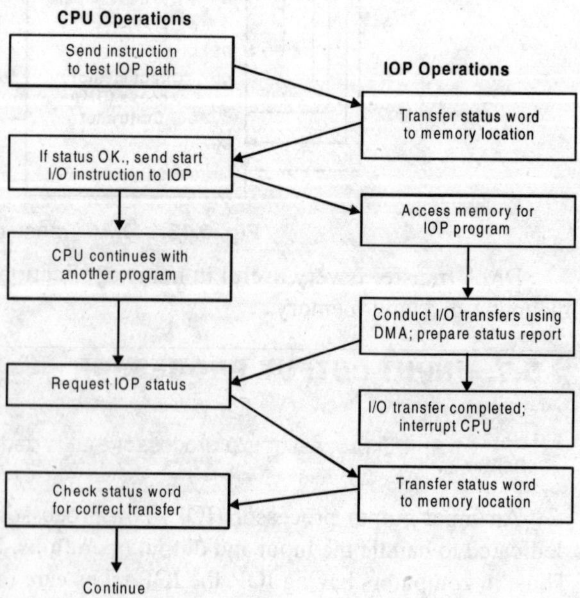

Fig. 5.15. CPU-IOP Communication.

interrupt. The IOP then responds by placing the status report at specified memory location. This status report contains whether the transfer has been completed without any errors or not.

5.8. SERIAL COMMUNICATION

In a communication network there are many devices which are communicating with each other, primarily computers. But there may be other types of devices like printer, interactive display devices, sensors etc. in a communication network. To send data each of these devices can use two types of data communication, parallel communication or serial communication. In parallel communication same number of communication lines/channels are required as of number of bits of data to be transferred from one device to another device. However, parallel communication is fast but if there are lesser number of communication lines or a single communication lines then data cannot be send parallely. In this situation serial communication is required. Transferring of information from one place to another place one bit at a time is known as serial communication.

In serial communication, all bits in the word are sent through a single line serially. The coming word is stored in the buffer on the sender side. Then, bit-wise each bit is taken and sent on the single communication channel in a serial manner. On the receiver end, again there is a buffer and coming bits are stored in the buffer in the same manner as they were stored in sender buffer. The advantage of the serial communication is that only one communication line is required to send any number of bits of the communication data. But the speed of the data transfer is decreased in serial communication.

There are three different modes in which serial communication can be implement. These are :

(i) **Simplex:** It allows transmission of data in only one direction *i.e.* data travel in only one direction. It is very rarely used in data communication because the receiver cannot communicate with the sender. Examples of simplex communication mode are radio and television broadcasting.

(ii) **Half-duplex:** It allows the transmission of data in either direction, but not at the same time. A pair of wires is needed for this mode. The time required to switch a half-duplex line from one direction to the other is called a turn around time.

(iii) **Full-duplex:** It allows the transmission of data in both direction simultaneouly. Full-duplex operation is possible on a four-wire link, with two-wire dedicated to each direction of transmission.

The communication lines which is used by the end devices is called data link and how the information will be transfer from one end to the other end is accomplished by a protocal. A data link control protocol is a set of rules that are followed by interconnecting computers and terminals to ensure the orderly transfer of protocols.

5.8.1. Asynchronous and Synchronous Transmission

There are two ways of implementing the serial communication. One method is asynchronous transmission and other method is synchronous.

5.8.1.1. Asynchronous Data Transfer

In asynchronous data transmission a special bit is inserted at both ends of the character code. Thus, each character will consists of three parts : a start bit, the character bits and the stop bits. The first bit

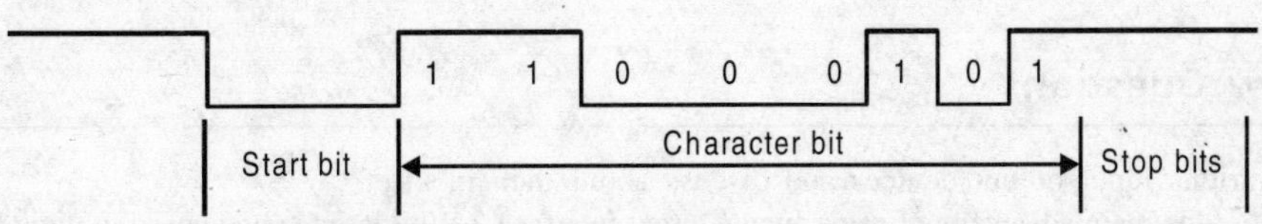

Fig. 5.16. Asynchronous Serial Transmission.

called the start bit, is always 0 and is used to indicate the begining of the character. The last bit of the character is always 1, called the stop bit. The Fig. 5.16 shows a format of asynchronous serial communication.

A receiver can detect the transmitted character having the knowledge of the transmission rules as follows :

(*a*) When a character is not being sent, the line is kept in the 1 state.

(*b*) The initiation of a character transmission is detected from the start bit, which is always 0.

(*c*) The character bit always follows the start bit.

(*d*) After the last bit of the character is transmitted, a stop bit is detected when the line returns to the 1- state for at least one bit time.

5.8.1.2. Synchronous Data Transfer

In this method, both sender and receiver share a common clock frequency. Bits are transmitted from one end to other continuously at the rate dictated by the clock pulse. Synchronous transmission does not use start and stop bits to frame character word. The modems are used for synchronous data transfer. These modem have internal clocks that are set to the frequency that bits are being transmitted in blocks consisting of several hundreds or thousands of bit each. The beginning and end of the each block are marked by appropriate codes and data within a block are organised according to the agreed set of rules. Modems requires a significant start-up time to complete such operations as transmitting and detecting carrier frequencies and establishing synchronization.

Summary

- Input-output system of a computer, provides an efficient way of communication between the central system and the outside world.
- Input-output devices attached to the computer are called peripherals.
- I/O interface is a hardware component.
- There are two types of asynchronous data transfer method known as strobe control and handshaking.
- Strobe control method uses a single control line to transfer data between two independent units.
- Handshaking method uses the control lines, one is generated by source unit and other by destination unit.
- The three ways by which the data can be sent from I/O device to the memory are programmed I/O, interrupt-initiated I/O and Direct memory Access.
- Interrupt is a signal sent by I/O devices to CPU when it is ready to send information to the memory or receive information from the memory.
- Direct memory access (DMA) is a technique for transferring data directly between main memory and peripherals without the intervention of CPU.
- The process where a memory cycle is taken by DMA from CPU is known as cycle stealing.
- The burst transfer method is implemented when data are being transferred by a fast devices like magnetic disk.
- An IOP is a processor with direct memory access capability that communicates with the I/O devices.
- When a computer system consisting of both CPU and IOP, CPU acts as the master processor and IOP is a slave processor.

Review Questions

1. List various input output devices and discuss about then in brief.
2. What are the main advantage of using Input/Output interface ? Why interfacing is used in digital computers ?
3. With the help of block diagram, discuss working of direct memory access ?
4. Discuss DMA transfer with the help of block diagram.

5. What are various modes of data transfer between CPU and I/O devices.
6. Discuss various Asynchronous data transfer.
7. Discuss :
   (*i*) I/O versus Memory Bus
   (*ii*) Isolated versus Memory Mapped
8. Discuss salient features of IOP (Input-Output Processor)
9. With the help of block diagram, discuss working of direct memory access (DMA). Also explain main features of an IOP. Give a brief comparison of DMA and IOP.
10. Explain how data are transferred serial. Why it is slower that parallel data transfer.
11. What is difference between I/O mapped input/output and memory mapped I/O ? What are the advantages and disadvantages of each ?
12. What is the basic advantage of using interrupt-initiated data transfer over transfer under program control without an interrupt ?
13. What is interrupt ? Discuss Daisy-Chaining Priority in details with the help of diagram.
14. What is an interrupt ? Describe its significance for digital computers. Also discuss various types of interrupts in brief.
15. Why are the read and write control lines in DMA controller bi-directional ? Under what condition and for what purposes are they used as inputs ? Under what condition and for what purposes are they used as outputs ?
16. Explain the working of a DMA controller.
17. What is the difference between I/O controlled transfer and DMA transfer ?
18. What do you understand be Serial Data Transfer ? Discuss different types of serial data transfer in brief. Also explain the following:
   (*i*) USART (*ii*) RS 232C
19. With neat block diagram, explain how DMA controller is initialised for DMA data transfer.
20. How data is transmitted in synchronous serial communication system?
21. How many characters per second can be transmitted over a 1200 baud line in asynchronous serial transmission in following modes. Assume a character code is of eight bits.
   (*i*) Synchronous Serial Transfer
   (*ii*) Asynchronous Serial Transfer with 2 Stop bits.
   (*iii*) Asynchronous Serial Transfer with one Stop bit.
22. Compare memory mapped I/O vs I/O mapped I/O.
23. Compare I/O versus memory bus.
24. Explain the use of time out mechanism in hand shaking data transfer scheme.
25. Discuss different techniques used for interfacing I/O units with the processor.
26. What do you mean by initialisation of DMA controller? How DMA controller works? Explain with suitable block diagram.
27. With neat diagram, explain the Strobe control data transfer method. State its disadvantages
28. What is the disadvantage of transferring data through strobe control method? Which method overcomes this disadvantage? Explain.
29. How is source Initiated Handshaking different from Destination Initiated Handshaking? Explain the two through Block and Timing diagrams.
30. Why do peripherals need special communication links to interface them with central processing unit ?

Objective Type Questions

1. In a vectored interrupt
   (*a*) The branch address is assigned to a fixed location in memory
   (*b*) The interrupting source supplies the branch information to the processor through an interrupt vector

(*c*) The branch address is obtained from a register in the processor
(*d*) None of the above

2. The communication between the components in a microcomputer takes place via the address and
(*a*) I/O bus (*b*) Data bus
(*c*) Interface (*d*) Control lines

3. An interface that provides I/O transfer of data directly to and from the memory unit and peripheral is termed as
(*a*) Transfer interface (*b*) Serial interface
(*c*) Bus request interface (*d*) Direct memory access (DMA)

4. An interface that provides a method for transferring binary information between internal storage and external devices is called
(*a*) I/O interface (*b*) I/O bus
(*c*) Input interface (*d*) Output interface

5. An interrupt for which hardware automatically transfers the program to a specific memory location is known as
(*a*) Software interrupt (*b*) Hardware interrupt
(*c*) Maskable interrupt (*d*) Vector interrupt

6. Cycle stealing technique is used in
(*a*) Interrupt based data transfer (*b*) Polled based data transfer
(*c*) DMA based data transfer (*d*) None of the above

7. DMA interface unit eliminates the need to use CPU registers to transfer data from
(*a*) MAR to MBR (*b*) MBR to MAR
(*c*) I/O units to memory (*d*) Memory to I/O units

8. In a non-vectored interrupt, the address of the interrupt service routine is
(*a*) Obtained from interrupt address table.
(*b*) Supplied by the interrupting I/O device.
(*c*) Obtained through vector address generator device.
(*d*) Assigned to a fixed memory location.

9. A control character is sent at the beginning as well as at the end of each block in the synchronous transmission, in order to
(*a*) Synchronise the clock of transmitter and receiver.
(*b*) Supply information needed to separate the incoming bits into individual character.
(*c*) Detect the error in transmission and received system.
(*d*) None of the above

10. The cycle stealing technique, allows the DMA controller to transfer
(*a*) Half of a data word at a time.
(*b*) One block of data at a time.
(*c*) One data byte at a time.
(*d*) One data word at a time.

11. During DMA acknowledgement cycle, CPU relinquishes
(*a*) Address bus only (*b*) Address bus and control bus
(*c*) Control bus and data bus (*d*) Data bus and address bus

ANSWERS

1. (*b*)	**2.** (*b*)	**3.** (*d*)	**4.** (*a*)	**5.** (*b*)
6. (*c*)	**7.** (*d*)	**8.** (*d*)	**9.** (*b*)	**10.** (*c*)
11. (*d*)				

❑❑❑

6

Chapter

Memory Organization

INSIDE THIS CHAPTER

6.1. BASIC MEMORY CHARACTERISTICS

The most modern computers are designed using the Von-Neumann Model, which is centered on Main Memory. The user programs and user data that perform the processing are stored initially on auxiliary memory devices from there they are to be brought to main memory by operating system. We know that memory is logically organized as a linear array of locations or addresses, with addresses from 0 to the maximum memory size the processor can address. This address is normally assigned by a programmer in OCTAL or Hexadecimal format.

The various different types of memory devices and technology used for fabrication for these devices for better performance and how each is part of the memory hierarchy system. We then look at cache memory (a special high-speed memory) and a method that utilizes memory to its fullest by means of virtual memory implemented via paging technique. The goal of Memory Hierarchy is to obtain the highest possible access speed while minimizing the total cost of the memory system.

Location of the Device

(*i*) CPU registers–we call sometime internal memory.

(*ii*) Cache memory–between CPU and Main memory or On chip Cache.

(*iii*) Main Memory–SRAM or DRAM Devices.

(*iv*) Secondary Storage (external or auxiliary device).

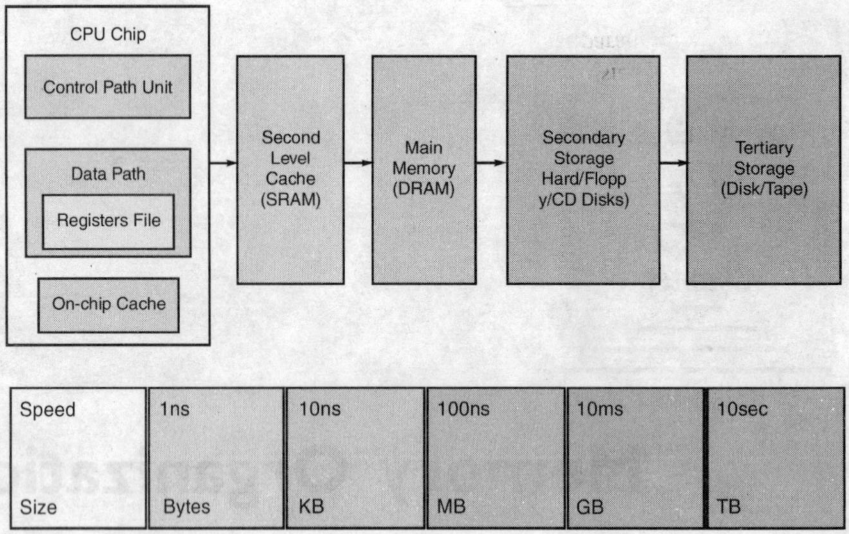

Fig. 6.1. CPU and Memory Interaction

Capacity of the Device—Unit of Data Transfer

The number of data lines into and out of the memory module. Internally it is usually governed by data bus width. Externally it is usually a block which is much larger than a word.

Access Methods For The Device

Sequential Access

We start at the beginning and read through in order and access time depends on location of data and previous location (tape)

Direct Access

Individual blocks have unique address in memory and Access is by jumping to vicinity plus sequential search. Access time depends on location and previous location.

Random Access

Individual addresses identify locations exactly. Access time is independent of location or previous access (RAM).

Associative

Data is located by a comparison with contents of a portion of the store. Access time is independent of location or previous access e.g. cache. The words are retrieved based on a portion of its contents rather than its address and constant access time.

Memory Device Performance

Access Time : The time it takes from the instant that an address is presented to the memory by CPU to the instant that data have been made available for use on data bus.

Memory Cycle Time

Time may be required for the memory to "recover" before next access. Cycle time is access + recover.

Data Transfer Rate or Memory Bandwidth

Rate at which data can be transferred into or out of a memory unit. Random access : 1/(Cycle time).

Non-random : $TN = TA + (N/R)$

TA = access time

N = number of bits

R = transfer rate (bps)

Physical Characteristics of a Device

(*i*) Volatility

(*ii*) Power Consumption

(*iii*) Erasable

(*iv*) Density

(*v*) Cost Per Bit Storage

Device Organisation and Trade-off for Performance

(*i*) Physical arrangement of bits into words.

(*ii*) Not always obvious e.g. interleaved unit.

Tradeoff among three main characteristics of memory (cost, size, access time)

(*i*) Smaller access time, greater cost per bit.

(*ii*) Greater capacity, smaller cost per bit.

(*iii*) Greater capacity, greater access time.

Different types of memory are required at different level of operation in a computer system. A classification is discussed further.

6.2. MEMORY HIERARCHY

The widening speed gap between CPU and main memory needs to be filled. Modern processor operations take of the order of 1 ns, while memory access requires 10s or even 100s of ns. The second parameter is memory bandwidth, which limits the instruction execution rate. Each instruction executed involves at least one memory access. Hence, a few to 100s of MIPS is the best that can be achieved. A fast buffer memory can help bridge the CPU-memory gap. The fastest memories are expensive and thus not very large.

One of the most important considerations in understanding the performance capabilities of a modern processor is the memory hierarchy. Unfortunately, as we have seen, not all memory is created equal, and some types are far less efficient and thus cheaper than others. To deal with this disparity, today's computer systems use a combination of memory types to provide the best performance at the best cost. This approach is called hierarchical memory. As a rule, the faster memory is, the more expensive it is per bit of storage. By using a hierarchy of memories, each with different access speeds and storage capacities, a computer system can exhibit performance above what would be possible without a combination of the various types. The base types that normally constitute the hierarchical memory system include registers, cache, main memory, and secondary memory.

At the bottom of the hierarchy are the relativity slow magnetic tapes used to store removable files. Magnetic disk used as back up storage. The main memory occupies a central position by being able to communicate directly with the CPU and with auxiliary memory device through an input/output processor when program not residing in main memory are needed by the CPU. They are in bought in from auxiliary memory. Program not currently needed in main memory are transferred into auxiliary memory to provided space for currently used programs and data.

As we move from top to bottom in hierarchy.

(*i*) Access Speed will Decrease
(*ii*) Storage Capacity will Increase
(*iii*) Cost Per Bit will Decrease

While the I/O processor manages data transfer between auxiliary memory and main memory, the cache organization is concerned with the transfer of information between main memory and CPU. Thus each is involved with a different level in memory hierarchy system. The reason for having two or three level of memory hierarchy is economics. As the storage capacity of the memory increase, the **cost per bit** for storing binary information decrease and the access time of the memory become longer.

6.3. CLASSIFICATION OF MEMORY

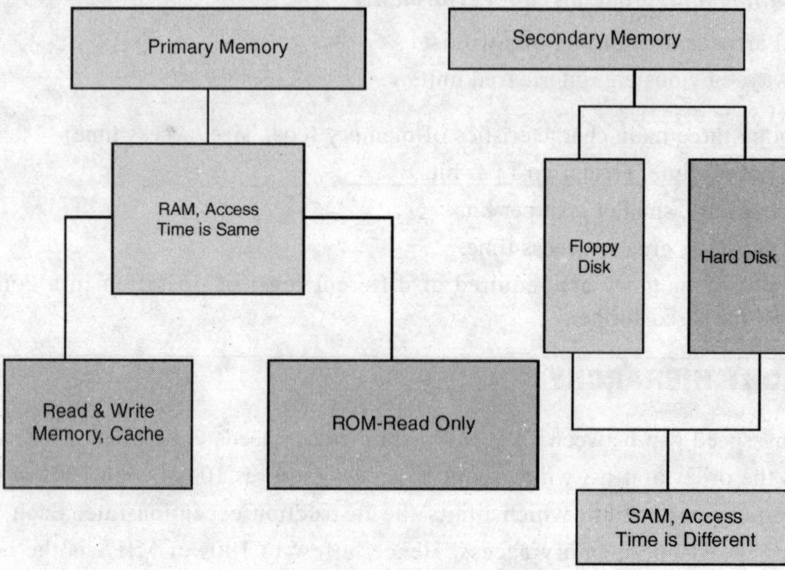

Fig. 6.2. Memory Classifications

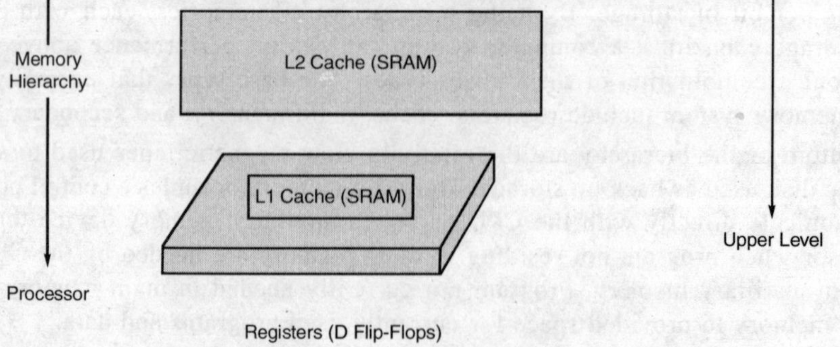

Fig. 6.3. Memory Hierarchy Levels

6.3.1. Main or Primary Memory

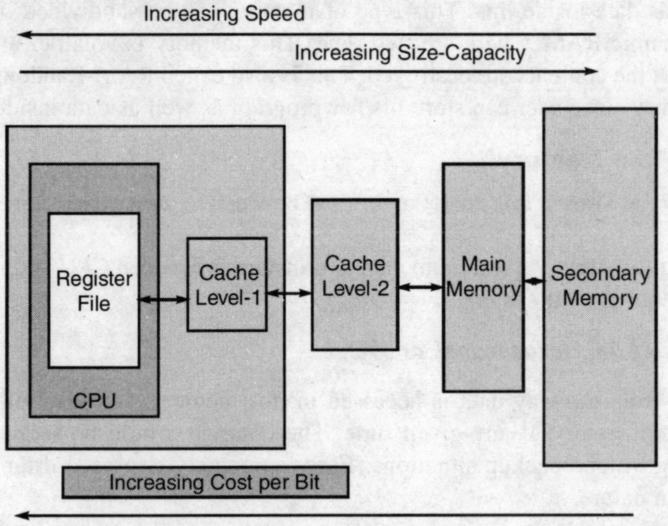

Fig. 6.4. Register File (CPU)

Internal or CPU Registers Memory

Every CPU has a register file or array-which is used by CPU as internal temporary memory for computation, therefore, sometimes this is called **internal scratch memory** of the CPU. A system programmer has access to this memory. Any register can be accessed by the system programmer for usage like system programming.

R/WM (Read Write Memory) or RAM Organisation

This memory is volatile, meaning that when the power is turned off, all the contents are destroyed.

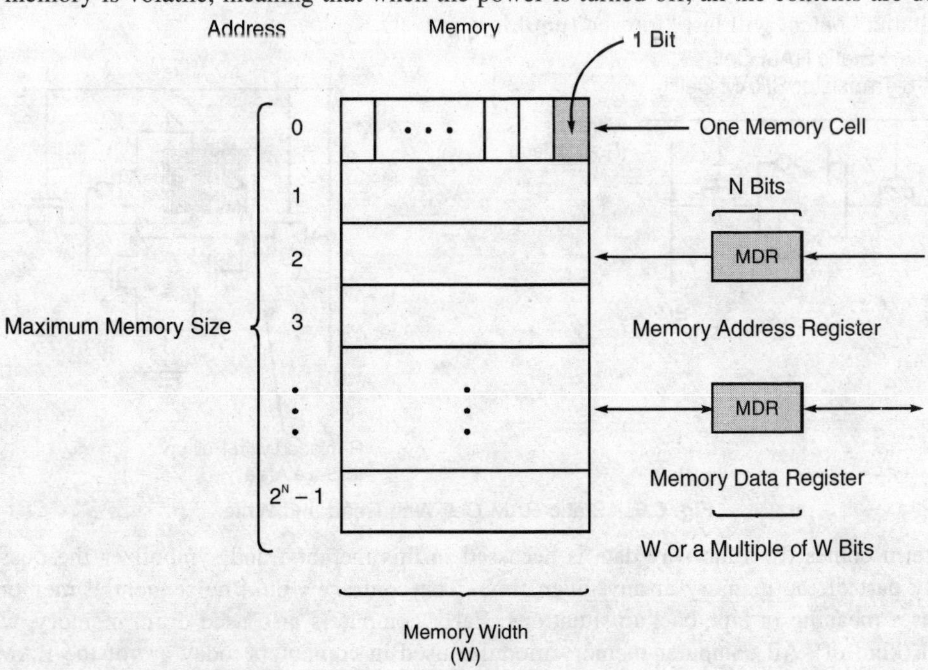

Fig. 6.5. Random Access Memory Addressing by CPU

This is also called RAM (Random Access Memory) and also called user memory since user can store his program as well as data inside this. This type of memory can be subdivided into two main groups: Static RAM and Dynamic RAM. There are two types.This memory is volatile, meaning that when the power is turned off, all the contents are destroyed. This is also called RAM-Random Access Memory and also called user memory since user can store his/her program as well as data inside it.

Performance of Main Memory

(*i*) **Memory Access Time :** It is the delay or time between request generation and arrival of word in the CPU.

(*ii*) **Memory Bandwidth :** At what rate data is transferred between CPU and main memory. Higher is the rate better is your system.

Address is Represented in Hexadecimal or Octal

The term comes from the way data is accessed in this memory, and symbolizes the possibility to access any part of the memory at any given time. The contrary would be sequential memory, which mostly has a meaning in tape backup situations. Early computers also used drum memory, which was sequential (kind of) in nature.

(*i*) **SRAM – Static Random Access Memory**

(*ii*) **DRAM – Dynamic Random Access Memory**

- FP – Fast Page RAM
- EDO – Extended Data Output
- SDRAM – Synchronous DRAM

Static RAM Organization and Main Features

(*i*) It is made of flip-flops.

(*ii*) Low density, high power, expensive, fast.

(*iii*) Static: content will last "forever"(until lose power).

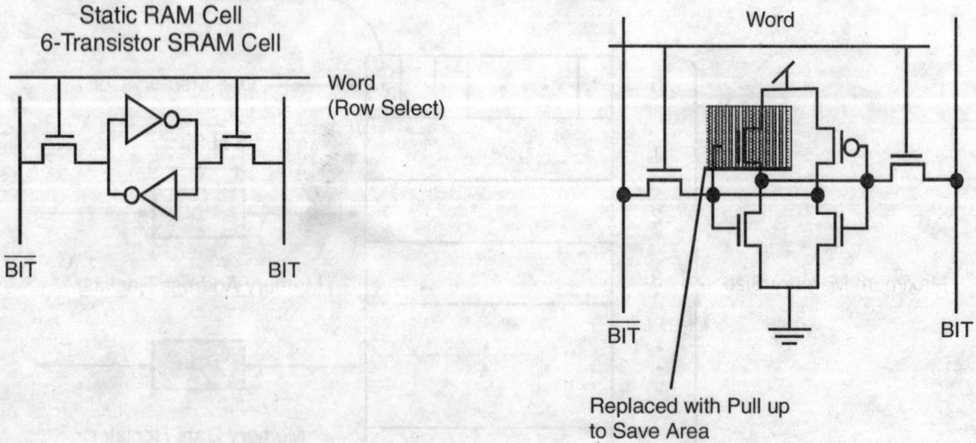

Fig. 6.6. Static RAM Cell With Read and Write

The term comes from the way data is accessed in this memory, and symbolizes the possibility to access any part of the memory at any given time. The contrary would be sequential memory, which mostly has a meaning in tape backup situations. Early computers also used drum memory, which was sequential (kind of). All computer memory modules used in computers today are of the RAM type. A

RAM is organized typically as bytes to store data and its address is represented in HEX-Format. Any memory configuration can be fabricated using basic SRAM Cell and address decoder circuit.

Write : (*i*) Drive Bit Lines (Bit = 1, Bit = 0) (*ii*) Select Row; Read : (*i*) Precharge Bit and Bit to V_{dd} or $V_{dd}/2 \Rightarrow$ Make Sure Equal! (*ii*) Select Row (*iii*) Cell Pulls One Line Low (*iv*) Sense Amp on Coloumn

(*i*) Based on flip-flops and requires several transistors.

(*ii*) Fast access time and bits stored as on/off switches.

(*iii*) No charges to leak and No refreshing needed when powered.

(*iv*) Used in Cache memory design.

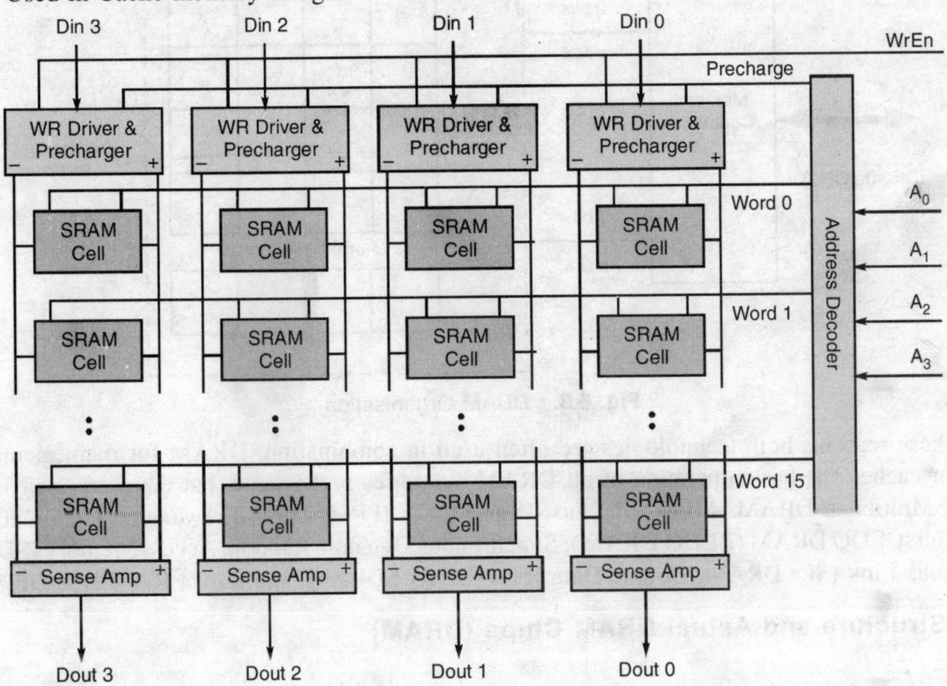

Fig. 6.7. SRAM Organisation 16-Works × 4 Bit

DRAM Organization

DRAM is made up of MOS Transistor Gates. It is a refreshing type memory. The difference between the two is that as long as power is maintained on the memory modules the DRAM will hold its information. The content of DRAM memory disappears from the memory within milliseconds, so in order to maintain its data it has to be refreshed periodically. This makes the DRAM memory much slower than the SRAM. The computer memories you usually see are a form of DRAM, like SDRAM and DDR-SDRAM.

(*i*) Requires one capacitor and one transistor.

(*ii*) But needs refresh every few milliseconds.

(*iii*) Need refresh circuits.

(*iv*) Slower access time and bits stored as charge in capacitors.

(*v*) Charges may leak.

(*vi*) Simpler construction and smaller per bit.

(*vii*) Less expensive.

Dynamic RAM is constructed of tiny capacitors that leak electricity. DRAM requires a recharge every few milliseconds to maintain its data. Static RAM technology, in contrast, holds its contents as long as power is available. SRAM consists of circuits similar to the D flip-flops. SRAM is faster and much more expensive than DRAM; however, designers use DRAM because it is much denser (can store many bits per chip), uses less power, and generates less heat than SRAM. The different types of SRAM include asynchronous **SRAM,** synchronous **SRAM**, and **pipeline burst SRAM.**

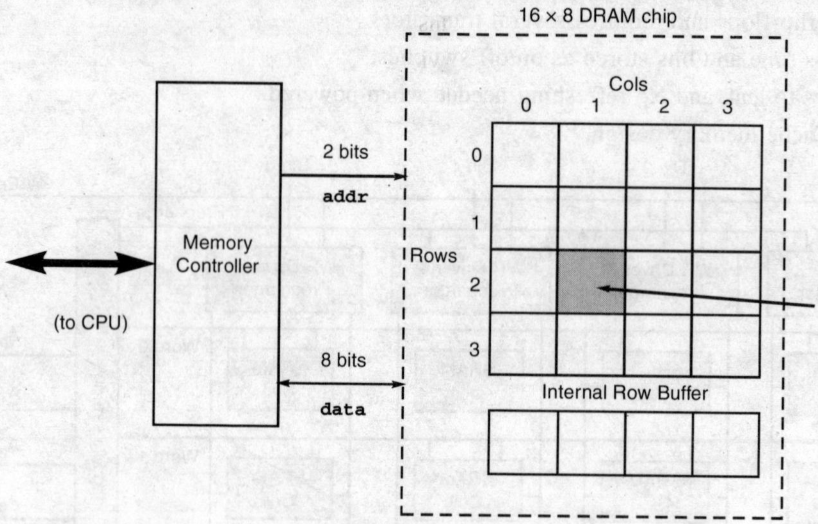

Fig. 6.8. DRAM Organisation

For these reasons, both technologies are often used in combination: DRAM for main memory and SRAM for cache. The basic operation of all DRAM memories is the same, but there are many flavors, including Multibank DRAM (MDRAM), Fast-Page Mode (FPM) DRAM, Extended Data Out (EDO) DRAM, Burst EDO DRAM (BEDO DRAM), Synchronous Dynamic Random Access Memory (SDRAM), Synchronous-Link (SL) DRAM, Double Data Rate (DDR) SDRAM, and Direct RAMBUS DRAM.

Circuit Structure and Actual DRAM Chips (DRAM)

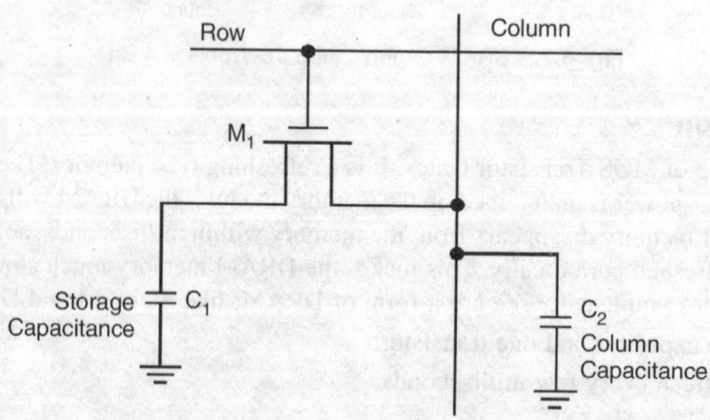

Fig. 6.9. DRAM Circuit

Read Only Memory (ROM)

Mask Programmable (Data "Written" During Device Fabrication")

Field Programmable (Data "Written" by Device Purchaser)

(*i*) PROM–Programmable Read Only Memory

(*ii*) EPROM–Erasable Programmable Read Only Memory

(*iii*) EEPROM–Electrically Erasable Programmable Read Only Memory.

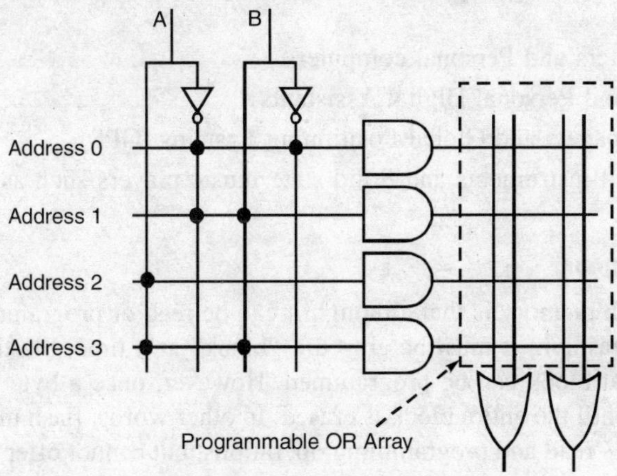

Fig. 6.10. PLD Architecture for PROM

PROM (Programmable Read-Only Memory)

It is a variation on ROM. PROM can be programmed by the user with the appropriate **Programming Device.** Whereas ROMs are hardwired, PROM has fuses that can be blown to program the chip. Once programmed, the data and instructions in PROM cannot be changed.

Flash Memory Applications and Limitations

The Flash memory was invented by Dr. Fujio Masuoka while working for Toshiba in 11084. According to Toshiba, the name 'Flash' was suggested by Dr. Masuoka's colleague, Mr. Shoji Ariizumi, because the erasure process of the memory contents reminded him of a flash of a camera.

Fig. 6.11. Flash Memory Device

Flash memory is a form of EEPROM (Electrically Erasable Programmable Read-Only Memory) that allows multiple memory locations to be erased or written in one programming operation. In other terms, it is a form of rewritable memory chip that, unlike a RAM (Random Access Memory) chip, power supply is not required to hold the contents. It is commonly used in memory cards, USB flash drives, MP3 players, digital cameras and mobile phones.

(*i*) High Density

(*ii*) Read/Write

(*iii*) Non Volatile – Data stored indefinitely without power

(*iv*) Combine best characteristics of traditional RAM/ROM.

Applications of Flash Memory

The Flash Memory technology has evolved into the preferred storage media for a variety of consumer and industrial devices.

1. Notebook computers and Personal computers
2. Digital cameras and Personal Digital Assistants
3. Cell phones and pagers and Global Positioning Systems (GPS)
4. Electronic musical instruments and Solid-state music players such as Television set-top boxes and MP3 players.

Limitations of Flash Memory

One limitation of flash memory is that although it can be read or programmed a byte or a word at a time in a random access fashion, it must be erased a "block" at a time. Starting with a freshly erased block, any byte within that block can be programmed. However, once a byte has been programmed, it cannot be changed again until the entire block is erased. In other words, flash memory (specifically NOR flash) offers random-access read and programming operations, but cannot offer random-access rewrite or erase operations.

(*i*) MRAM - MAGNETIC RAM

(*ii*) Infineon, IBM show prototype 16Mb magnetic RAM. Prototype MRAM chip brings power-saving technology one step closer

(*iii*) MRAM is expected to first be used in mobile applications as a replacement for flash.

(*iv*) At some point in the future, MRAM could also replace SRAM and DRAM, which is cheap to manufacture but is slower than MRAM and requires a constant power supply to retain stored data.

(*v*) Using MRAM instead of DRAM in notebooks PCs could help extend battery life, according to Infineon and IBM.

6.3.2. Auxiliary Memory Devices

(*i*) Floppy Disks

(*ii*) Hard Disks

(*iii*) Magnetic Disks

(*iv*) Magnetic Tapes

(*v*) CD-ROM Disks

(*vi*) Optical Disk

(*vii*) DVD-digital Video Disk

Floppy Disks

A soft magnetic disk is called *floppy* because it flops if you wave it (at least, the 5¼-inch variety does). Unlike most hard disks, floppy disks (often called *floppies* or *diskettes*) are portable, because you can remove them from a disk drive. Disk drives for floppy disks are called floppy drives. Floppy disks are slower to access than hard disks and have less storage capacity, but they are much less expensive. And most importantly, they are portable.

Floppies come in three basic sizes :

1. **8-inch :** The first floppy disk design, invented by IBM in the late 11060s and used in the early 11070s as first a read-only format and then as a read-write format. The typical desktop/laptop computer does not use the 8-inch floppy disk.

2. **5¼-inch :** The common size for PCs made before 11087 and the predecessor to the 8-inch floppy disk. This type of floppy is generally capable of storing between 100 K and 1.2 MB of data. The most common sizes are 360 K and 1.2 MB.
3. **3½-inch :** *Floppy* is something of a misnomer for these disks, as they are encased in a rigid envelope. Despite their small size, microfloppies have a larger storage capacity than their cousins — from 400K to 1.4MB of data. The most common sizes for PCs are 720K (double-density) and 1.44MB high-density.

Hard Disks

Do You Know

A hard disk (commonly known as a HDD (hard disk drive) or hard drive and formerly known as a fixed disk.

Hard disk is a non-volatile storage device which stores digitally encoded data on rapidly rotating **platters** with **magnetic** surfaces. Strictly speaking, "drive" refers to an entire unit containing multiple platters, a read/write head assembly, driver electronics, and motor while "hard disk" (sometimes "platter") refers to the storage medium itself. It is connected to main memory through the bus and a disk controller.

The processor only occasionally interacts with secondary memory.

1. Secondary memory is where programs and data are kept on a long-term basis. Common secondary storage devices are the hard disk and floppy disks. The hard disk has enormous storage capacity compared to main memory.
2. The hard disk is usually contained in the systems unit of a computer. Data and programs on the hard disk are organized into files—named sections of the disk.

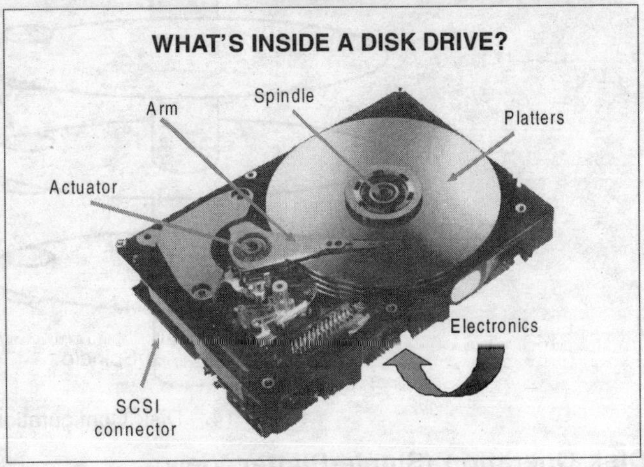

Fig. 6.12. Hard Disk-Components

3. Hard disk might have a storage capacity of 40 gigabytes. This is about 300 times the amount of storage in main memory (assuming 128 megabytes of main memory). A hard disk is very slow compared to main memory.

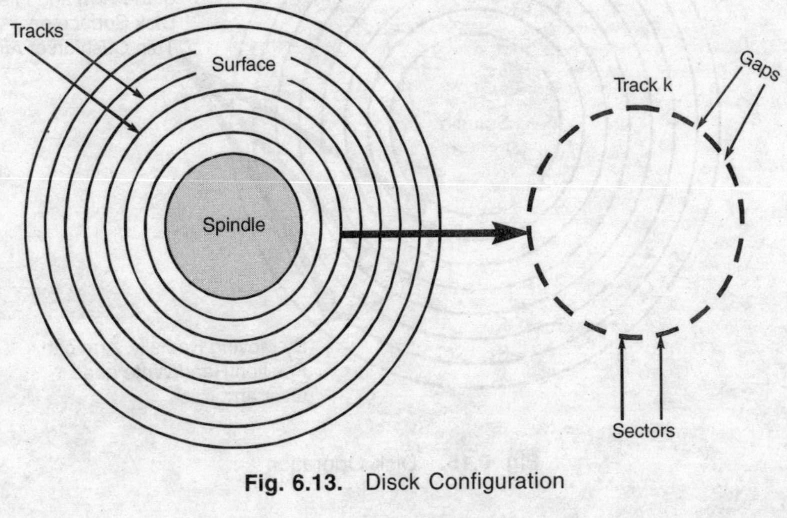

Fig. 6.13. Disck Configuration

Hard disks were originally developed for use with computers. In the 21st century, applications for hard disks have expanded beyond computers to include digital video recorders, digital audio players, personal digital assistants, and digital cameras. The need for large-scale, reliable storage, independent of a particular device, led to the introduction of configurations such as RAID, hardware such as network attached storage (NAS) devices, and systems such as storage area networks (SANs) for efficient access to large volumes of data.

Disks consist of platters, each with two surfaces. Each surface consists of concentric rings called tracks. Each track consists of sectors separated by gaps.

Disk Operation (Multiplotter View)

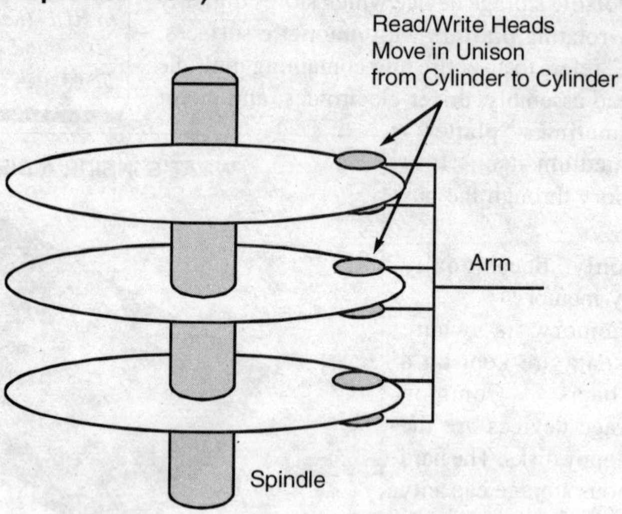

Fig. 6.14. Disk Configuration-Multi-Platters

Disk Operation (Single-Platter View)

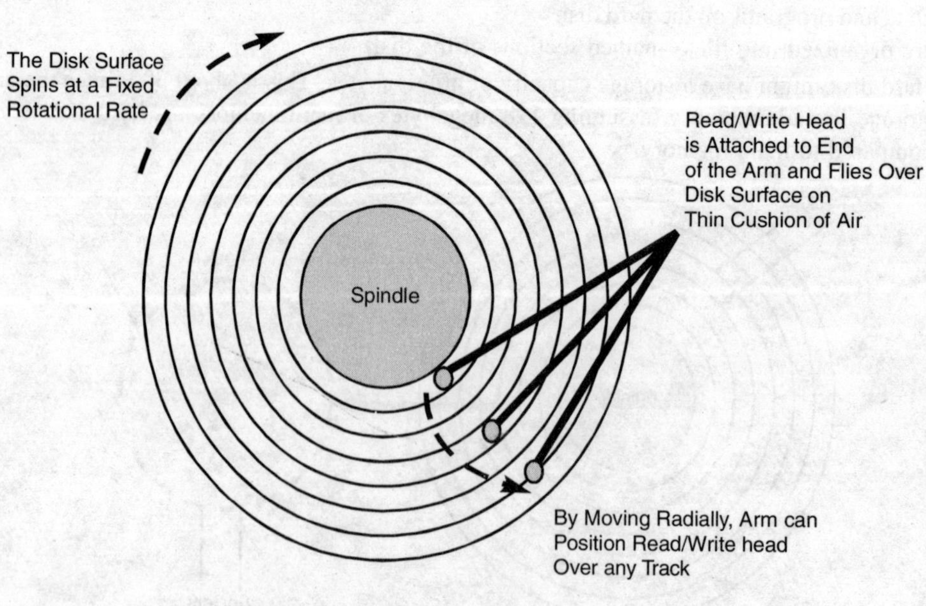

Fig. 6.15. Disk Operation

Disk Electronics

Quantum Viking (circa 1997)

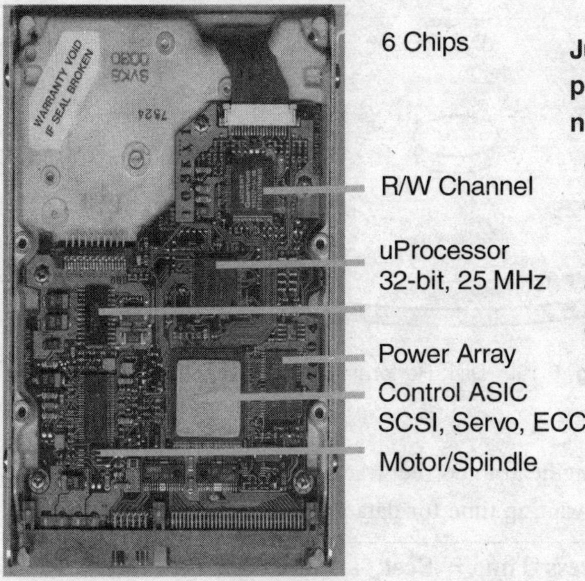

Just like a small computer– processor, memory, network

- **CONNECT TO DISK**
- **CONTROL PROCESSOR**
- **CACHE MEMORY**
- **CONTROL ASIC**
- **CONNECT TO MOTOR**

Fig. 6.16. Disk Electronics Components

How Data is Organised In Hard-Disk?

Organizing Data on Disk

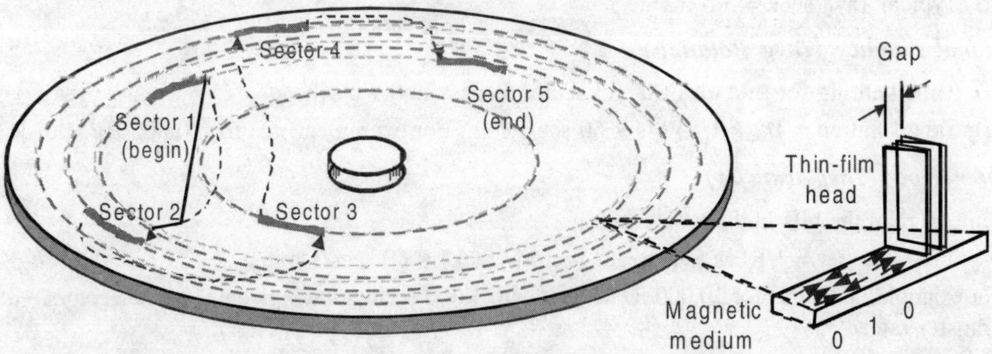

Magnetic recording along the tracks and the read/write head.

	0	16	32	48	1	17	33	49	2	
	30	46	62	15	31	47	0	16	32	
	60	13	29	45	61	14	30	46	62	
	27	43	59	12	28	44	60	13	29	

Logical numbering of sectors on several adjacent tracks.

Fig. 6.17. Data Organisation on Disk

$$\text{Seek time} = a + b(c-1) + b(c-1)^{1/2}$$

Reducing Average Seek Time and Rotational Latency by Performing Disk Accesses out of Order.

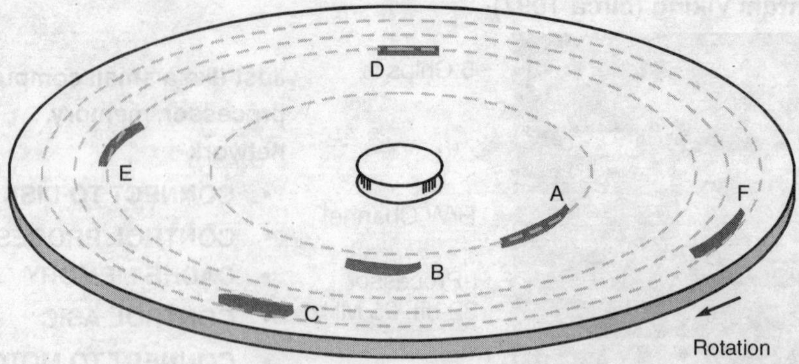

Fig. 6.18. Disk Performance

Disk Speed

(*i*) **Seek Time :** The time in moving head to correct track.

(*ii*) **Rotational Latency :** It is the waiting time for data to rotate under head.

Access Time = Seek + Latency

Disk Access Time

Average time to access a specific sector approximated by:

Taccess = Tavg seek + Tavg rotation + Tavg transfer Seek time (Tavg seek)

(*i*) Time to position heads over cylinder containing target sector

(*ii*) Typical Tavg seek = 3-5 ms

Rotational Latency (Tavg Rotation)

(*i*) Time waiting for first bit of target sector to pass under r/w head.

(*ii*) Tavg rotation = 1/2 × 1/RPMs × 60 sec/1 min. For example, 3 ms for 10,000 RPM disk.

Transfer Time (Tavg transfer)

Time to read the bits in the target sector

Tavg transfer = 1/RPM x 1/(avg # sectors/track) × 60 secs/1 min

For example, 0.006ms for 10,000 RPM disk with 1,000 sectors/track given 512-byte sectors, ~85 MB/s data transfer rate.

Disk Access Time Example

Given :

(*i*) Rotational rate = 7,200 RPM

(*ii*) Average seek time = 5 ms

(*iii*) Avg # sectors/track = 1000

Derived average time to access random sector :

(*i*) Tavg rotation = 1/2 x (60 secs/7200 RPM) × 1000 ms/sec = 4 ms.

(*ii*) Tavg transfer = 60/7200 RPM × 1/400 secs/track × 1000 ms/sec = 0.008 ms.

- Taccess = 5 ms + 4 ms + 0.008 ms = 9.008 ms.
- Time to second sector: 0.008 ms.

TIPS

Access time dominated by seek time and rotational latency. First bit in a sector is the most expensive, the rest are free. SRAM access time is about 4 ns/doubleword, DRAM about 60 ns.

6.3.3. Disk Caching–RAID Idea

Same idea as processor cache: bridge main-disk speed gap. Read/write an entire track with each disk access : Access one sector, get 100s free, assume hit rate around 100%. Rotational latency eliminated; can start from any sector. Need back-up power so as not to lose changes in disk cache.

Placement options for disk cache in the disk controller: It suffers from bus and controller latencies even for a cache hit closer to the CPU. Avoids latencies and allows for better utilization of space-Intermediate or multilevel solutions

1. Several improvements in disk-use techniques involve the use of multiple disks working cooperatively.
2. Disk striping uses a group of disks as one storage unit.
3. RAID schemes improve performance and improve the reliability of the storage system by storing redundant data.
   (*a*) *Mirroring* or *shadowing* keeps duplicate of each disk.
   (*b*) *Block interleaved parity* uses much less redundancy.

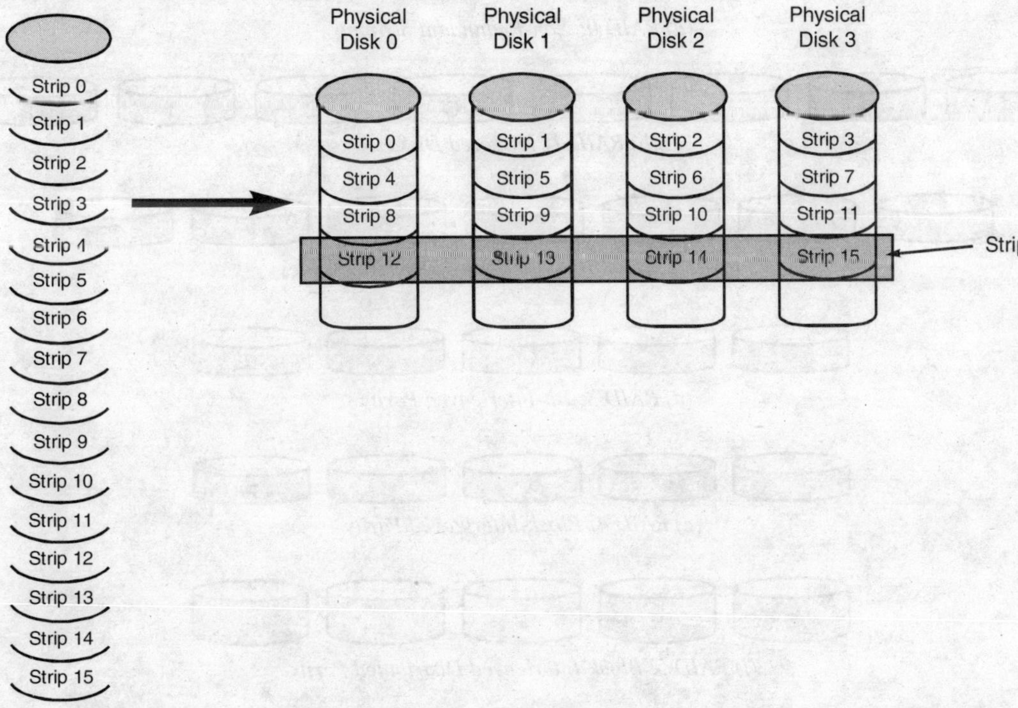

Fig. 6.19. Disk Cache-Raid Stripping

4. Disk arrays can be made fault-tolerant by redundantly storing information in various ways.
5. Five types of array architectures, RAID-1 through RAID-5, were defined by the Berkeley paper, each providing disk fault-tolerance and each offering different trade-offs in features and performance.

6. In addition to these five redundant array architectures, it has become popular to refer to a non-redundant array of disk drives as a RAID-0 array.

RAID Common Characteristics

1. A set of physical disk drives viewed by the OS as a single logical drive.
2. Data are distributed across the array of disk drives.
3. Redundant disk capacity is used to store parity information, which guarantees data recoverability in case of a disk failure.

Data Striping

(*i*) Fundamental to RAID is "striping", a method of concatenating multiple drives into one logical storage unit.

(*ii*) Striping involves partitioning each drive's storage space into strips which may be as small as one sector (512 bytes) or as large as several megabytes.

RAID Structure

(*i*) **RAID** – provides **reliability** via **redundancy**.

(*ii*) RAID is arranged into six different levels:

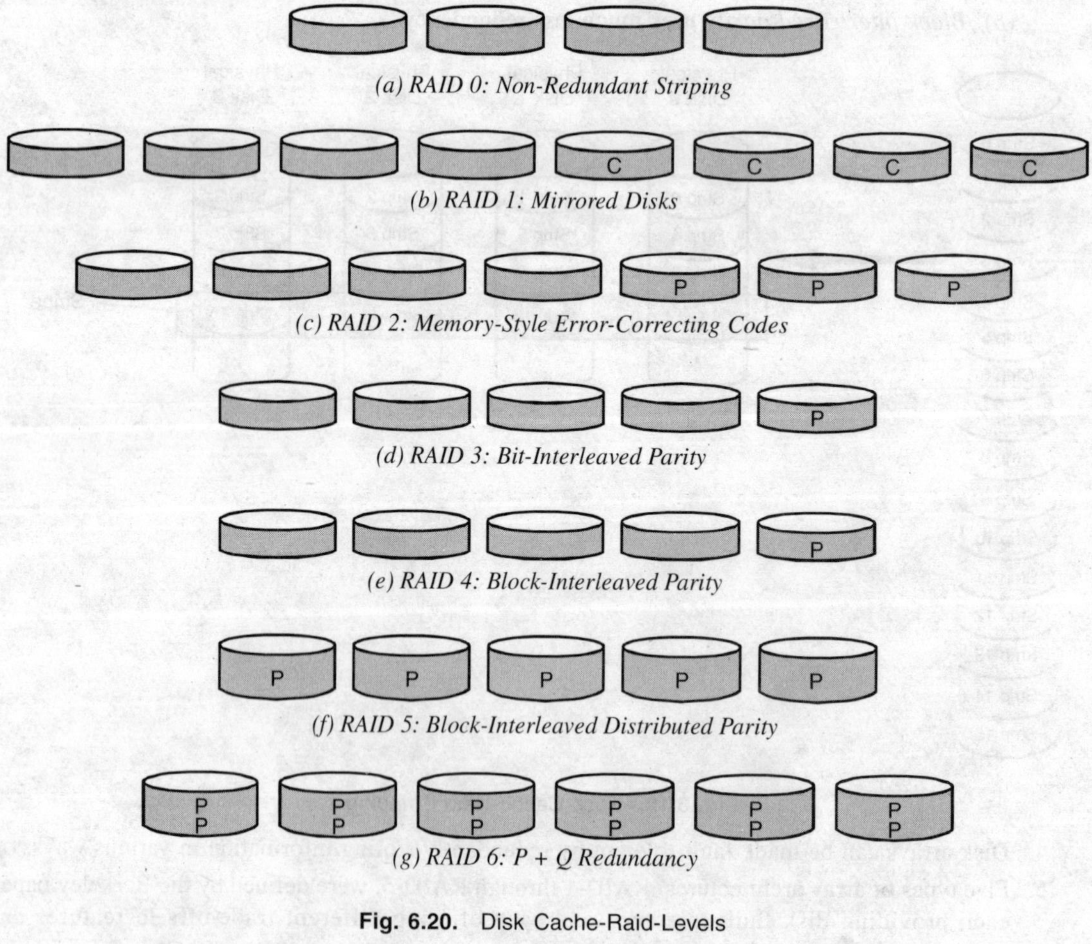

(a) RAID 0: Non-Redundant Striping

(b) RAID 1: Mirrored Disks

(c) RAID 2: Memory-Style Error-Correcting Codes

(d) RAID 3: Bit-Interleaved Parity

(e) RAID 4: Block-Interleaved Parity

(f) RAID 5: Block-Interleaved Distributed Parity

(g) RAID 6: P + Q Redundancy

Fig. 6.20. Disk Cache-Raid-Levels

Category	Level	Description	I/O Request Rate (Read/Write)	Data Transfer Rate (Read/Write)	Typical Application
Striping	0	Nonredundant	Large strips: Excellent	Small strips: Excellent	Applications requiring high performance for noncritical data
Mirroring	1	Mirrored	Good/Fair	Fair/Fair	System drives; critical files
Parallel access	2	Redundant via Hamming code	Poor	Excellent	
	3	Bit-interleaved parity	Poor	Excellent	Large I/O request size applications, such as imaging, CAD
Independent access	4	Block-interleaved parity	Excellent/Fair	Fair/Poor	
	5	Block-interleaved distributed parity	Excellent/Fair	Fair/Poor	High request rate, read-intensive, data lookup
	6	Block-interleaved dual distributed parity	Excellent/Poor	Fair/Poor	Applications requiring extremely high availability

RAID-0 Level

(*i*) RAID Level 0 is not redundant.

(*ii*) In level 0, data is split across drives, resulting in higher data throughput.

(*iii*) Performance is very good, but the failure of any disk in the array results in data loss.

(*iv*) This level is commonly referred to as striping.

RAID-1 Level

(*i*) Provides redundancy by writing all data to two or more drives.

RAID 1 (Mirrored)

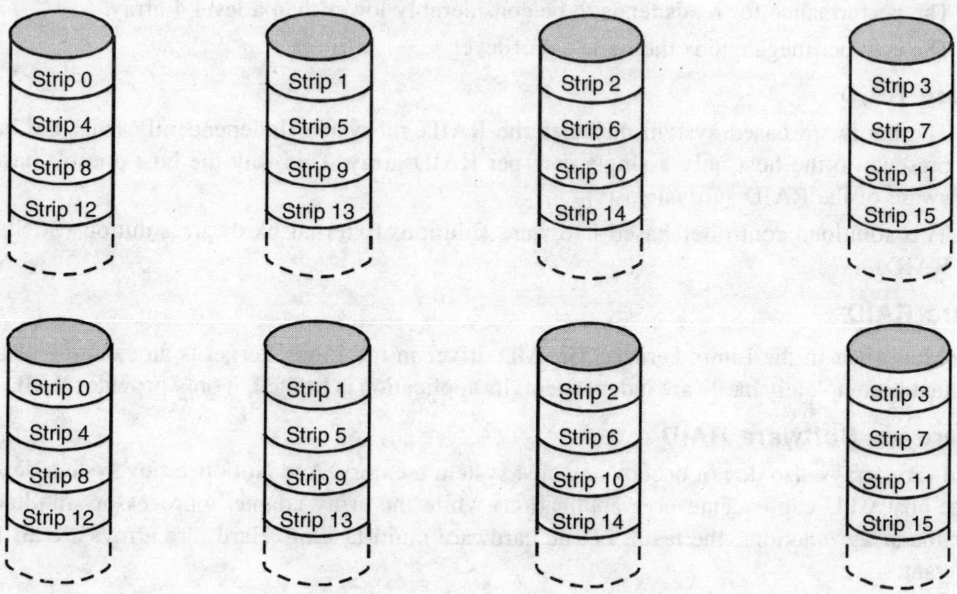

Fig. 6.21. Disk Cache-Raid-Level-1

(*ii*) The performance of a level 1 array tends to be faster on reads and slower on writes compared to a single drive, but if either drive fails, no data is lost.

(*iii*) The cost per megabyte is high.

(*iv*) This level is commonly referred to as mirroring.

RAID-2

(*i*) RAID Level 2 uses Hamming error correction codes.

(*ii*) Is intended for use with drives which do not have built-in error detection.

(*iii*) All SCSI drives support built-in error detection, so this level is of little use when using SCSI drives.

RAID-3

(*i*) RAID Level 3 stripes data at a byte level across several drives, with parity stored on one drive. It is otherwise similar to level 4.

(*ii*) Byte-level striping requires hardware support for efficient use.

RAID-4

(*i*) RAID Level 4 stripes data at a block level across several drives, with parity stored on one drive.

(*ii*) The parity information allows recovery from the failure of any single drive.

(*iii*) The performance of a level 4 array is very good for reads (the same as level 0).

(*iv*) Writes require that parity data be updated each time. This slows small random writes, but large writes or sequential writes are fairly fast.

(*v*) Because only one drive in the array stores redundant data, the cost per megabyte of a level 4 array can be fairly low.

RAID-5

(*i*) RAID Level 5 is similar to level 4, but distributes parity among the drives.

(*ii*) Can speed small writes in multiprocessing systems, since the parity disk does not become a bottleneck.

(*iii*) The performance for reads tends to be considerably lower than a level 4 array.

(*iv*) The cost per megabyte is the same as for level 4.

Hardware RAID

(*i*) The hardware based system manages the RAID subsystem independently from the host and presents to the host only a single disk per RAID array. This way the host doesn't have to be aware of the RAID subsystems(s).

(*ii*) Two solutions controller based hardware solution. External hardware solution (SCSI—SCSI RAID).

Software RAID

The MD driver in the Linux kernel : The MD driver in the Linux kernel is an example of a RAID solution that is completely hardware independent. Its application is limited, it only provides RAID level 0.

Hardware vs. Software RAID

Hardware arrays also do not occupy any host system memory. And Not operating system dependent. Since the host CPU can execute user applications while the array adapter's processor simultaneously executes the array functions, the result is true hardware multi-tasking. Hardware arrays are also highly fault tolerant.

Magnetic Disks

(i) Disk substrate coated with magnetically material (iron oxide…rust).

(ii) Substrate used to be aluminium.

(iii) Now glass is used because it has Improved surface uniformity and Increases reliability. Reduction in surface defects which reduced read/write errors. Better stiffness and better shock/damage resistance.

Magnetic Tape

Magnetic tape is a non-volatile storage medium consisting of a magnetic coating on a thin plastic strip. Nearly all recording tape is of this type, whether used for video, audio storage or general purpose digital data storage using a computer. Magneto-optical and optical tape storage products have been developed using many of the same concepts as magnetic storage, but have achieved little commercial success.

CD-ROM Disks

CD-ROM is a Compact Disc that contains data accessible by a computer. While the Compact Disc format was originally designed for music storage and playback, the format was later adapted to hold any form of binary data. CD-ROMs are popularly used to distribute computer software, including games and multimedia applications, though any data can be stored (up to the capacity limit of a disc). Some CDs hold both computer data and audio with the latter capable of being played on a CD player, whilst data (such as software or digital video) is only usable on a computer. These are called **Enhanced CDs.**

Data is stored on the disc as a series of microscopic indentations **("pits",** with the gaps between them referred to as **"lands").** A laser is shone onto the reflective surface of the disc to read the pattern of pits and lands. Because the depth of the pits is approximately one-quarter to one-sixth of the wavelength of the laser light used to read the disc, the reflected beam's phase is shifted in relation to the incoming beam, causing destructive interference and reducing the reflected beam's intensity. This pattern of changing intensity of the reflected beam is converted into binary data.

CD-ROM Format

A CD-ROM sector contains 2352 bytes, divided into 108 24-byte frames. The CD-ROM is, in essence, a data disk, which cannot rely on error concealment, and therefore requires a higher reliability of the retrieved data. In order to achieve improved error correction and detection, a CD-ROM has a third layer of Reed-Solomon error correction. A Mode-1 CD-ROM, which has the full three layers of error correction data, contains a net 2048 bytes of the available 2352 per sector. In a Mode-2 CD-ROM, which is mostly used for video files, there are 2336 user-available bytes per sector. The net byte rate of a Mode-1 CD-ROM is 44.1k×2048/(6×108) = 153.6 kB/s. The playing time is 74 minutes, or 4440 seconds, so that the net capacity of a Mode-1 CD-ROM is 682 MB.A 1x speed CD drive reads 75 consecutive sectors per second.

CD Sector Contents : A standard 74 min CD contains 333,000 blocks or sectors.

Optical Storage CD-ROM

Originally optical CD-ROM was used for audio system. The capacity used was 650Mbytes giving over 70 minutes audio. The Polycarbonate coated with highly reflective coat, usually aluminium material was used for fabrication. Data stored in the form of pits. Read operation is done by reflecting laser. A constant packing density and constant linear velocity

An **optical disc** is a flat, circular, usually polycarbonate disc whereon data is stored in the form of pits (or bumps) within a flat surface, usually along a single spiral groove that covers the entire recorded surface of the disc. This data is generally accessed when a special material on the disc (often aluminum) is illuminated with a laser diode. The pits distort the reflected laser light. Most optical discs, with the exception of a few such as black CD-ROMs designed for the original Sony Play station, have a characteristic prismatic or iridescent appearance created by the grooves in the reflective layer.

Other Optical Storage

(*i*) CD-Recordable (CD-R) WORM and compatible with CD-ROM drives

(*ii*) CD-RW

- Erasable
- Getting cheaper
- Mostly CD-ROM drive compatible
- Phase change-Material has two different reflectivities in different phase states

CD Operation

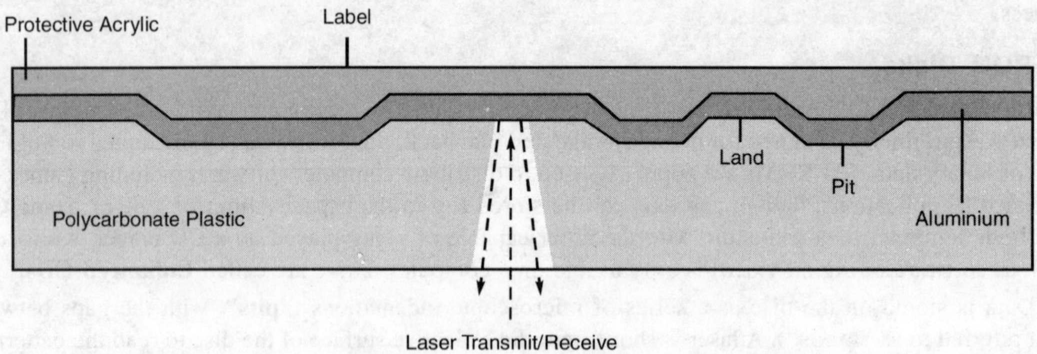

Fig. 6.22. CD Operation Diagram

CD-ROM Drive Speeds

Audio is single speed.

(*i*) Constant linier velocity

(*ii*) 1.2 ms-1

(*iii*) Track (spiral) is 5.27km long

(*iv*) Gives 4391 seconds = 73.2 minutes

Other speeds are quoted as multiples as 24x

CD-ROM Format

(*i*) Mode 0=blank data field

(*ii*) Mode 1=2048 byte data error correction

(*iii*) Mode 2=2336 byte data

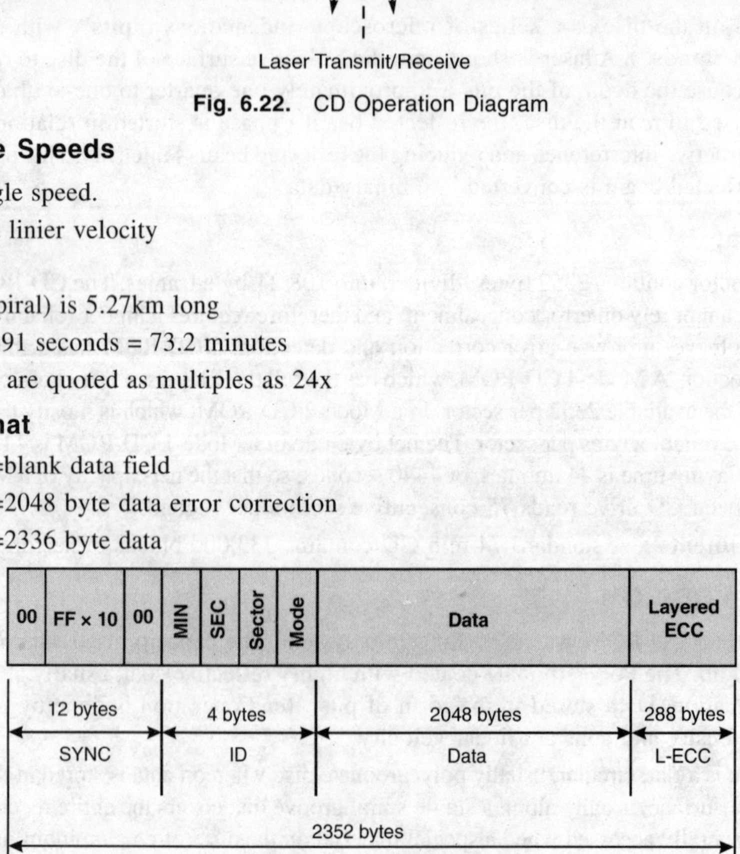

Fig. 6.23. CD-ROM Format

Digital Video Disks (DVD)

DVD commonly "**Digital Versatile Disc**" or "**Digital Video Disc**" is an optical disc storage media format that can be used for data storage including movies with high video and sound quality. DVDs resemble compact discs as their diameter is the same (120 mm (4.72 inches) or occasionally 80 mm (3.15 inches) in diameter), but they are encoded in a different format and at a much higher density.

> **Do You Know**
>
> *DVDs with properly formatted and structured audio are DVD-Audio discs.*

All DVD discs, regardless of type, are *DVD-ROM* discs. This includes replicated (factory pressed), recorded (burned), video, audio, and data DVDs. A DVD with properly formatted and structured video content is a *DVD-Video*. Everything else (including other types of DVD discs with video) is referred to as a *DVD-Data* disc. However, many refer to a "DVD-ROM" as a pressed data.

DVD—Technology

(*i*) Multi-layer technology is used and available in very high capacity (4.7G per layer).

(*ii*) Full length movie on single disk using MPEG compression.

(*iii*) Finally standardized. Movies carry regional coding and players only play correct region films.

DVD—Writable

(*i*) Lots of trouble with standards.

(*ii*) First generation DVD drives may not read first generation DVD-W disks.

(*iii*) First generation DVD drives may not read CD-RW disks.

HD DVD

High Density DVD or High-Definition DVD is a high-density optical disc format designed for the storage of data and high-definition video. High-definition video generally refers to any video system of higher resolution than standard-definition (SD), i.e. NTSC, PAL and SECAM. It is important to note the difference between general purpose high-definition video as discussed in this article, and its specific applications in television HDTV.

6.3.4. Associative Memory

In an associative memory any stored item can be accessed by using the contents of the item in question, generally some specified sub field, as an address. The sub field chosen to address the memory is called the key. Items stored in an associative memory can be viewed as having the two-field format.

> **Do You Know**
>
> *Associative memories are also commonly known as Content Addressable Me-mories (CAMs).*

Key, Data

Where KEY is the stored address and DATA is the information to be accessed. For example, if a page table of the kind shown in Figure 6.24 is placed in an associate memory, the page address can be selected as the key, while the page frame, presence bit, change bit and access rights from the data. Such a memory can then be accessed with a request such as: Read the page frame number corresponding to page address E.

An associative cache employs a tag, that is, a block address, as the key. At the start of a memory access, the incoming tag is compared simultaneously to all the tags stored in the cache's tag memory. If a match (cache hit) occurs, a match-indicating signal triggers the cache to service the requested memory access. A no-match signal identifies a cache miss, and the memory access requested is forwarded to main memory for service. A cache block containing the target address is then sent from main memory to the cache, and at the same time, a data word is sent to the CPU or transferred from CPU to the cache, in response to the original access request.

Figure 6.24 shows the general structure of an **associative memory.** Each unit of stored information is a fixed-length word. Any sub field of the word can be chosen as the key. Here the desired key is specified by *mask register*, whose contents identify the bit positions (which need not be adjacent) that define the key. The current key is compared simultaneously with all stored words; those that match the key output a match signal, which enter a select circuit, which enables the data field to be accessed. If several entries have the same key, then the select circuit, which enables the data field to be accessed. If several entries have the same key, then the select circuit determines which data field is to be read out. It can, for example, read out all matching entries in some predetermined order. Since all words in the memory are required to compare their keys with the input key simultaneously, each needs its own match circuit. The match and select circuits make associative memories much more complex and expensive than conventional memories.

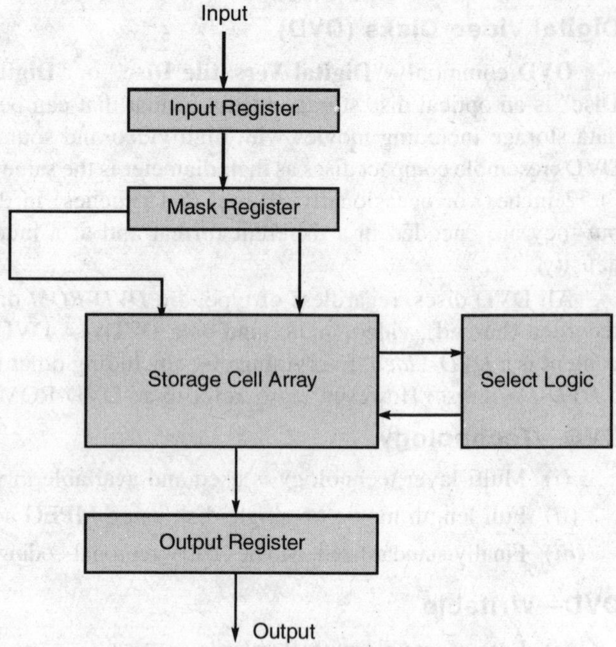

Fig. 6.24. Content Addressable Memory

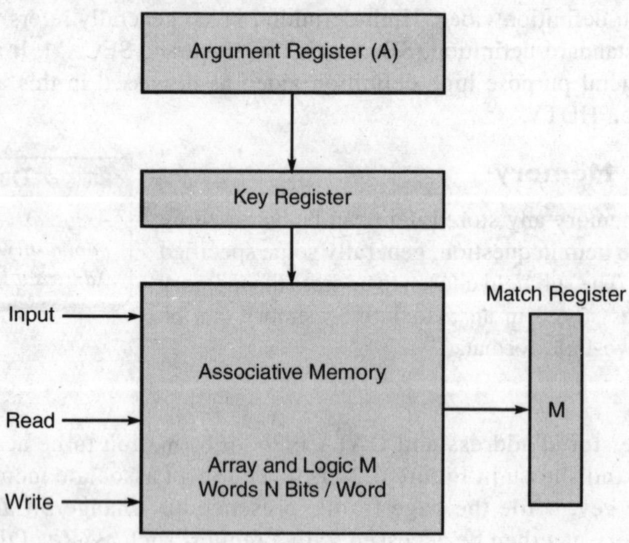

Fig. 6.25. AM Diagram

This is the type of memory where we access data by searching or matching the contents and not by address value. Accessed by the content of the data rather than by an address. This is also called content addressable memory (CAM).

In this the data we need to search is kept in Argument register. This is of the same length as of word size. Since we have m words of n bits length we will have argument register of length n bits. Also we have M as match register which gives us the matching result in terms of those particular bits high. This is equal to the number of words in memory, so match register is of length m bits.

For example if we have to search 1011 and you have words

Like 1011
0111
1000
1100
0010
1011
0111
1011

In this we have occurrence of 1011, 3 times i.e for 1st, 6th and 8th word. Thus the value of match register will be high at 1st, 6th and 8th place else it will be low.

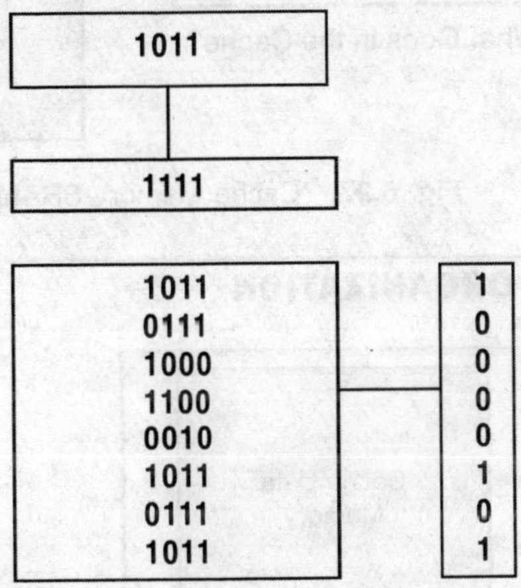

Fig. 6.26.

In this the value of key register is 1111 which represents that it is matching all the bits of argument register to every word of associative memory. In case we need to choose only some bits for checking as we want all words ending with one. The value of key register will be 0001 (as we are matching only last bit)

And the value of Match Register will be 11000111.

6.4. CACHE MEMORY

Cache is a very small and fast memory that is inserted between the larger, slower main memory and the CPU. This memory holds the currently active segments of a program and its data.

(*i*) 'Cache' refers to a fast intermediate memory within a larger memory system. It serves as a buffer between CPU and main memory. (Fig. 6.27)

(*ii*) It provides the CPU with fast single cycle access to its external memory and an efficient way to place a small portion of the memory on the same chip as a micro-processor.

Here memory words are stored in a cache data memory and are grouped into small pages, called cache blocks or lines. Each block is marked with a block address called the tag. The collection of tag addresses is stored in a special memory called cache tag memory. (Fig. 6.28)

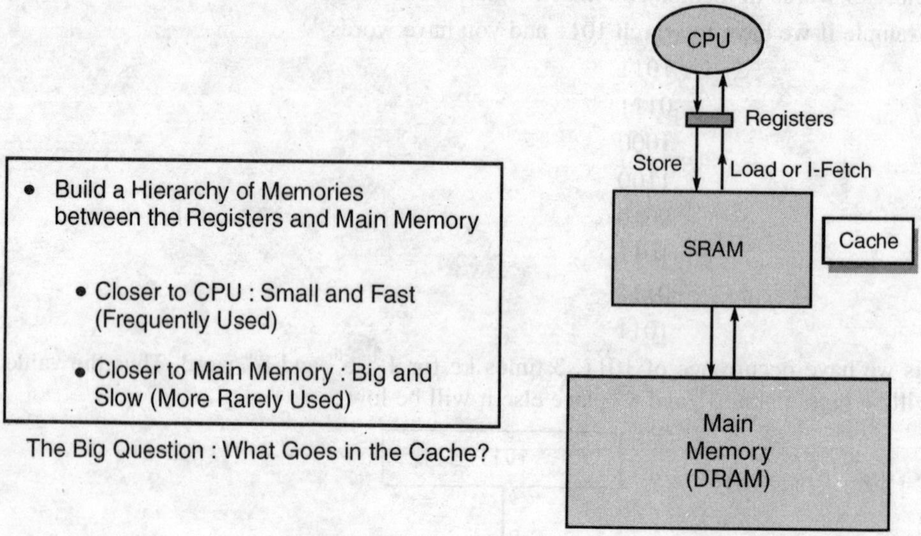

Fig. 6.27. Cache Memory-SRAM

6.5. CACHE MEMORY ORGANIZATION

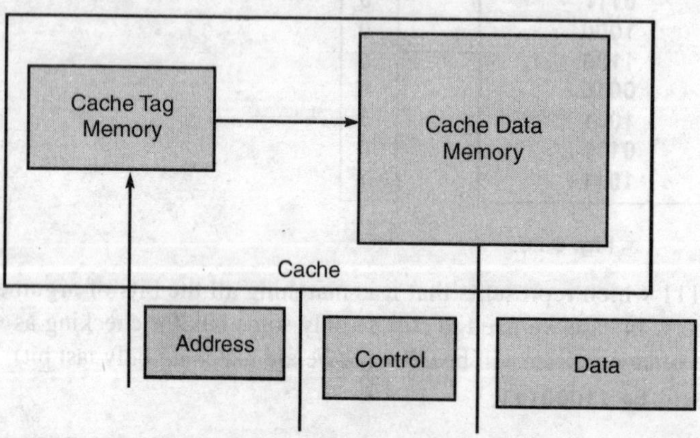

Fig. 6.28. Basic Structure of a Cache

> **Do You Know**
> *The "content" that is addressed in content addressable cache memory is a subset of the bits of a main memory address called a field.*

The fields into which a memory address is divided provide a many-to-one mapping between larger main memory and the smaller cache memory. Many blocks of main memory map to a single block of cache. A *tag* field in the cache block distinguishes one cached memory block from another. The size of each field into which a memory address is divided depends on the size of the cache. Suppose our memory consists of 2^{14} words, cache has $16 = 2^4$ blocks, and each block holds 8 words.

Thus memory is divided into $2^{14} / 2^3 = 2^7$ blocks. For our field sizes, we know we need 4 bits for the block, 3 bits for the word, and the tag is what's left over :

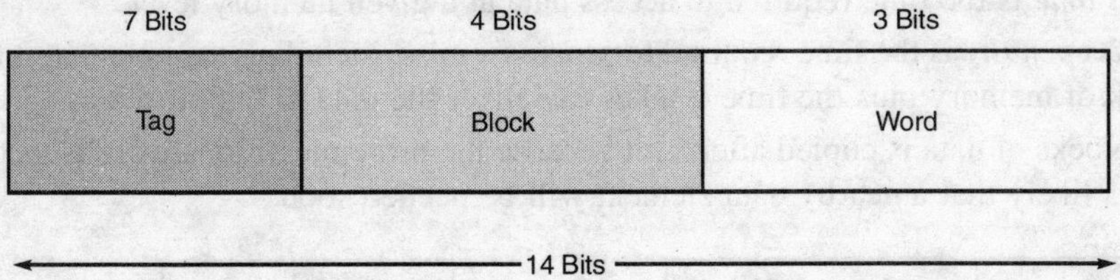

As an example, suppose a program generates the address **1AA**. In 14-bit binary, this number is: **00000110101010**.The first 7 bits of this address go in the tag field, the next 4 bits go in the block field, and the final 3 bits indicate the word within the block.

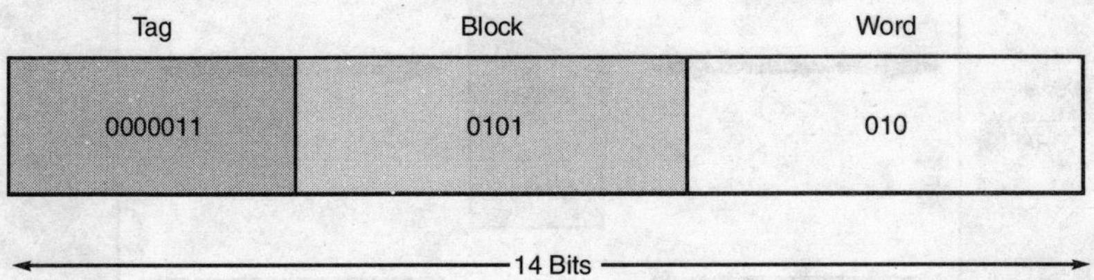

Cache Working Concept

To access a particular piece of data, the CPU first sends a request to its nearest memory, usually cache. If the data is not in cache, then main memory is queried. If the data is not in main memory, then the request goes to disk. Once the data is located, then the data, and a number of its nearby data elements are fetched into cache memory.

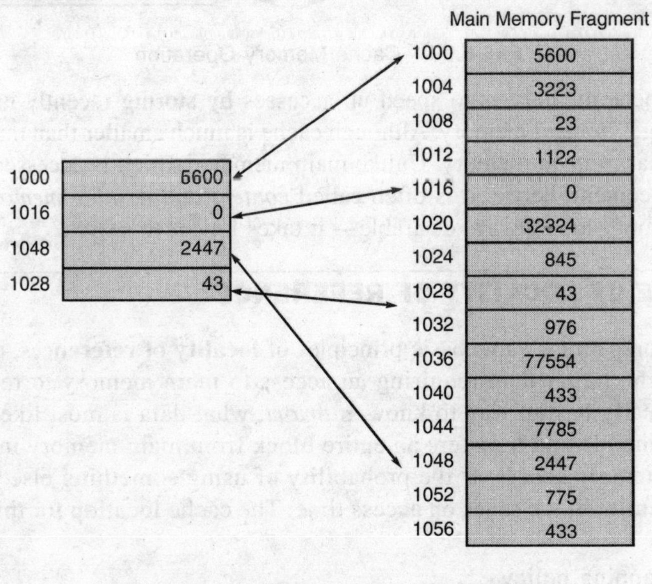

Fig. 6.29. Cache And Main Memory Data

This leads us to some definitions.

1. A *hit* is when data is found at a given memory level.
2. A *miss* is when it is not found.
3. The *hit rate* is the percentage of time data is found at a given memory level.
4. The *miss rate* is the percentage of time it is not.
5. Miss rate = 1 – hit rate.

6. The *hit time* is the time required to access data at a given memory level.
7. The *miss penalty* is the time required to process a miss, including the time that it takes to replace a block of memory plus the time it takes to deliver the data to the processor.

An entire blocks of data is copied after a hit because the principle of locality tells us that once a byte is accessed, it is likely that a nearby data element will be needed soon.

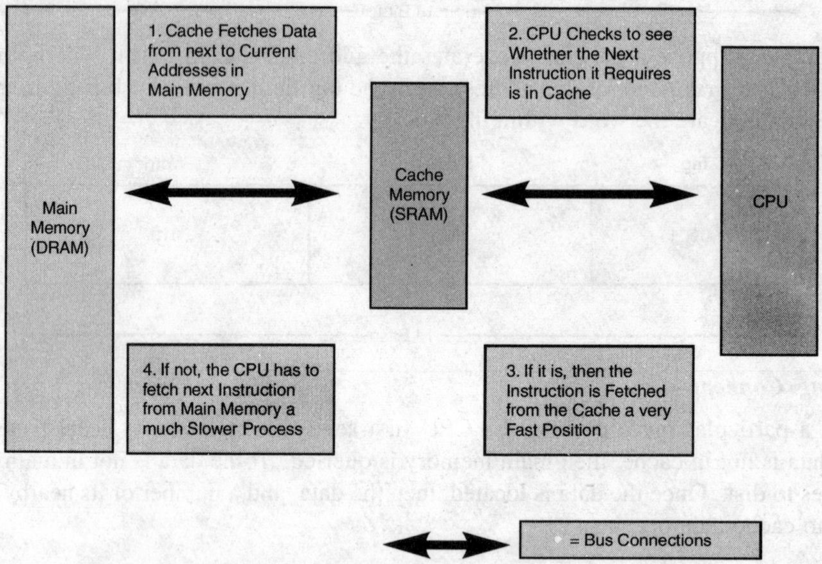

Fig. 6.30. Cache Memory Operation

The purpose of cache memory is to speed up accesses by storing recently used data closer to the CPU, instead of storing it in main memory. Although cache is much smaller than main memory, its access time is a fraction of that of main memory. Unlike main memory, which is accessed by address, cache is typically accessed by content; hence, it is often called *content addressable memory*. Because of this, a single large cache memory isn't always desirable— it takes longer to search.

6.6. PRINCIPLE OF LOCALITY OF REFERENCE

Cache memory works on the same basic principles of **locality of references,** by copying frequently used data into the cache rather than requiring an access to main memory to retrieve the data every time. The computer really has no way to know, *a priori*, what data is most likely to be accessed, so it uses the locality principle and transfers an entire block from main memory into cache whenever it has to make a main memory access. If the probability of using something else in that block is high, then transferring the entire block saves on access time. The cache location for this new block depends on two things:

(*i*) The cache mapping policy

(*ii*) The cache size

Which affects whether there is room for the new block or no? : According to the property, over a short interval of time the addresses generated by a typical program refer to a few localized areas of memory repeatedly. The memory hierarchy was developed based on a program behavior known as locality of reference, memory references are generated by the CPU for either instruction or data access. These accesses tend to be clustered in certain regions in time, space and ordering.

These are three dimensions of the locality of reference property:

(*a*) Temporal Locality

(*b*) Spatial Locality

(*c*) Sequential Locality

(*a*) **Temporal Locality :** Recently referenced items (Inst. Or data) are likely to be referenced again in the near future. This is often caused by special program constructs such as iterative loops, process stacks, and temporary variable or sub routines. Once a loop is entered or a subroutine is called, a small code segment will be referenced repeatedly many times, thus temporal locality tends to cluster the access in the recently used areas.

(*b*) **Spatial Locality :** This refers to the tendency for a process to access items whose addresses are near one another. For example operations on tables or array involve access of a certain clustered area in the address space.

(*c*) **Sequential Locality :** In a typical program the execution of instructions follows a sequential order (or the program order) unless branch instructions create out-of-order execution. The ratio of in-order execution to out of order execution is roughly 5 to 1 in ordinary programs. The sequentially in program behavior also contributes to the spatial locality because sequentially coded instructions and array elements are often stored in adjacent locations each type of locality affect the design of memory hierarchy. The spatial locality assists us in determining the size of unit data transfers between adjacent memory levels. The temporal locality also helps to determine the size of memory at successive levels. The principal of locality will guide us the design of cache, main memory and even virtual memory organization.

Principle of Locality

Programs tend to use data and instructions with addressed near or equal to those they have used recently. Temporal Locality items are likely to be referenced again in the near future. Spatial locality items with nearby addressed tend to be referenced close together in time.

Locality Example

Data : Reference array element in succession (stride 1 reference pattern) : Spatial locality reference sum each iteration temporal locality. Reference instructions in sequence. Spatial locality cycle through loop repeatedly : temporal locality

```
Sum = 0;
  for (i = 0; i < n; i++)
      sum += a[i];
  return sum;
```

There are three forms of locality to summarize;

(*i*) *Temporal locality* - Recently-accessed data elements tend to be accessed again.

(*ii*) *Spatial locality* - Accesses tend to cluster.

(*iii*) *Sequential locality* - Instructions tend to be accessed sequentially.

6.7. CACHE MAPPING SCHEMES

Address Mapping

When a tag address is presented to the cache, it must be quickly compared to the stored tags to determine whether a matching tag is currently assigned to the cache. The obvious approach of scanning all the tags in sequence is unacceptably slow. The fastest technique for implementing tag comparison is *associative* or content addressing, which permits the input tag to be compared simultaneously to all tags in the cache-tag memory. Pure associative memories are very expensive, however so it is only feasible to use them in small caches and TLBs. Various less costly techniques have been developed to solve this problem, some of which make limited use of associative addressing.

Block	Tag	Data	Valid
0	00000000	Words A, B, C, ...	1
1	11110101	Words L, M, N, ...	1
2	--------		0
3	--------		0

Direct Mapped Cache

A simpler, address-mapping technique for cache is known as direct mapping. Let M_1 be divided into $s_t = 2^s$ regions M_1 (0), M_1 (1),......, M_1 (s_1 – 1) called sets, each of which stores a block of an

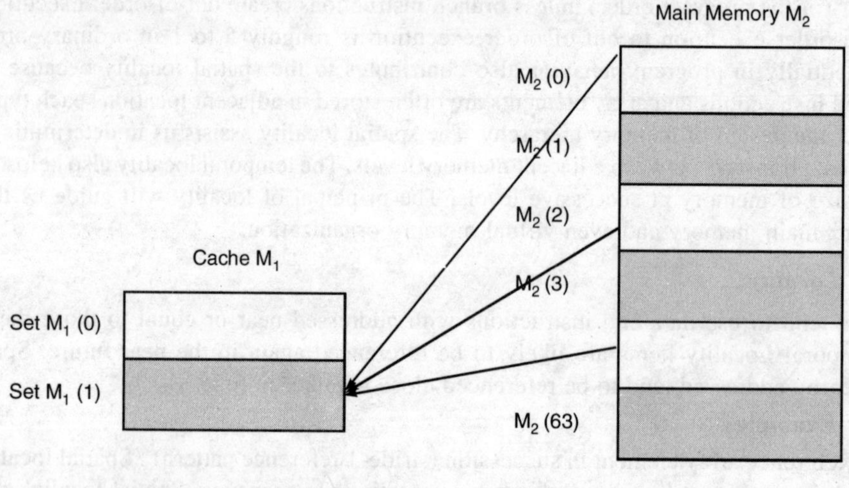

Fig. 6.31. Direct-Mapped Cache With Block Capacity of Two

consecutive words. Main memory M_2 is similarly divided into one-block regions M_2 (0), M_2 (1), ..., ... M_2 (s_2 – 1). With direct mapping, each block M_2 (i) in M_2 is mapped into one specific M_1 (j) in M_1. The set address j is determined from I by the rule

$$\mathbf{J = i \quad (modulo\ s_1)}$$

For example, if $s_1 = 2$ as in Fig. 6.31, every even-address (unshaded) block in M_2 is mapped into M_1 (0) and every odd-address (shaded) block in M_2 is mapped into M_1 (1). The hardware needed to implement direct mapping is fairly simple. The low-order s bits of each block address A form a set address that identifies the unique cache set that can store the block in question. The remaining t high-order bits of A now constitute the tag, and only these bits need be stored in the cache's tag memory.

Consequently, the cache tag memory can be an ordinary RAM that is addressed by the s-bit set-address part of an incoming memory address A. If there are 2^d words per set, then the low-order d bits of A form the displacement address of the word in question within its block. Thus an incoming address has three parts: a t-bit tag, an s-bit set address, and a d-bit displacement.

The main drawback of direct mapping is that the cache's hit ratio drops sharply if two or more frequently used blocks happen to map onto the same region in the cache. This possibility is minimized by the fact that such blocks are relatively for apart in memory-address space. For example, $s_1 = 2^6 = 64$, then only the blocks with addresses i, i + 64, i + 128, i + 1102 can be mapped into the same cache set M_1 (i).

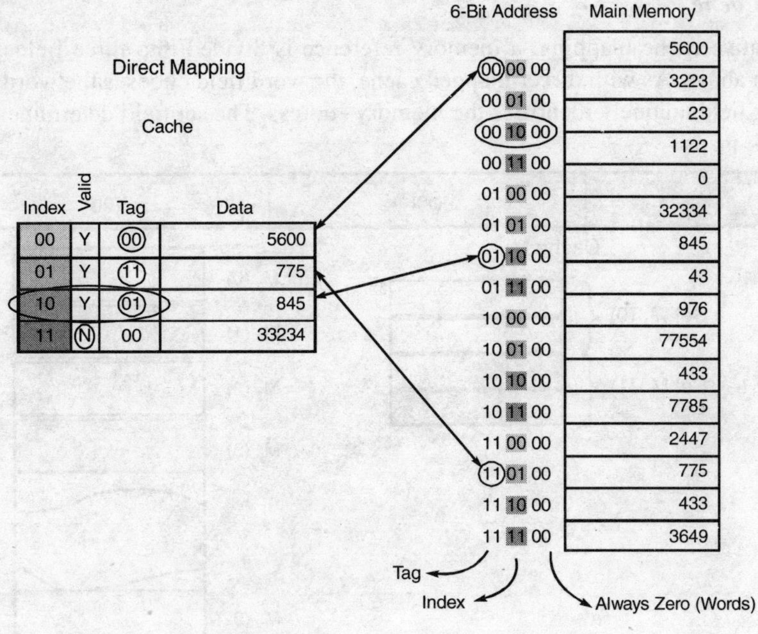

Fig. 6.32. Cache Data Transfer

Accessing a Direct-Mapped Cache

Example 6.1. Show cache addressing for a byte-addressable memory with 32-bit addresses. Cache line W = 16 B. Cache size L = 4096 lines (64 KB).

Solution. Byte offset in line is $\log_2 16 = 4$b. Cache line index is $\log_2 4096 = 12$ b. This leaves $32 - 12 - 4 = 16$ b for the tag.

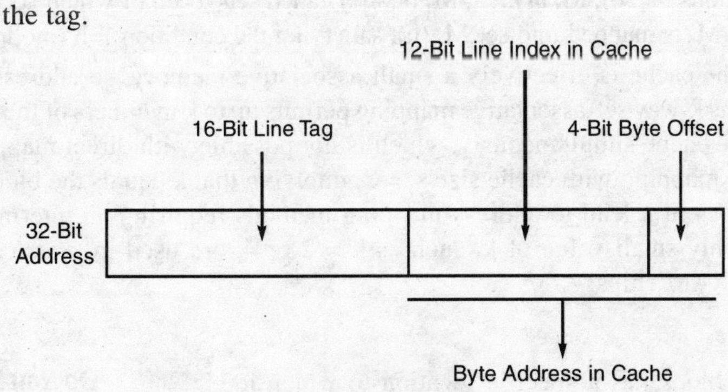

Fig. 6.33

Components of the 32-bit address is an example direct-mapped cache with byte addressing.

Hit and Misses

When the CPU reads from memory, calculate the index and tag is the data in the cache ? Yes – a hit, you're done! Data not in cache ? This is a miss. Read the word from memory, give it to the CPU. *Update* the cache so we won't miss again. Write the data and tag for this memory location to the cache. (Exploits temporal locality). The hit rate and miss rate are the fraction of memory accesses that are hits and misses. Typically, hit rates are around 95%. Many times instructions and data are considered separately when calculating hit/miss rates.

Set Associative Cache

In set associative cache mapping, a memory reference is divided into three fields: TAG, SET, and WORD, as shown above. As with direct-mapped cache, the word field chooses the word within the cache block, and the tag field uniquely identifies the memory address. The set field determines the set to which the memory block maps.

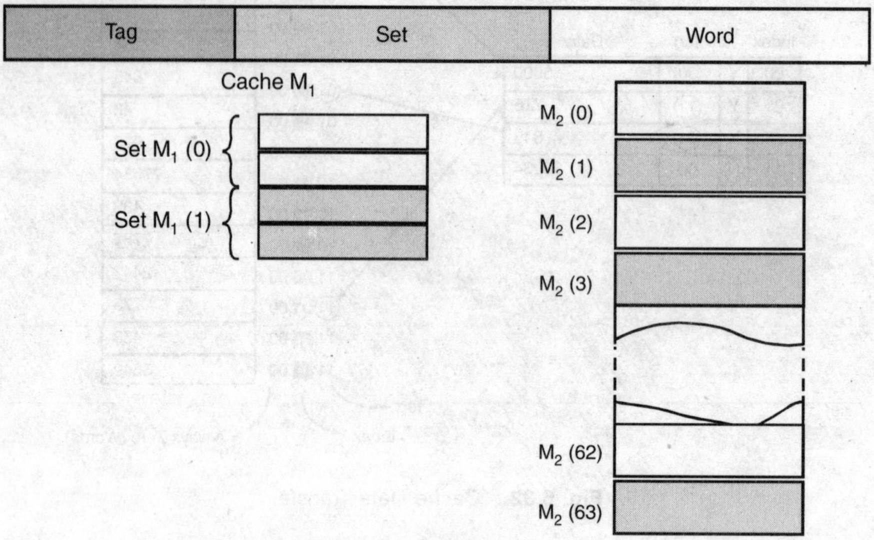

Fig. 6.34. Cache With Two-way Set-associative Addressing

A more general address mapping method for caches, called **set associative,** includes pure associative and direct mapping as special cases. As in direct mapping, blocks in main memory M_2 are grouped into equivalence classes determined by their addresses.M_2 (i), and M_2 (j) are in same equivalence class E if I=j (modulo s_1 multiblock regions M_1 (0), M_1 (1),...., M_1 ($s_1 - 1$) called sets, each of which accommodates $k = 2^h$ blocks. A block M_2 (i) in M_2 is mapped into set M_1 (h), satisfying the condition I=h (modulo s_1). (Fig. 6.34)

Each set M_1 (h) in the cache is effectively a small associative memory, so address mapping within each set is associative. This k-way set-associative mapping permits up to k members of the same equivalence class E to be stored in the cache simultaneously, which is not possible with direct mapping. Figure 6.34 illustrates set-associative mapping with cache size $s_1 = 2$, implying that k equals the block capacity of the cache. Intermediate values of k lead to address-mapping methods requiring an intermediate amount of associative hardware. Only small value of k, such as k = 2 or 4, are used in practice which makes it feasible to use low-cost RAM chips.

Fully Associative Cache

> **Do You Know**
>
> *Direct mapped cache is not as expensive as other caches because the mapping scheme does not require any searching.*

Each main memory block has a specific location to which it maps in cache; when a main memory address is converted to a cache address, the CPU knows exactly where to look in the cache for that memory block by simply examining the bits in the block field. This requires the entire cache to be built from *associative memory* so it can be searched in parallel. That is, a single search must compare the requested tag to *all* tags in cache to determine whether the desired data block is present in cache. Associative memory requires special hardware to allow associative searching, and is, thus, quite expensive.

With direct mapping, if a block already occupies the cache location where a new block must be placed, the block currently in cache is removed (it is written back to main memory if it has been modified or simply overwritten if it has not been changed). With fully associative mapping, when cache is full, we need a replacement algorithm to decide which block we wish to throw out of cache (we call this our *victim block*). A simple first-in, first-out algorithm would work, as would a least-recently used algorithm. Instead of placing memory blocks in specific cache locations based on memory address, we could allow a block to go anywhere in cache. In this way, cache would have to fill up before any blocks are evicted. This is how *fully associative* cache works. A memory address is partitioned into only two fields: the tag and the word.

Suppose, as before, we have 14-bit memory addresses and a cache with 16 blocks, each block of size 8. The field format of a memory reference is:

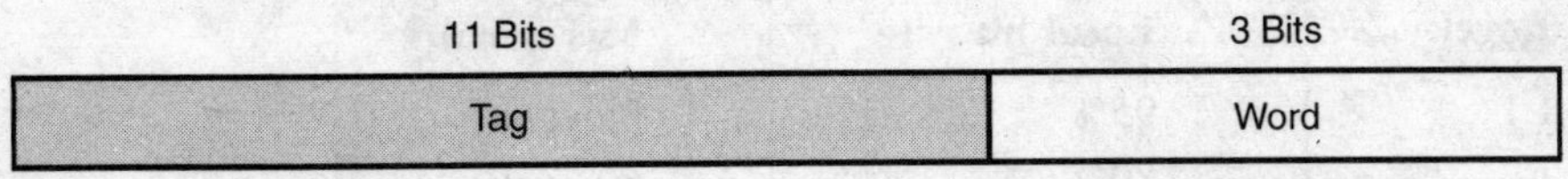

When the cache is searched, all tags are searched in parallel to retrieve the data quickly. This requires special, costly hardware.

Cache Types

Caches are distinguished by the kinds of information they store.

1. Instruction I-Cache Stores Instructions Only,
2. Data or D-cache Stores Data Only.

Separating the stored data in this way recognizes the different access behavior patterns of instructions and data. For example, programs tend to involve few write accesses, and they often exhibit more temporal and spatial locality than the data they process. A cache that stores both instructions and tat is referred to as unified. A split cache, on the other hand, consists of two associated but largely independent units: an I cache for instructions and D-cache for data. While a unified cache is simpler, a split cache makes it possible to access programs and data concurrently. A split cache can also be designed to manage its I- and D-cache components differently.

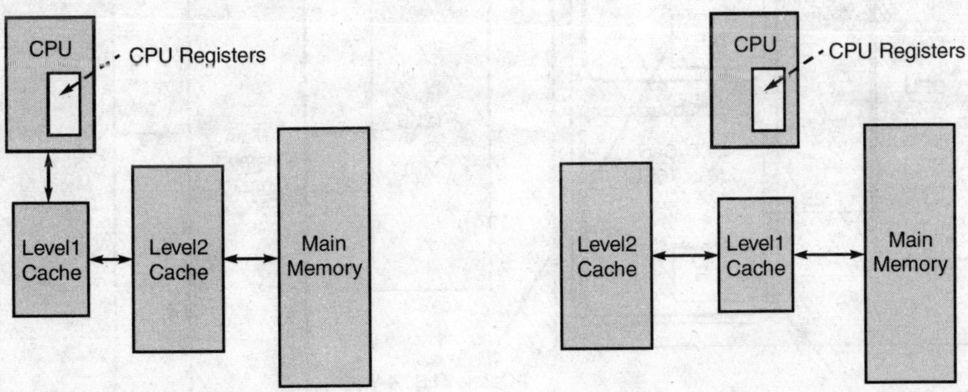

(a) Level 2 between Level 1 and Main *(b) Level 2 Connected to "Backside" Bus*

Fig. 6.35. Multi-Level Cache

Caches are also classified by the level they occupy in the memory hierarchy. Early computers employed a single, multichip cache that occupied one level of the hierarchy between the CPU and main memory. Two developments made it desirable to introduce two or more cache levels in high-performance systems: the feasibility of including part of the real memory space on a microprocessor chip and growth in the size (but not the speed) of main memory in typical computers. A level 1 (L1) or primary cache is an efficient way to implement an on-chip memory. An additional memory level can be introduced via an off-chip, level 2 (L2) or secondary cache. The desirability of an L2 cache increases with the size of main memory, assuming that the size of the on-chip, L1 cache is fixed. As main-memory size increases further, even more cache levels may be desirable.

Performance of a Two-Level Cache System

Example 6.2. A system with L1 and L2 caches has a CPI of 1.2 with no cache miss. There are 1.1 memory accesses on average per instruction. What is the effective CPI with with cache misses factored in ? What are the effective hit rate and miss penalty overall if L1 and L2 caches are modeled as a single cache ?

Level	Local hit rate	Mis penalty
L1	95%	8 cycles
L2	80%	60 cycles

Solution. $C_{eff} = C_{fast} + (1 - h_1)\,[C_{medium} + (1 - h_2)C_{slow}]$

Becase C_{fast} is included in the CPI of 1.2, we must account for the rest

$CPI = 1 + 2 + 1.1(1 - 0.95)\,[8 + (1 - 0.8)60] = 1.2 + 1.1 \times 0.05 \times 20 = 2.3$

Overall : hit rate 99% (95% + 80% of 5%), miss penalty 60 cycles.

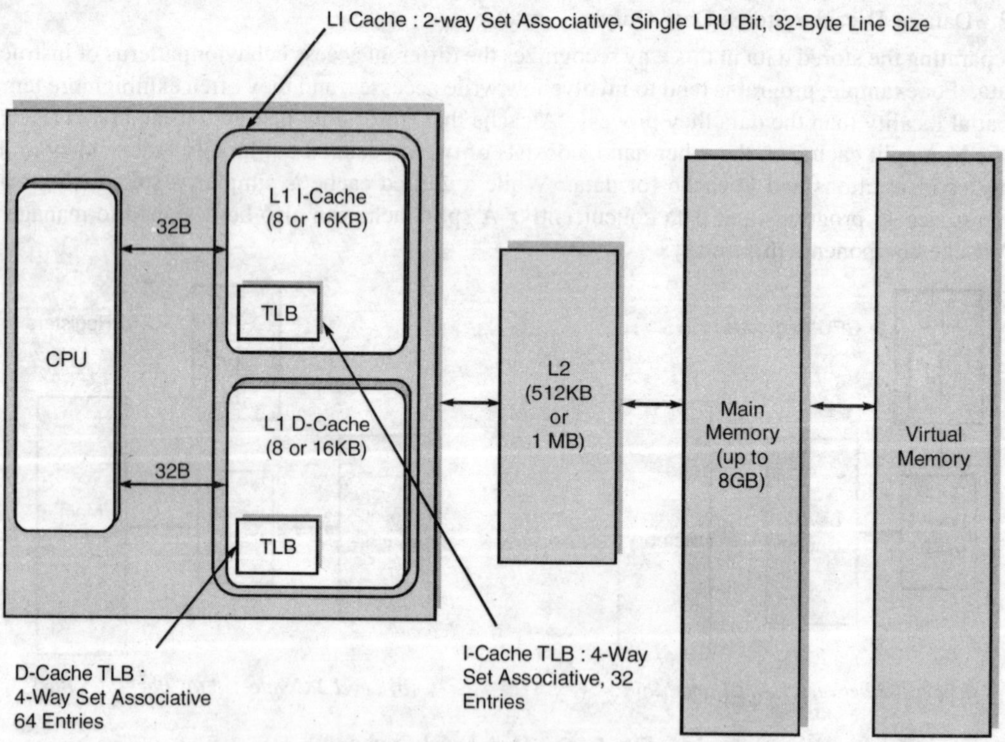

Fig. 6.36. A Pentium Processor-Two-Level Cache

The Pentium architecture supports both paging and segmentation, and they can be used in various combinations including unpaged unsegmented, segmented unpaged, and unsegmented paged. The processor supports two levels of cache (L1 and L2), both having a block size of 32 bytes. The L1 cache is next to the processor, and the L2 cache sits between the processor and memory. The L1 cache is in two parts: and instruction cache (I-cache) and a data cache (D-cache).

6.8. CACHE REPLACEMENT POLICIES

> **Do You Know**
> *Direct mapping is no need for a replacement policy because the location for each new block is predetermined.*

In a direct-mapped cache, if there is contention for a cache block, there is only one possible action: The existing block is kicked out of cache to make room for the new block. This process is called *replacement*. However, with fully associative cache and set associative cache, we need a replacement algorithm to determine the "victim" block to be removed from cache. When using fully associative cache, there are K possible cache locations (where K is the number of blocks in cache) to which a given main memory block may map. With N-way set associative mapping, a block can map to any of N different blocks within a given set. How do we determine which block in cache should be replaced?

The algorithm for determining replacement is called the *replacement policy*.

There are several popular replacement policies. One that is not practical but that can be used as a benchmark by which to measure all others is the *optimal* algorithm. We like to keep values in cache that will be needed again soon, and throw out blocks that won't be needed again, or that won't be needed for some time. An algorithm that could look into the future to determine the precise blocks to keep or eject based on these two criteria would be best. This is what the optimal algorithm does. We want to replace the block that will not be used for the longest period of time in the future. For example, if the choice for the victim block is between block 0 and block 1, and block 0 will be used again in 5 seconds, whereas block 1 will not be used again for 10 seconds, we would throw out block 1.

We need algorithms that best approximate the optimal algorithm. We have several options. For example, we might consider temporal locality. We might guess that any value that has not been used recently is unlikely to be needed again soon. We can keep track of the last time each block was accessed (assign a timestamp to the block), and select as the victim block the block that has been used least recently. This is the **least *recently used (LRU)*** algorithm. Unfortunately, LRU requires the system to keep a history of accesses for every cache block, which requires significant space and slows down the operation of the cache.

FIRST IN, FIRST OUT (FIFO) is another popular approach. With this algorithm, the block that has been in cache the longest (regardless of how recently it has been used) would be selected as the victim to be removed from cache memory.

Another approach is to select a victim at *random*. The problem with LRU and FIFO is that there are degenerate referencing situations in which they can be made to *thrash* (constantly throw out a block, then bring it back, then throw it out, then bring it back, repeatedly.

Performance of Cache Memory System

The cache is the fastest component in the memory hierarchy, so it is desirable to make the average memory access time Te seen by the CPU as close as possible to access time Tc of the cache.

To achieve this goal, M_1 should satisfy a very high percentage of all memory reference; that is, the cache-hit ratio h should be almost one. A high hit ratio is possible because of the locality-of-reference property discussed earlier.

Hit Ratio - % of memory accesses satisfied by Cache memory system

Let Te: Effective memory access time in Cache memory system

Tc: Cache access time

Tm: Main memory access time

$$Te = Tc + (1 - h)\,Tm$$

For example, $Tc = 0.4\ \mu s,\ Tm = 1.2\ \mu s,\ h = 0.85\%$

Hence: $Te = 0.4 + (1 - 0.85) * 1.2 = 0.58\ \mu s$

Consider a k-way set associative cache M_1 defined by the following parameters; the number of sets s_1, the number of blocks (lines) per set k, and the number of bytes per block (also called the line size) p_1. Recall that the cache is fully associative when $s_1 = 1$ and is direct-mapped when k = 1. The number of bytes stored in the cache's data memory, usually referred to as the cache size s_1, is given by the following formula :

$$S_1 = ks_1p_1$$

Or, in words, Cache size = Number of blocks (lines) per set + Number of sets × Number of bytes per block

Although other factors, such as the tag memory (directory) size, influence the over-all cost C_1 of the cache. We can calculate cache impact on processor performance

$$\text{CPU time} = \text{IC} \times (\text{CPI}_{\text{execution}} + \text{CPI}_{\text{mem\_stall}}) \times \text{Cycle Time}$$

$$\text{CPI}_{\text{mem\_stal}} = \text{Memory Inst Frequency} \times \text{Miss Rate} \times \text{Miss Penalty}$$

We can calculate cache impact on processor performance. Calculate average memory access time (AMAT)

$$\text{AMAT} = \text{Hit time} + \text{Miss rate} \times \text{Miss penalty}$$

6.9. CACHE MEMORY DESIGN PARAMETERS

1. Cache Size (In Bytes Or Words) : A larger cache can hold more of the program's useful data but is more costly and likely to be slower.

2. Block or Cache-Line Size (A unit of data transfer between cache and main) : With a larger cache line, more data is brought in cache with each miss. This can improve the hit rate but also may bring low-utility data in.

3. Placement Policy : Determining where an incoming cache line is stored. More flexible policies imply higher hardware cost and may or may not have performance benefits (due to more complex data location).

4. Replacement Policy : Determining which of several existing cache blocks (into which a new cache line can be mapped) should be overwritten. Typical policies: choosing a random or the least recently used block.

5. Write Policy : Determining if updates to cache words are immediately forwarded to main (*write-through*) or modified blocks are copied back to main if and when they must be replaced (*write-back* or *copy-back*).

Improving Cache Performance

For a given *cache size,* the following design issues and tradeoffs exist :

Line width (2^W). Too small a value for W causes a lot of main memory accesses; too large a value increases the miss penalty and may tie up cache space with low-utility items that are replaced before being used.

Set size or associativity (2^S). Direct mapping (S = 0) is simple and fast; greater associativity leads to more complexity, and thus slower access, but tends to reduce conflict misses. More on this later.

Line replacement policy. Usually LRU (least recently used) algorithm or some approximation thereof; not an issue for direct-mapped caches. Somewhat surprisingly, random selection works quite well in practice.

Write policy. Modern caches are very fast, so that write-through is seldom a good choice. We usually implement write0back or copy-back, using write buffers to soften the impact of main memory latency.

Compulsory, Capacity and Conflict Misses

1. ***Compulsory misses* :** With *on-demand fetching*, first access to any item is a miss. Some "compulsory" misses can be avoided by prefetching.
2. ***Capacity misses* :** We have to oust some items to make room for others. This leads to misses that are not incurred with an infinitely large cache.
3. ***Conflict misses* :** Occasionally, there is free room, or space occupied by useless data, but the mapping/placement scheme forces us to displace useful items to bring in other items. This may lead to misses in future.

Write Through

When writing into memory. If Hit, both Cache and memory is written in parallel. If Miss, Memory is written for a read miss, missing block may be overloaded onto a cache block. Memory is always updated -> Important when CPU and DMA I/O are both executing. Slow, due to the memory access time.

Write-Back (Copy-Back)

When writing into memory if hit, only Cache is written. If Miss missing block is brought to Cache and write into Cache for a read miss, candidate block must be written back to the memory. Memory is not up-to-date, i.e., the same item in Cache and memory may have different value.

6.10. VIRTUAL MEMORY

Virtual memory was developed in approximately 1959-1962, at the University of Manchester for the Atlas Computer, completed in 1962. In 1961; Burroughs released the B5000, the first commercial computer with virtual memory. When the program is actually executed, the logical/virtual addresses are converted into real memory addresses.

> **Do You Know**
>
> *Virtual memory as an alternate set of memory addresses. Programs use these virtual addresses rather than real addresses to store instructions and data.*

Virtual memory is a concept used in some large computer systems that permit the user to construct programs as though a large memory space is available equal to the totality of the auxiliary memory. Each address that is referenced by the CPU goes through an address mapping from the so-called virtual address to a physical address in main memory. Virtual memory is used to give programmers the illusion that they have a very large memory at their disposal, even though the computer actually has a relatively small main memory.

Before the development of the virtual memory technique, programmers in the 1940s and 1950s had to manage directly two-level storage such as main memory or RAM and secondary memory in the form of hard disks or earlier, magnetic drums. The VM concept enlarges the address space, the set of addresses a program can utilize. Virtual memory might contain twice as many addresses as main memory.

Objective : When a computer is executing many programs at the same time, Virtual memory make the computer to share memory efficiently. Eliminate a restriction that a computer works in memory which is small and be limited. When many programs is running at the same time, by distributing each suitable memory area to each program, VM protect programs to interfere each other in each memory area.

Working Principle of VM

(*i*) To facilitate copying virtual memory into real memory, the operating system divides virtual memory into **pages**, each of which contains a fixed number of addresses. Each page is stored on a disk until it is needed. When the page is needed, the operating system copies it from disk to main memory, translating the virtual addresses into real addresses. A *physical address* is the actual memory address of physical memory.

(*ii*) Programs create *virtual addresses* that are *mapped* to physical addresses by the memory manager. *Page faults* occur when a logical address requires that a page be brought in from disk. *Memory fragmentation* occurs when the paging process results in the creation of small, unusable clusters of memory addresses.

A virtual memory system provides a mechanism for translating program-generated addresses into correct main memory locations. This is done dynamically while programs are being executed in the CPU. The translation or mapping is handled automatically by the hardware by means of a mapping table.

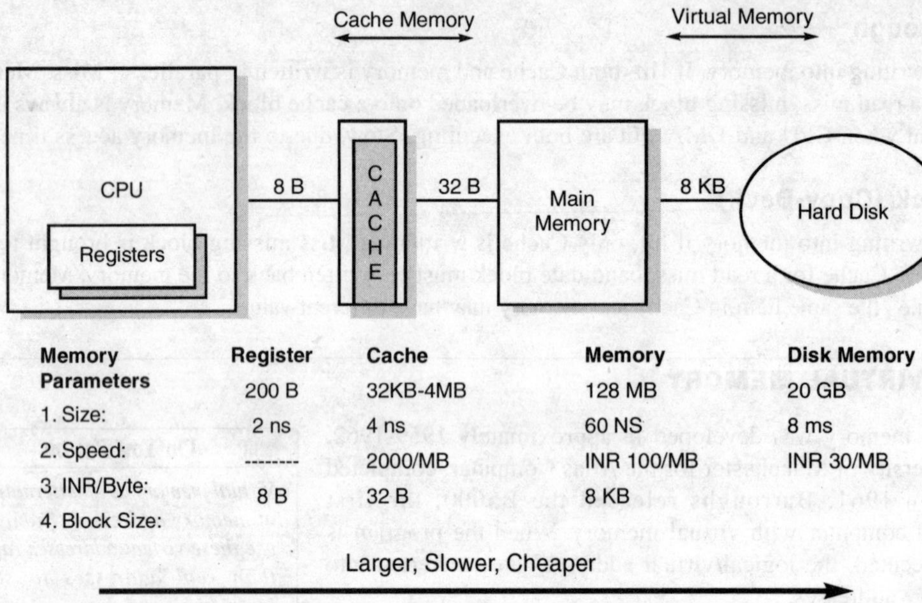

Fig. 6.37. Virtual Memory in Memory Hierarchy

6.11. MEMORY MANAGEMENT HARDWARE

Definition Memory management is the hardware base that makes a virtual memory system possible and which allows software to reference physical memory by virtual addresses, quite often more than once. It accomplishes this through the use of page and page tables.

Use a section of memory to translate virtual addresses into physical addresses via a series of table lookups.

The software that handles the page fault is generally part of an operating system and the hardware that detects this situation.

The concept of a logical address space that is bound to a separate **physical address space** is central to proper memory management. Logical Address generated by the CPU; also referred to as virtual address. Physical address seen by the memory unit.

Logical and physical addresses are the same in compile-time and load-time address-binding schemes; logical (virtual) and physical addresses differ in execution-time address-binding scheme. Main memory and virtual memory are divided into equal sized pages. The entire address space required by a process need not be in memory at once. Some parts can be on disk, while others are in main memory.

Further, the pages allocated to a process do not need to be stored contiguously— either on disk or in memory. In this way, only the needed pages are in memory at any time, the unnecessary pages are in slower disk storage.

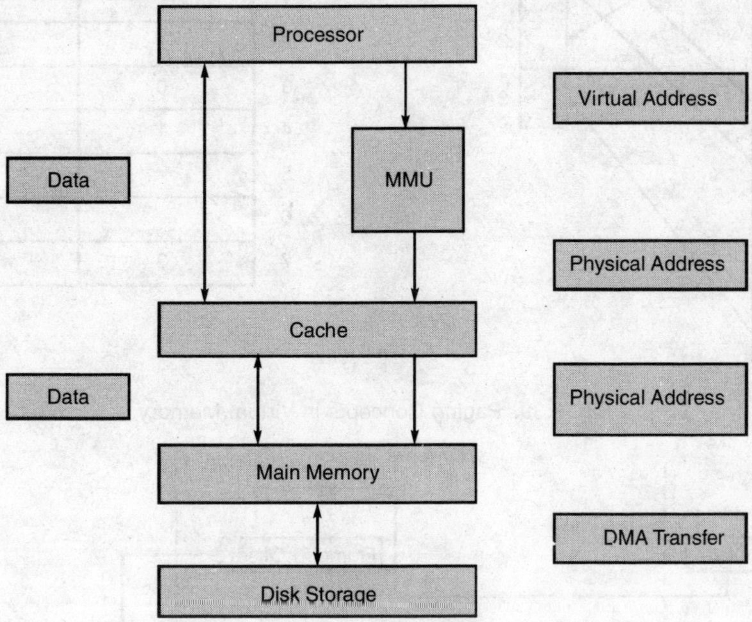

Fig. 6.38. Virtual Memory Concept Diagram

Page Tables : One for kernel addresses or more for user space processes. Page Table Entry (PTE) one per virtual page. When a process generates a virtual address, the operating system translates it into a physical memory address. To accomplish this, the virtual address is divided into two fields: A *page* field, and an *offset* field. The page field determines the page location of the address, and the offset indicates the location of the address within the page.The logical page number is translated into a physical page frame through a lookup in the page table.

Address generated by CPU is divided into :

Page number (*p*) – used as an index into a *page table* which contains base address of each page in physical memory.

Page offset (d) – combined with base address to define the physical memory address that is sent to the memory unit.

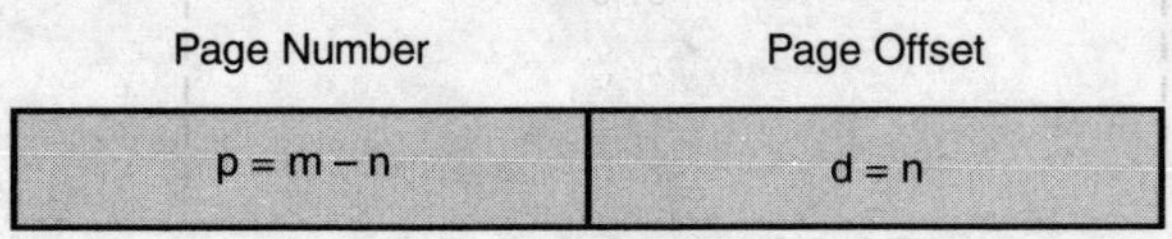

For given logical address space 2^m *and page size* 2^n

Paging provides a somewhat easier interface for programs, in that its operation tends to be more automatic and thus transparent. Each unit of transfer, referred to as a page, is of a fixed size and swapped by the virtual memory manager outside of the program's control. Instead of utilizing a segment/offset addressing approach, as seen in segmentation, paging uses a linear sequence of virtual addresses which are mapped to physical memory as necessary.

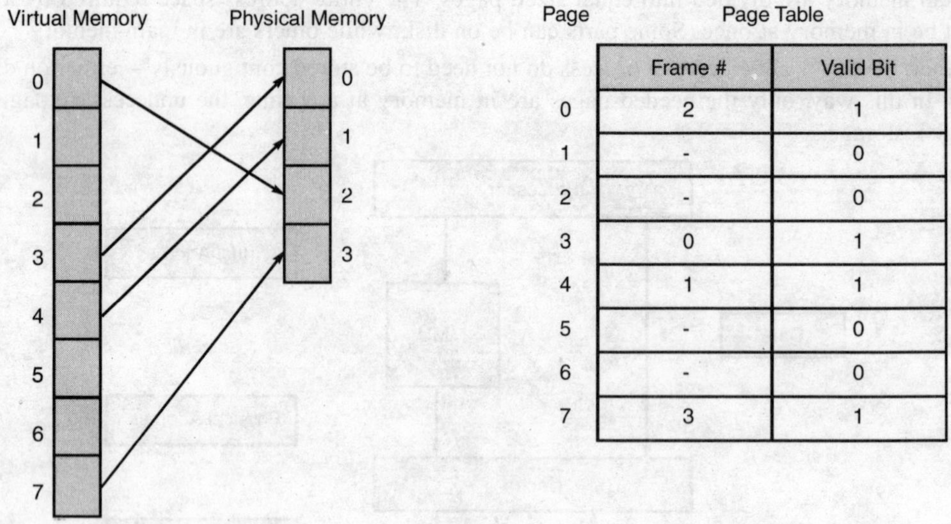

Fig. 6.39. Paging Concepts in Virtual Memory

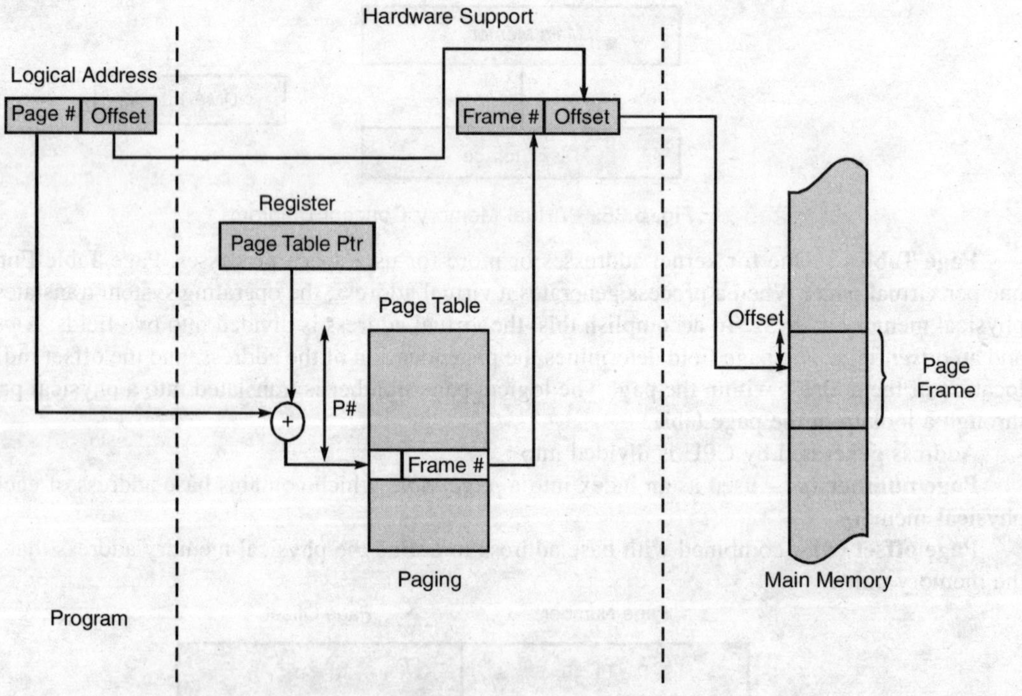

Fig. 6.40. Hardware For Paging

Due to this addressing approach, a single program may refer to series of many non-contiguous segments. Although some internal fragmentation may still exist due to the fixed size of the pages, the approach virtually eliminates external fragmentation.

Page table maintains mapping of process pages onto frames. Hardware support is needed to support translation of relative addresses within a program (logical addresses) into the memory addresses. Paging makes virtual memory possible. The Logical to physical address mapping is dynamic.

Page Fault

An interrupt to the software raised by the hardware when a program accesses a page that is not mapped in physical memory. When a program accesses a memory location in its memory and the page corresponding to that memory is not loaded. When a program accesses a memory location in its memory and the program does not have privileges to access the page corresponding to that memory.

Segmentation

1. Segmentation involves the relocation of variable sized segments into the physical address space. Generally these segments are contiguous units, and are referred to in programs by their segment number and an offset to the requested data.
2. Efficient segmentation relies on programs that are very thoughtfully written for their target system. Since segmentation relies on memory that is located in single large blocks, it is very possible that enough free space is available to load a new module, but can not be utilized.
3. Segmentation may also suffer from internal fragmentation if segments are not variable-sized, where memory above the segment is not used by the program but is still **"reserved"** for it.

Benefits

(*a*) More processes may be maintained in the main memory hence, better system utilization and throughput.

(*b*) The process size is not restricted by the physical memory size: the process memory is *virtual.* Less disk I/O to swap/load programs.

Paging Replacement Algorithms

(*i*) OPT (MIN) is eliminates the page that be not expected to be used.

(*ii*) FIFO (First Input First Output) is rather than choosing the victim page at random, the oldest page is the first to be removed.

(*iii*) LRU (Least Recently Used) to move out the page that is the least rarely used.

(*iv*) LFU (Least Frequently Used) to move out the page that is not used often in the past.

Now see some brief introduction about multiprocessors and Dataflow architecture.

Multiprocessors

Multiprocessors are architectures composed of multiple computing units which operate in a synchronous mode. Generally, both shared and local memory can be available for these processors. The communication between the processors takes place either through the shared memory area or through the interprocessor messages.

Dataflow Architectures

These architectures are functionally distributed architectures, in which the operations are triggered with the arrival of data at these processors. They may be viewed as very general MIMD parallel architectures.

Characteristics of Multiprocessors

A multiprocessor system is an interconnection of two or more CPU, with memory and input-output equipment. As defined earlier, multiprocessors can be put under MIMD category. The term multiprocessor is some times confused with the term multicomputers. Though both support concurrent operations, there is an important difference between a system with multiple computers and a system with multiple processors. In a multicomputers system, there are multiple computers, with their own operating systems, which communicate with each other, if needed, through communication links. A multiprocessor system, on the other hand, is controlled by a single operating system, which coordinates the activities of the various processors, either through shared memory or interprocessor messages.

The advantages of multiprocessor systems are:

- Increased reliability because of redundancy in processors
- Increased throughput because of execution of
- Multiple jobs in parallel
- Portions of the same job in parallel

A single job can be divided into independent tasks, either manually by the programmer, or by the compiler, which finds the portions of the program that are data independent, and can be executed in parallel. The multiprocessors are further classified into two groups depending on the way their memory is organized. The processors with shared memory are called tightly coupled or shared memory processors. The information in these processors is shared through the common memory. Each of the processors can also have their local memory too. The other class of multiprocessors is loosely coupled or distributed memory multi-processors. In this, each processor have their own private memory, and they share information with each other through interconnection switching scheme or message passing.

The principal characteristic of a multiprocessor is its ability to share a set of main memory and some I/O devices. This sharing is possible through some physical connections between them called the interconnection structures.

Interconnection Structures

The principal characteristic of a multiprocessor is its ability to share a set of main memory and some I/O devices. This sharing is possible through some physical connections between them called the interconnection structures. Some of these schemes are:

- Time-shared common bus
- Multiport memory
- Crossbar switch
- Multistage switching network
- Hypercube Interconnection
- Let us examine each of them in details.

Time-Shared Common Bus

This is the simplest interconnection structure, in which all the processors along with the memory are connected on a time shared common bus (refer Fig. 6.41).

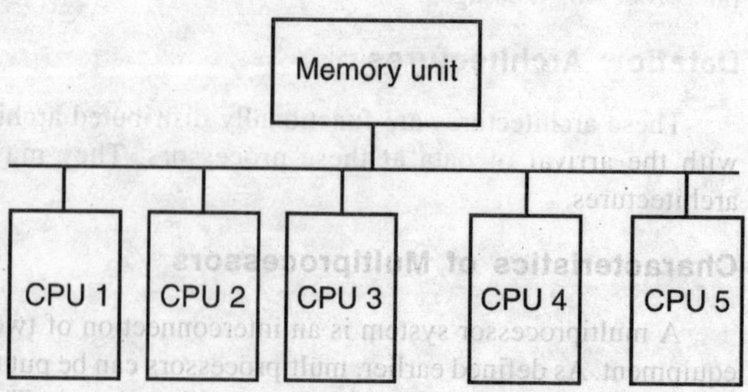

Fig. 6.41. Time - shared common bus organisation.

As all of them are sharing the same bus, only one processor can communicate with the memory at a time. The processor that is at any time communicating with the memory has the conrol over the bus. The other processor requesting the services of the bus waits for the bus to become free. A command is issued to the destination unit, indicating what operation is to be performed. The address of the source is recognised by the destination unit, which gets activated, after which the transfer is initiated. There could be more than one units requesting the services of the bus at any one time, in which case the conflict is resolved by the bus controller, which does it by assigning the priorities to various processors. The configuration resolution methods include static or fixed priorities like first-in first-out queues and daisy chaining.

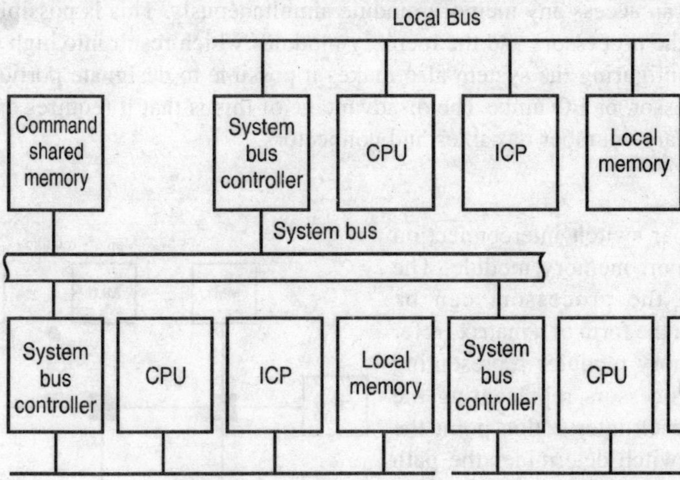

Fig. 6.42. Multibus structure for Multiprocessors.

As only one transfer can take place at any one time on the bus, the overall performance of the system is limited by the bandwidth of the bus. While one processor is communicating with the memory through the bus, the processors can only be doing some local task. The performance of the system can be increased by increasing the number of buses, but this increases the complexity and the cost of the system. The other alternative is to have a number of local buses, each connecting a processor to its local memory (refer to Fig. 6.42). Then there is some common memory to all the processors. This memory is called shared memory, and contains the global information, which needs to be shared by more than one processors. While one processor is accessing the shared memory, the other processors could be accessing their local memory or doing some other local task.

Multiport Memory

This, configuration is suitable and used for both uniprocessor and multiprocessor systems. In this configuration, there are separate buses connecting each CPU to each memory module (refer to Fig. 6.43).

Each processor bus consists of the address, data and the appropriate control lines used to communicate the CPU with the memory module. At any time there could be more than one processor requesting the services of a specific memory module. This conflict is

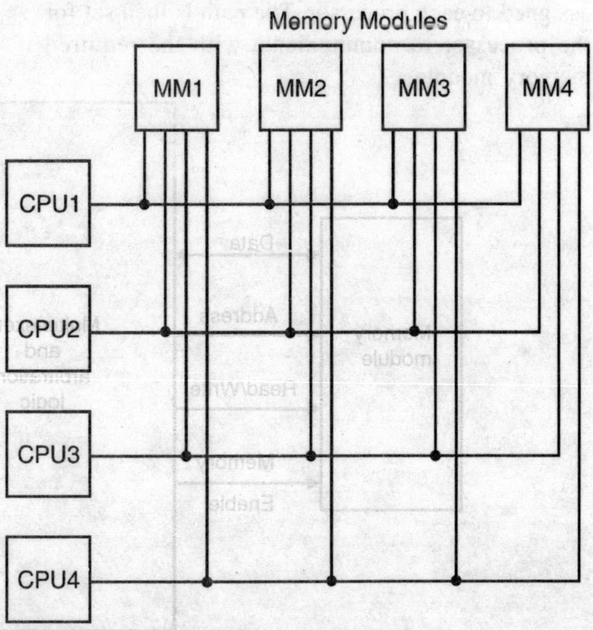

Fig. 6.43. Multiport Memory Organization.

resolved by assigning fixed priorities to each of the memory port. The systems can then be conFig.d as necessary at each installation. In Fig. 6.43, for example, the priorities can be allocated depending upon the physical placement of the processor module with respect to the memory module. Which means CPU will have priority over CPU2,CPU2 gets priority over CPU3,and CPU4 has the lowest priority.

You would notice in this case, as long as two processors are not trying to access the same memory module, the processors can access any memory module simultaneously. This is possible, because of the multiple paths between the processors and the memory modules which result into high transfer rate. The flexibility possible in configuring the system also makes it possible to designate portions of memory as private to a certain processor, or I/O units. The disadvantage of this is that it requires expensive memory and control logic and a large number of cables and connectors.

Crossbar Switch

Functionally, crossbar switch interconnection is very similar to multiport memory module. The memory modules and the processors can be visualized to be placed in the form of a matrix (refer to Fig. 6.44), with memory modules representing the columns and the processors representing the rows of the matrix. At each intersection point the switch is placed. This switch determines the path from the processor to the memory module. The switch has a control logic which determines the path from the processor to the memory module. The processor places the address of the memory to access on the bus. This address is captured by the switch, which checks for the conflicts. In case of a conflict in the form of multiple requests, the switch resolves the conflict depending upon the priority assigned to each processor. The path is then set for the processor to communicate with the required memory module.

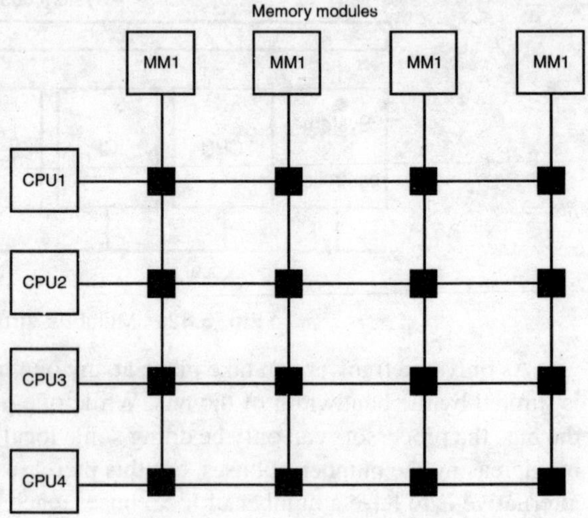

Fig. 6.44. Crossbar Switch.

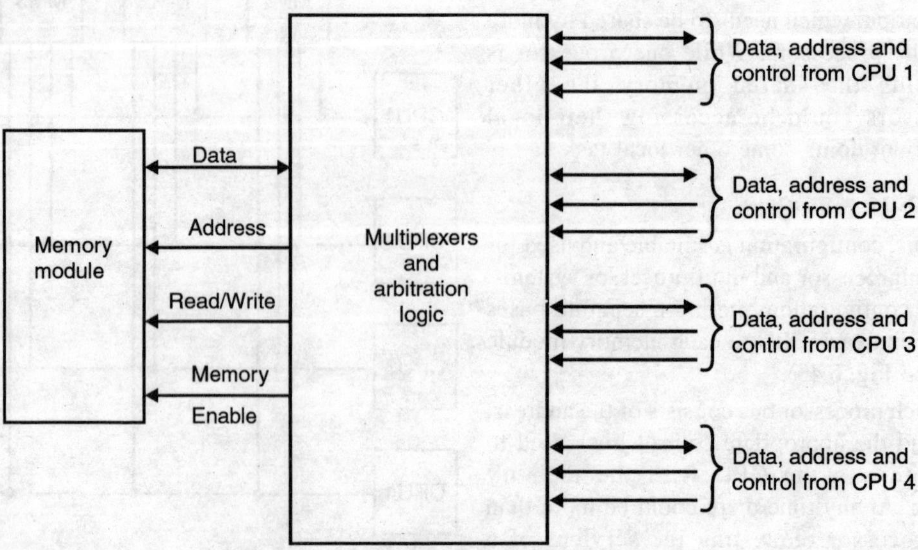

Fig. 6.45. Crossbar Switch Interconnection.

The switch can thus be imagined to be some sort of a combination of multiplexers and arbitration logic (refer to Fig. 6.45). The crossbar switch, organization supports simultaneous transfers from all memory modules, because there is a separate path associated with each module. However, the hardware required to support the crossbar switch is very large and complex and thus difficult to maintain.

Multistage Switching Network

Before starting the discussion on the multistage switching network, let us first try and understand the principles involved in the construction of a crossbar switch. Consider the Fig. 6.46, a 2X2 crossbar switch. The switch shown has 2 inputs, labelled A and B and 2 outputs labelled 0 and 1. This switch has the capability to connect its inputs A and B to either outputs 0 or 1,depending on the value of some control bit attached to the switch. Say, if the control bit is zero, the input A is connected to the output 0, and when the control bit is equal to 1, the input A is connected to output 1. Because there are no buffers in the switch, the process of resetting up the switch after a request rejection is very slow. To speedup the process, some buffers are inserted within the switch (Please refer to Fig. 6.47).

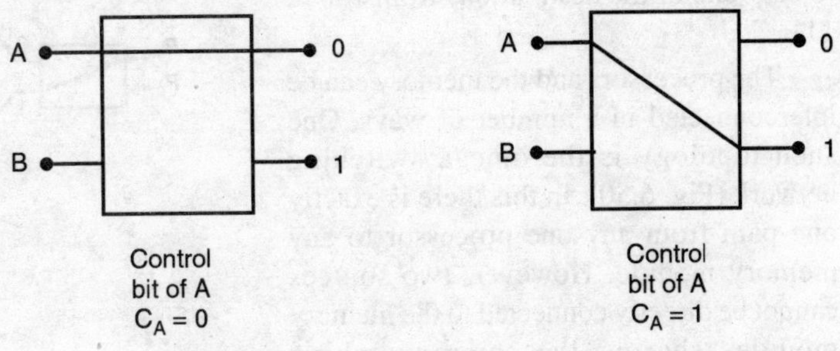

Fig. 6.46. A 2 X 2 Crossbar switch

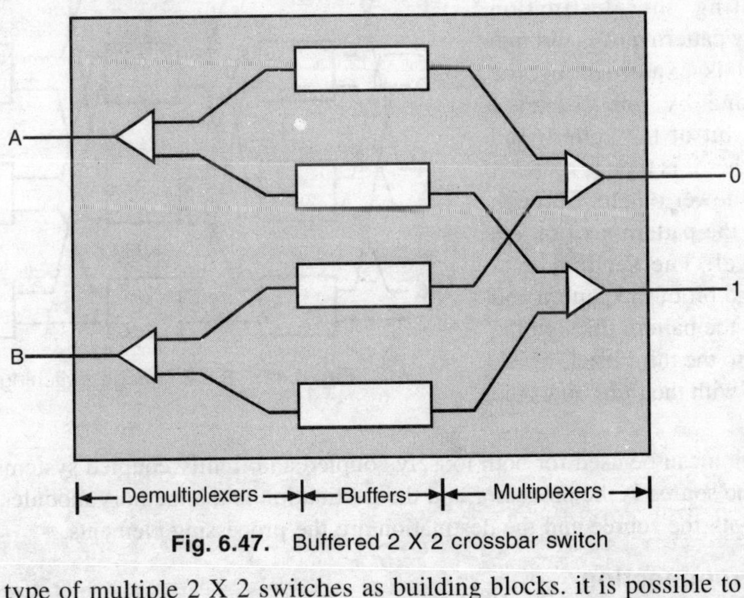

Fig. 6.47. Buffered 2 X 2 crossbar switch

Using such type of multiple 2 X 2 switches as building blocks. it is possible to build a multistage. network to control the communication between multiple sources and destinations. These can be connected to each other, either in the form of a binary tree (Fig. 6.48) or omega network (Fig. 6.48). To understand the working of such type of systems, consider the binary tree of Fig. 6.49. The two processors P1 and P2 are connected through switches to eight memory modules marked in binary from 000 to 111. The path from the source to the destination is determined by the binary bits of the destination. The first bit of the destination number determines the switch output in the first level. The second bit specifies the output of the switch in the second level and so on. For example, to connect P1 to memory 101, it is necessary to form a path from P1 to output 1 in the first level switch, output 0 in the second level switch, and output 1 in the third level switch. P1 and P2 can be connected to any of the eight memory modules. Certain request patterns cannot be satisfied simultaneously. For example, if P1 is connected to one of the destinations 000 through 011, then P2 can be connected to only one of the destinations from 100 to 111.

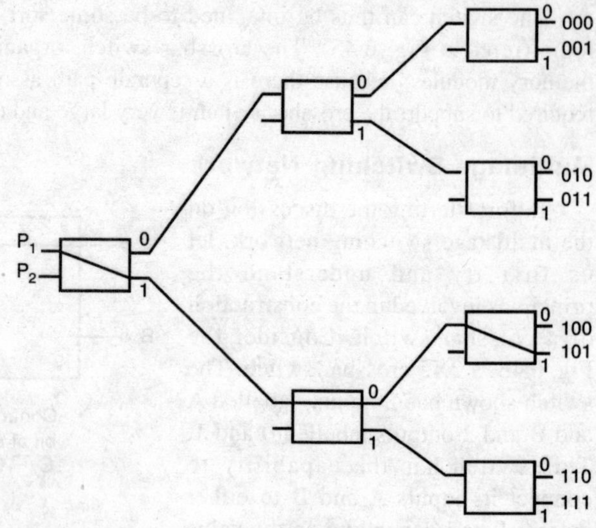

Fig. 6.48. Binary tree with 2 × 2 switches.

The processors and the memory can be interconnected in a number of ways. One such topology is the omega switching network (Fig. 6.50). In this there is exactly one path from any one processor to any memory module. However, two sources cannot be directly connected to the memory modules sharing the same switching network. Example, two sources cannot be connected simultaneously to destinations 000 and 001 or 010 and 011 and so on. The path from the source to the destination is determined in the same way as in the previous case. The source sends a 3-bit pattern representing the destination number. This binary pattern moves through the network, and is examined by the junction boxes one by one. Level 1 examines the first bit of the pattern and depending on whether it is 0 or 1, routes it to the upper or the lower output. Through one of the outputs the pattern now moves to the second level. The second level examines the second bit of the pattern and accordingly moves the pattern through the appropriate output to the third level, which repeats the process with the third bit of the pattern.

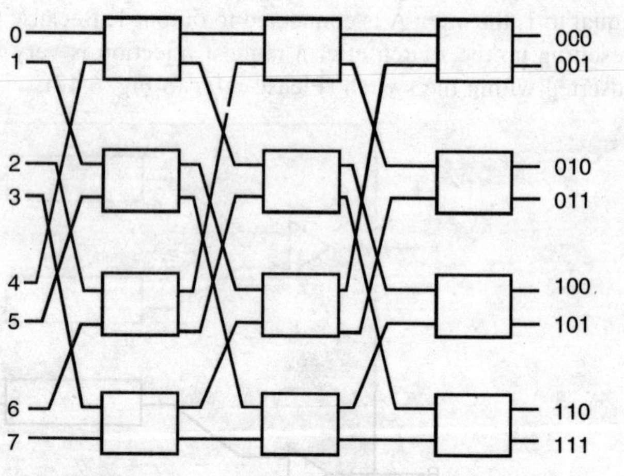

Fig. 6.49. 8 x 8 Omega switching network.

The configuration can be used for both loosely coupled and tightly coupled systems. For the tightly coupled systems, the source is the processor and the destination is the memory modules. For the loosely coupled systems, both the source and the destination are the processing elements.

Hypercube Interconnection

The hypercube or the binary n-cube multiprocessor structure is a loosely coupled system composed of N = (2) processors connected in a n-dimensional binary cube. Each processor forms a node of the cube. These processors could be the I/O processors, the CPUs or even the memory modules. In an n-dimensional hypercube, each node has a direct link with 'n' other nodes. For example,consider Fig. 6.50 (*a*), (*b*) and (*c*). Fig. 6.50(*a*) represents a 1 dimensional hypercube. As you would notice, the hypercube has 21 = 2 nodes, and each node is connected directly to 1 neighbouring node. The nodes can be numbered using 1 bit.

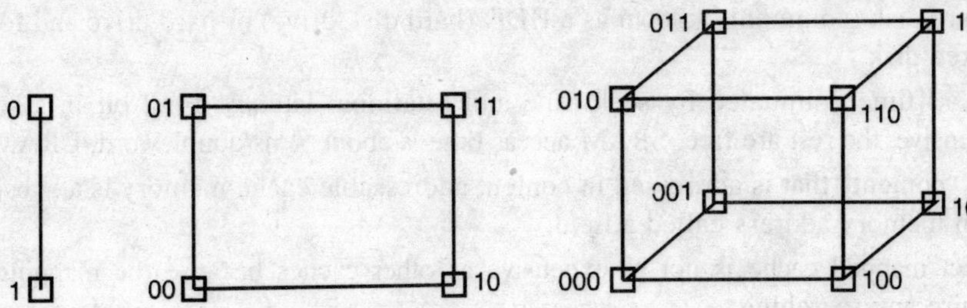

(a) One-cube *(b) Two-cube* *(c) Three-cube*

Fig. 6.50. Hypercube structures for n = 1, 2 and 3.

Similarly Fig. 6.50(*b*) shows a 2-dimensional hypercube, which has 2^2 = 4 nodes, and each node is directly connected to 2 other nodes. Each node can be represented using 2 bits (00, 01, 10, 11). The same description can be extended for three dimensional hypercube and so on. Please note that in the nomenclature of the nodes a direct neighbour of a node can be located on changing only one bit of a given node. For example, the direct neighbour of node 001 are: 011 or 101 or 050.

If any node has to communicate with the other node, to which it is not directly connected, it routes through the directly connected nodes, forming a path. The following example illustrates this for a 3-dimensional hypercube [Fig. 6.50(*c*)]. Node 010 can communicate directly with node 01 1, but to communicate with node 101 it has to go through atleast three links. The path could be 010 —> 110 => 111 => 101 or alternately, it could even be 010 => 000 => 100 => 101. In a similar fashion a number of other possible routes can also be derived.

A general routing procedure can thus be devised:

Take an exclusive OR of the source node address with the destination node address. The resulting binary value will have 1-bits corresponding to the axes on which the two nodes differ.

Send the message along with any one of these axes.

Summary

- Basic memory characteristics and memory structure.
- Memory classifications are described as Main memory and Auxiliary memory-Floppy Disks, Hard-disks, CD-ROM etc.
- Different types of memory devices and their organizations is discussed.-DISK GEOMETRY AND DISK SPEED is explained.
- Disk caching techniques and various levels of RAID are presented.
- RAID-0 is the fastest and most efficient array type but offers no fault-tolerance.
- RAID-1 is the array of choice for performance-critical, fault-tolerant environments. In addition, RAID-1 is the only choice for fault-tolerance if no more than two drives are desired.
- RAID-2 is seldom used today since ECC is embedded in almost all modern disk drives.
- RAID-3 can be used in data intensive or single-user environments which access long sequential records to speed up data transfer. However, RAID-3 does not allow multiple I/O operations to be overlapped and requires synchronized-spindle drives in order to avoid performance degradation with short records.
- RAID-4 offers no advantages over RAID-5 and does not support multiple simultaneous write operations.
- RAID-5 is the best choices in multi-user environments which are not write performance sensitive. However, at least three and more typically five drives are required for RAID-5 arrays.

- A hard disk (commonly known as a HDD (hard disk drive) or hard drive and formerly known as a fixed disk.
- Access time dominated by seek time and rotational latency. First bit in a sector is the most expensive, the rest are free. SRAM access time is about 4 ns/doubleword, DRAM about 60 ns.
- The "content" that is addressed in content addressable cache memory is a subset of the bits of a main memory address called a field.
- Direct mapped cache is not as expensive as other caches because the mapping scheme does not require any searching.
- Direct mapping is no need for a replacement policy because the location for each new block is predetermined.
- Virtual memory as an alternate set of memory addresses. Programs use these virtual addresses rather than real addresses to store instructions and data.
- Memory management is the hardware base that makes a virtual memory system possible and which allows software to reference physical memory by virtual addresses, quite often more than once. It accomplishes this through the use of page and page tables.
- One for kernel addresses or more for user space processes. Page Table Entry (PTE) one per virtual page. When a process generates a virtual address, the operating system translates it into a physical memory address.
- Operations on tables or array involve access of a certain clustered area in the address space.
- The sequentially in program behavior also contributes to the spatial locality because sequentially coded instructions and array elements are often stored in adjacent locations each type of locality affect the design of memory hierarchy.

Review Questions

1. How do you classify memory Devices?
2. How memory decoder is used to connect CPU with RAM?
3. What is cache memory?
4. Explain the characteristics of memory devices.
5. How a Cache memory is Organized?
6. What are RAID-3 and RAID-5 levels?
7. How RAM/ROM is organized?
8. Write short notes on Optical memory Devices with their applications.
9. How hard disk is organized? What is disk geometry?
10. Explain the speed of Hard Disk. What is CD-ROM?
11. How memory Hierarchy helps in Computer Performance improvements?
12. What is HIT and MISS ratio in Cache?
13. What are different Addressing schemes?
14. What is two-level cache?
15. Explain Set-Associative Mapping with example.
16. What is the CPU effective time due to Cache?
17. What are Cache Design parameters?
18. How memory management is done?
19. Explain Virtual memory.

❑❑❑

7 Chapter

Advanced Concepts of Computer Architecture

INSIDE THIS CHAPTER

7.1. INTRODUCTION TO PIPELINING

Pipelining is the most common implementation technique used today to increase the performance of computer systems. Pipelining takes advantage of instruction level parallelism found in everyday programs. In other words it increases performance by doing multiple things at the same time. Think of a pipeline as an assembly line. Instead of building cars or computers an instruction is slowly executed a piece at a time.

Each stage of the pipeline is connected to the next stage. Each stage must complete its portion of the execution before the instruction moves to the next stage. The machine cycle time is the time required by the slowest pipe stage just as an assembly line can only run as fast as the slowest worker.

The MIPS Pipeline

There are five steps of execution

IF	–	instruction fetch
ID	–	instruction decode
EX	–	execute
MEM	–	memory access
WB	–	write back

We can pipeline MIPS by fetching a new instruction each clock cycle.

Example 7.1. A nonpipelined machine has execution steps that require 50, 50, 60, 50, 50 ns respectively and that 5ns is required at each stage of a pipeline due to latch and skew overhead.

The original nonpipelined machine has an average execution time of 260 ns.

The pipelined machine must have a cycle time as long as the slowest stage plus the overhead or 65 ns for each stage.

The average execution time for an instruction has been decreased to 65ns giving a speedup of 4 with the ideal being 5.

7.1.1. Pipelining

Previously we saw that a datapath can be divided up so that execution proceeds in multiple cycles. The advantage of this division is that instructions that do not use the entire datapath can complete in fewer cycles. In addition, the clock rate can be increased because we do not have to extend the clock cycle to handle the longest instruction. However, each instruction must still wait for the one ahead of it to complete before it can begin to execute. This approach could be likened to a washing machine: different types of loads require different sequences of steps, but each load of clothes must be completely washed and removed before we can put in another load.

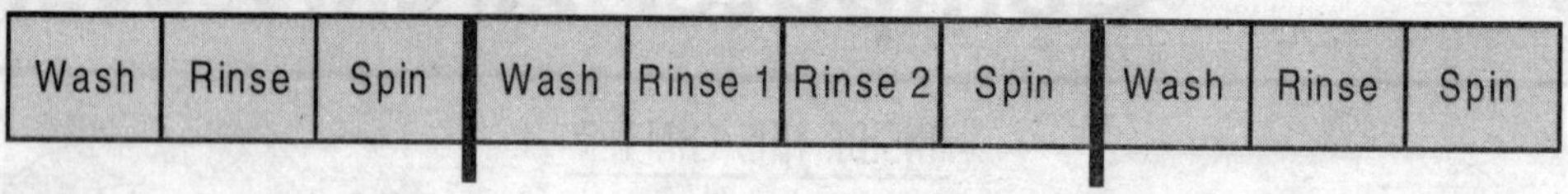

Fig. 7.1.

Suppose we wanted to be able to wash more clothes in a given time. Notice that, at any given time, the machine could be doing one of four different steps. Why can't it do these all at once for multiple loads?

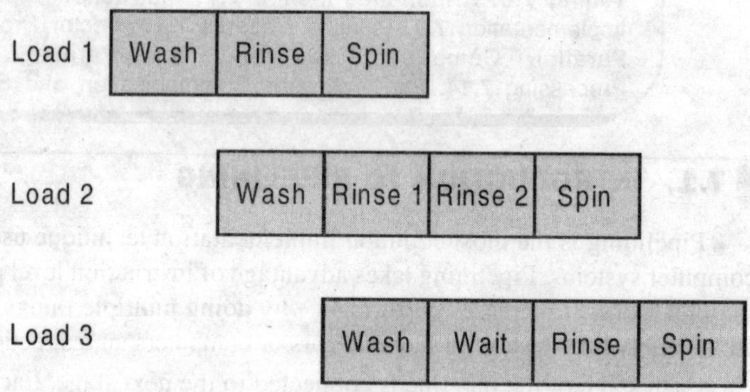

Fig. 7.2.

Because it can only hold one load at a time.

So how can we enable it to perform multiple steps at once?

If we have separate units for each of the steps, each with its own drum (storage), we can transfer the clothes from one step to the next and start a new load of clothes right away.

What is it about this change that enables us to run multiple loads at once?

The steps are independent, and they each can hold (store) a load of clothes.

In a datapath, the same basic idea can be used to process multiple instructions at once. It requires that the hardware for each of the steps be independent—able to operate separately from the hardware of the other steps—and that it have storage for intermediate state information.

There is one other aspect of this approach that isn't obvious at first. The loads of clothes, and the instructions, can only start if they are independent of each other. For example, if you know you need to run a sweatshirt through twice to get rid of a stain, you can't start the second load until the first one is completed. To make this approach worthwhile, you also need to have enough loads of wash to keep all of the units busy. Fortunately, in a datapath, there are usually plenty of instructions, and we take different (independent) instructions from memory in a simple sequential order.

Notice that the first example requires 10 cycles to complete while the second example takes only 6 cycles. This is a 40 percent speed increase. Notice also that if more and more instructions were to be processed, the average time per instruction remains the average number of steps in an instruction (in this case, something between 3 and 4) but the rate at which instructions are completed is much closer to one per cycle (roughly a 60 to 75 percent improvement).

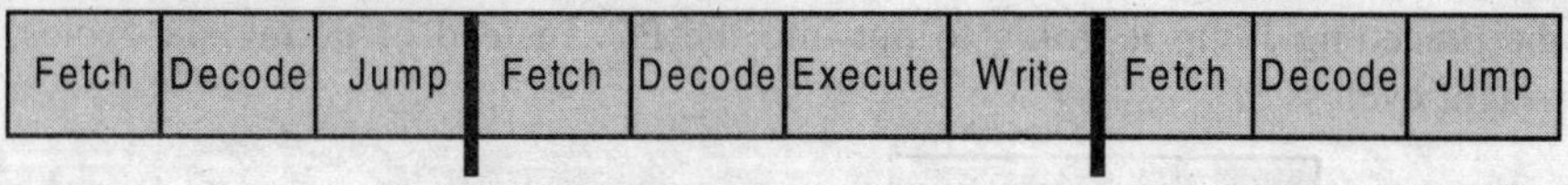

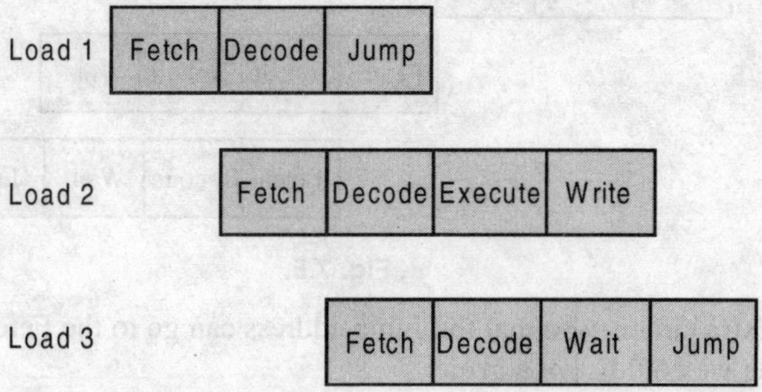

Fig. 7.3.

Thus, instructions flow through the datapath in a steady stream, something like fluid flowing through a pipe. Hence this style of datapath is called a pipeline.

When we pour fluid into a pipe, it takes some time (proportional to the length of the pipe) for the first drops to emerge from the other end. But if we pour steadily, then fluid continues to emerge from the pipe at whatever rate it enters.

For an infinite sequence of instructions that keep the pipeline full, the rate at which instructions are executed is one per cycle, where a cycle is a pipeline stage time. Thus, if we were to compare the instruction execution rate of a pipelined vs. a non-pipelined version of the same datapath on such an instruction stream, the pipelined processor would execute at N times the rate of the non-pipelined processor, where N is the number of stages in the pipe.

For a finite sequence of S instructions, the effective execution time is S/(N + S – 1) because the first instruction takes N cycles and the remaining S – 1 instructions each take another cycle to be completed. The speedup is thus SN/(N + S – 1). As you can see, if S approaches infinity, the execution time approaches 1 cycle per instruction and the speedup approaches N. Unfortunately, we can't keep a pipeline full forever.

If we stop pouring fluid for a while, then air enters the pipe and a bubble forms in the fluid flow. If we start pouring again, the other end of the pipe will have a pause in its output that equals the length of the pause in the pouring.

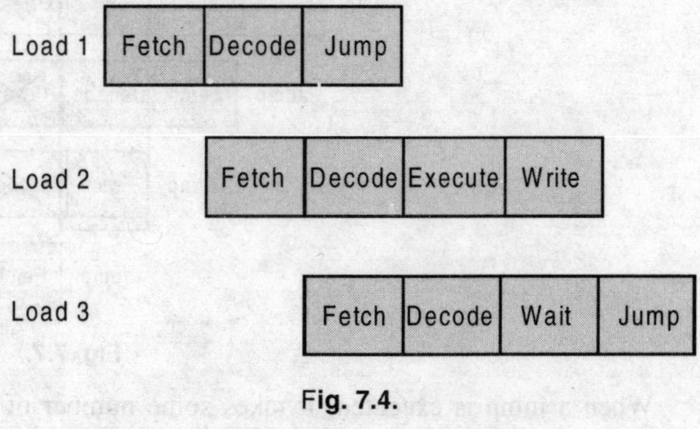

Fig. 7.4.

The pipelined datapath in a computer behaves the same way—it shows the same initial delay in emitting the first result, and any gaps in the input produce corresponding gaps in the output that are referred to as "bubbles".

Gaps in the instruction stream due to memory delays.

Jumps —notice that the jump occurs at the end of an instruction, but it invalidates the fetches that went on after it was fetched. The jump operation is not an independent instruction—the instruction that follows it depends on its execution. Let's look again at the preceding example. What's wrong with this picture?

The second instruction can't start until the first one is finished because it's Fetch step depends on the value that the preceding Jump is going to put into the PC. Instead of taking six cycles, this sequence really takes eight cycles.

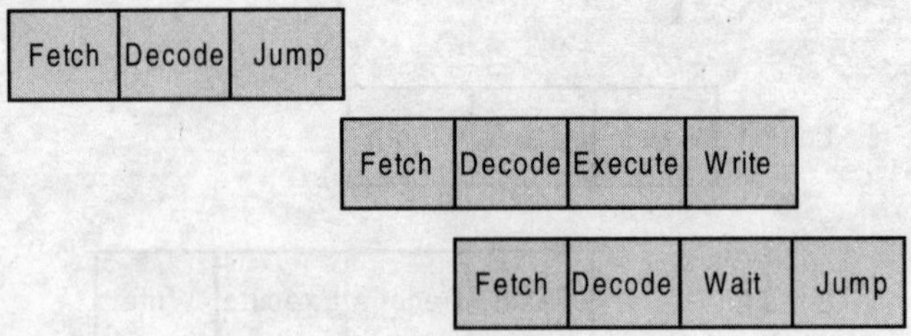

Fig. 7.5.

If we add some extra circuitry so that the Jump address can go to the Fetch unit at the same time that it goes to the PC, then we could save a cycle:

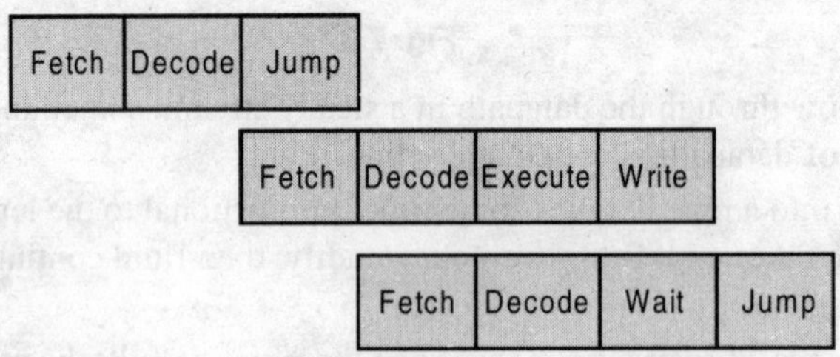

Fig. 7.6.

Even so, our pipeline is still less efficient in the presence of Jumps. Jumps account for a large fraction of the inefficiency seen in actual pipelines. In some programs, especially those written in object-oriented languages, jumps account for 25% to 30% of the instructions executed. How does this affect the efficiency of a pipeline?

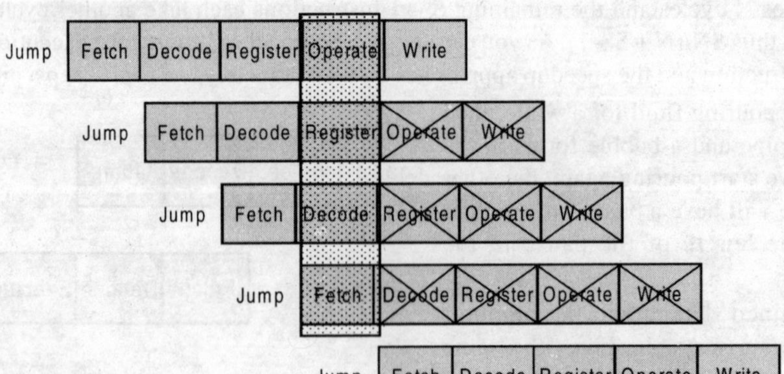

Fig. 7.7.

When a jump is executed, it takes some number of cycles for the control unit to detect that the instruction is a jump. With an unconditional jump, it can immediately switch to fetching from the target location, and the gap is just one or two pipeline stages (whatever stage the decode occurs plus one more to transfer the target address to the PC and simultaneously start the next fetch).

With a conditional jump (or branch), the number of cycles before it is detected is typically greater. Even though the control unit can identify the instruction as a branch, it can't tell whether it will be taken until some computation is performed. Thus, the delay can be almost the full length of the pipeline. For example, in a 5-stage pipeline, a branch might not be determined until the fourth stage. Thus, the three stages behind it in the pipeline are invalid and another instruction must enter the pipe at that point. The three invalid stages represent three instructions that were partially executed on the speculation that the branch would not be taken. They also represent three wasted pipeline cycles that are called the jump penalty.

If the jump penalty in a pipeline is P, the probability that an instruction is a jump is J, and the probability that the jump is taken is T, then the average instruction time (in cycles) for the pipeline is 1 + PJT. The efficiency is the inverse of this quantity: 1/(1 + PJT).

The probabilities in this formula are dependent on the specific program being executed. Tight loops and frequent subroutine calls increase the probabilities while long stretches of straight-line code decrease the probabilities.

Let's look at some examples to see the magnitude of this effect. Suppose we have a pipe as above, with a penalty of 3 cycles, and a program has 30% branch instructions and 25% of them are taken. What is the average instruction time and the efficiency?

$$\text{Average time} = 1 + 3 \times 0.3 \times 0.25 = 1.225$$
$$\text{Efficiency} = 1/\text{average time} = 81.6\%$$
$$\text{Speedup} = \text{Efficiency} \times \text{Number of stages} = 4.08$$

If the jumps are taken 50% of the time

$$\text{Average time} = 1 + 3 \times 0.3 \times 0.5 = 1.45$$
$$\text{Efficiency} = 1/\text{average time} = 68.9\%$$
$$\text{Speedup} = \text{Efficiency} \times \text{Number of stages} = 3.45$$

So, for fairly conservative scenarios, we see reductions in performance of nearly a third. Suppose the probability that jump is taken rises to 70%.

$$\text{Average time} = 1 + 3 \times 0.3 \times 0.7 = 1.63$$
$$\text{Efficiency} = 1/\text{average time} = 61.3\%$$
$$\text{Speedup} = \text{Efficiency} \times \text{Number of stages} = 3.07$$

Branch Prediction and Changing the Detection Point

What simple change could we make in this case to reach an 81.6% level of efficiency?

Always assume that a branch is taken.

Making such an assumption (one way or the other) is called static branch prediction. We are using statistical analysis of our programs to make a design decision regarding the treatment of branches in order to reduce the loss of efficiency due to branches. Of course, if our sample set of programs isn't representative, then our static strategy fails. On the other hand, once we make such a decision, we can often have a compiler provide some help.

Let's look at what happens if the branch detection point changes. What would you expect to be the effect if the detection point occurs later? Let's say it happens in stage 7 of an 8-stage pipe (that is, the penalty is 6, and the maximum speedup is 8), and we have 30% branches with a 50% probability of being taken.

$$\text{Average time} = 1 + 6 \times 0.3 \times 0.5 = 1.90$$
$$\text{Efficiency} = 1/\text{average time} = 52.6\%$$
$$\text{Speedup} = \text{Efficiency} \times \text{Number of stages} = 4.21$$

As we could guess, moving the detection point later increases the penalty and decreases the efficiency. With a penalty of 3, the efficiency is 68.9% but it drops to 52.6% when the penalty is 6. Note however that, because we increased the length of the pipeline, the speedup still increases. If the penalty is reduced to 2 in a 5-stage pipe, then

$$\text{Average time} = 1 + 2 \times 0.3 \times 0.5 = 1.30$$
$$\text{Efficiency} = 1/\text{average time} = 76.9\%$$
$$\text{Speedup} = \text{Efficiency} \times \text{Number of stages} = 3.85$$

So earlier detection is better, as we would expect.

Several techniques can be used to improve the efficiency of a pipeline in the presence of jumps. One is to tag jumps as being likely to be taken or not. For example, in a FOR loop, the jump that returns to the top of the loop could be tagged as likely to be taken. That is, a bit in the opcode could be reserved to indicate the likelihood of the jump being taken. This allows us to move the detection point up to the decode stage (perhaps reducing the penalty to one). Of course, when we guess wrong (the loop does eventually exit) then we pay a larger penalty, but hopefully we do so infrequently.

With a true FOR loop, we can even use a special counter that signals when the loop is about to exit, and thus we pay no penalty at all for the jump at the end of the loop. Unfortunately, C, C++, and Java do not contain true FOR loops (they can have additional conditions) so this applies mostly to Pascal, Ada and Fortran programs.

Another approach that is used is to tag instructions as being independent of the branch and have the compiler move them after the branch. For example, if we have some calculation to perform prior to the branch, but the branch does not depend on its outcome, we can move its instructions to follow the branch and tag them as not having to be discarded if the branch is taken. Thus, the penalty gap can be filled with useful work no matter what the branch does. Unfortunately, in typical programs it's difficult to find more than one or two such independent instructions. This "delay slot" technique, as it is called, also presents the problem that it we want to make the pipeline longer in the next generation of the processor design, we have to be careful to keep the branch penalty the same (which may not be easy to do) or else recompile our software to try to fill a larger gap.

These are all static methods as they rely on analysis of the code prior to execution. At execution time we can also use the past behavior of a branch to help predict its future actions. Most individual branches exhibit a behavior in which there is a typical action and an exceptional case. Thus, once a branch is executed, we can guess that it will behave the same way again.

A processor can be built with a table that stores the locations and actions of branches as it encounters them (called a branch history table). When a branch is fetched, its address causes the history table to recall its prior action and the control unit can assume that it will behave the same way and take the appropriate action.

Of course, if the branch takes the exceptional action, then the full penalty must be paid. In addition, the first time a branch is encountered, the full branch penalty must be taken. This approach is called dynamic branch prediction because it takes place at run time.

By combining static and dynamic methods, it is quite possible to achieve a 90% prediction rate so that the effective probability of a branch being taken is just 10%. With our previous example, we have:

$$\text{Average time} = 1 + 3 \times 0.3 \times 0.1 = 1.09$$
$$\text{Efficiency} = 1/\text{average time} = 91.7\%$$

So, for a 5-stage pipeline, we manage to get effectively a speedup of 4.6 over a non-pipelined system for an infinite instruction stream. Note that this is better performance from a 5-stage pipe than we got from our example of an 8-stage pipe with less aggressive prediction.

Because all of the stages of a pipeline advance in lock-step, the basic cycle time of a pipe is determined by the slowest stage. There is thus a significant incentive for the designer to equalize the time through all of the stages. Otherwise, speed in some stages is wasted. As we've seen, speed is often related to power consumption. Thus, power may be wasted in an imbalanced pipeline.

There are several ways that we could go about balancing the timing of individual stages.

Use faster or slower components as needed.

Shift some functionality between stages.

Split long stages into two shorter stages.

Combine two fast stages into one slower stage.

If, instead of having the stages operate in lockstep, we have them pass data to each other as they finish with it and the successor unit is available, then we have a different kind of system called a "dataflow" processor. Dataflow designs can have pipelines with nonlinear topologies (that is, branches), and they are perhaps most commonly employed in signal processing, where a long and complex calculation must be performed on a stream of data. The calculation is broken into stages and these are activated on a demand basis as data enters the system.

7.2. HOW PIPELINING WORKS

The basic principle behind pipelining is to utilize the time cycle of the instruction execution unit most efficiently. In one of the earliest instruction execution unit belonging to the 8085, a four step instruction cycle consisting of instruction fetch, decode, data fetch and execution was used (Figure 7.8). However it can be seen that in such a simple instruction cycle if we assume that each step takes one machine cycle than the executiori of a instruction will take a minimum of four cycle. Thus it is a clear wastage of time as only one unit of the whole instruction cycle is used at a time. In order to use the machine efficiently we can try to utilize every unit of the instruction cycle thereby executing instructions every cycle.

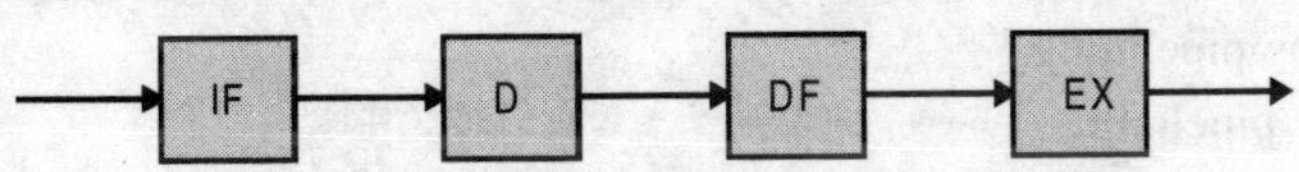

Fig. 7.8. Simple pipeline and its stages.

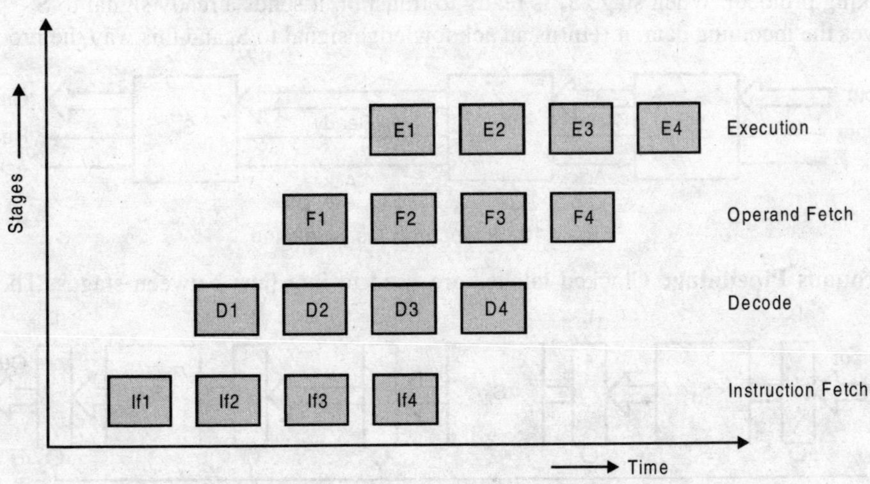

Fig. 7.9. Pipeline space time diagrams.

Pipelining, a standard feature in RISC processors, is much like an assembly line (Figure 7.9). Because the processor works on different steps of the instruction at the same time, more instructions can be executed in a shorter period of time.

A useful method of demonstrating this is the laundry analogy. Let's say that there are four loads of dirty laundry that need to be washed, dried, and folded. We could put the first load in the washer for 30 minutes, dry it for 40 minutes, and then take 20 minutes to fold the clothes. Then pick up the second load and wash, dry, and fold, and repeat for the third and fourth loads. Supposing we started at 6 PM and worked as efficiently as possible, we would still be doing laundry until midnight.

However, a smarter approach to the problem would be to put the second load of dirty laundry into the washer after the first was already clean and whirling happily in the dryer. Then, <vhile the first load wa* being folded, the second load would dry, and a third load could be added to the pipeline of laundry. Using this method, the laundry would be finished by 9:30.

With a pipelined processor, we get the best of both: the low CPI of the single cycle CPU, and the low cycle time of the multiple cycles CPU, the cost, as usual, is increased complexity. Note that pipelining does not improve the performance of a single instruction. Rather, it improves the performance of groups of instructions. This is one of the classic issues in computer science: latency vs. throughput. A pipelined processor can't execute a single load instruction any faster than a multi-cycle CPU. However, a pipelined processor can execute five load instructions faster than a multi-cycle CPU.

7.3. LINEAR PIPELINE PROCESSORS

A linear Pipeline processor is a cascade of processing stages which are linearly connected to perform a fixed function over a stream of data flowing from one end to other. In modern computers, linear pipelines are applied for instruction execution, arithmetic computation, memory access operations.

A linear pipeline processor is constructed with the processing stages. External inputs are fed into the pipeline at the first stage S_1. The processed results are passed from stage S_i to stage S_{i+1} for all $i = 1, 2, \ldots K - 1$. The final result emerges from the pipeline at the last stage S_k.

Depending on the control of data flow along the pipeline, linear pipelines are divided into 2 categories

- Asynchronous pipelining
- Synchronous pipelining

Asynchronous Pipelining: Data flow between adjacent stages in asynchronous pipeline is controlled by handshaking protocol. When stage S_1 is ready to transmit, it sends a ready signal to S_{i+1}. After stage $S_i + 1$ receives the incoming data, it returns an acknowledge signal to S_i and this way the process goes on.

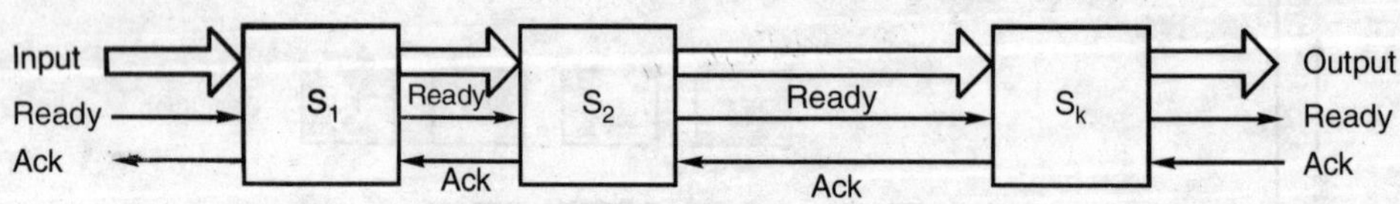

Fig. 7.10. Asynchronous pipelining

Synchronous Pipelining: Clocked latches are used to interface between stages. The latches are

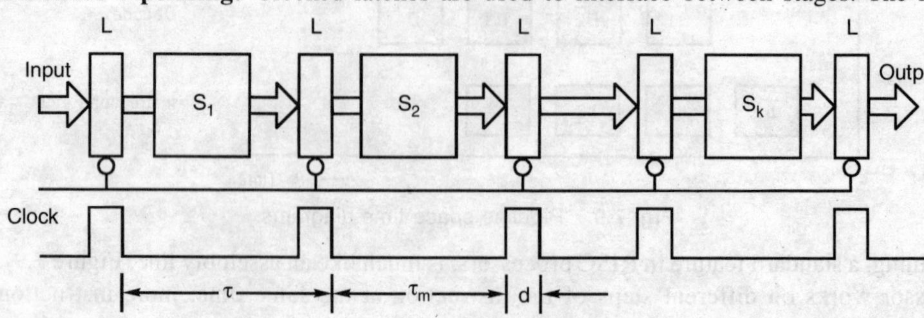

Fig. 7.11. Synchronous Pipelining

made with master slave flip flops, which can isolate inputs from outputs upon the arrival of a clock pulse, all latches transfer data to the next stage simultaneously.

7.4. NON-LINEAR PIPELINE PROCESSORS

A dynamic pipeline can be reconfigured to perform variable functions at different times. The traditional linear pipelines are static pipelines because they are used to perform fixed functions. A dynamic pipeline allows feed forward and feedback connections in addition to the streamline connections. For this reason, we call such a structure as non-linear pipeline.

This pipeline has three stages. Besides the streamline connections from S_1 to S_2 and from S_2 to S_3, there is feed forward connection from S_2 to S_3 and two feedback connections from S_3 to S_2 and from S_3 to S_1.

These feed forward and feedback connections make the scheduling of successive event into the pipeline a non trivial task. With these connections, the output of the pipeline is not necessarily from the last stage. In fact, following different dataflow patterns, one can use the same pipeline to evaluate different functions.

7.5. PIPELINE CLOCK AND TIMING

Clock cycle of the pipeline: τ

Latch delay: **d**

Maximum stage delay: τ_m

$$\tau = \max\{\tau_m\} + d$$

Pipeline frequency: **f**

$$f = 1/\tau$$

K-stage pipeline processes n tasks in k + (n – 1) clock cycles

K cycles for the first task and $n - 1$

Cycles for the remaining $n - 1$ tasks

Total time to process n tasks

$$T_k = [k + (n-1)]\tau$$

For the non-pipelined processor

$$T_1 = nk\tau$$

Speedup factor
$$S_k = \frac{T_1}{T_k} = \frac{nk\tau}{[k+(n-1)]\tau} = \frac{nk}{k+(n-1)}$$

7.6. ARITHMETIC PIPELINE

The principles used in instruction pipelining can be used in order to improve the performance of computers in performing arithmetic operations such as add, subtract and multiply. In this case, these principles will be used to realize the arithmetic circuits inside the ALU. In this section, we will elaborate on the use of arithmetic pipeline as a means to speed up arithmetic operations. We will start with fixed-point arithmetic operations and then discuss floating-point operations.

7.6.1. Fixed-Point Arithmetic Pipelines

The basic fixed point arithmetic operation performed inside the ALU is the addition of two n-bit operands $A– a_{n-1}\, a_{n-2}, ..., a_2a_1a_0$ and $B–b_{n-1}\, b_{n-2}, ..., b_2b_1b_0$.

Addition of these two operands can be performed using a number of techniques.

These techniques differ in basically two attributes: degree of complexity and achieved speed. These two attributes are somewhat contradictory; that is, a simple realization may lead to a slower circuit while a complex realization may lead to a faster circuit.

7.6.2. Floating-Point Arithmetic Pipelines

Using a similar approach, it is possible to pipeline floating-point (FP) addition/sub-traction. In this case, the pipeline will have to be organized around the operations needed to perform FP addition. The main operations needed in FP addition are exponent comparison (EC), exponent alignment (EA), addition (AD), and normalization (NZ). Therefore, a possible pipeline organization is to have a four-stage pipeline each performing an operation from EC, EA, AD, and NZ.

Pipeline arithmetic units are usually found in very high speed computers. They are used to implement floating point operation, multiplication of fixed point numbers and similar computations encountered in scientific problems.

Arithmetic pipeline for addition and subtraction

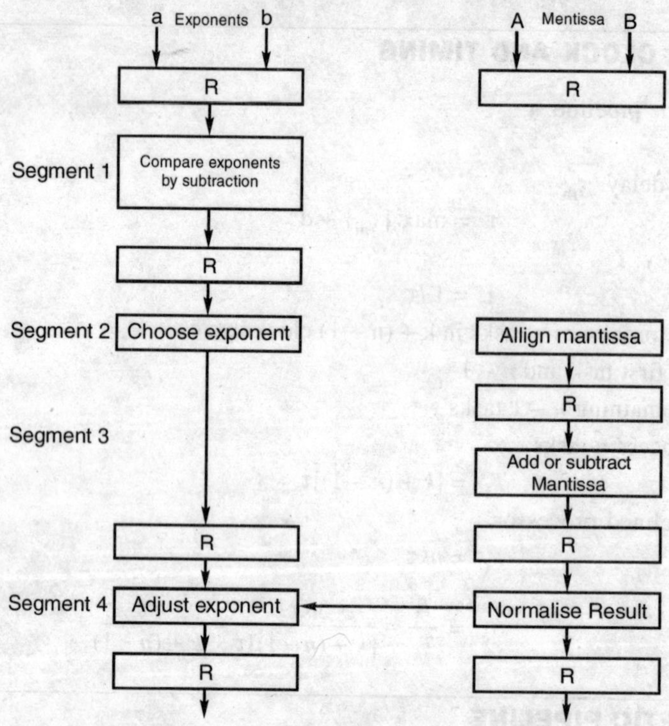

Fig. 7.12

- The exponent are compared by subtracting them to determine their difference.
- Exponent difference determine how many times the mantissa associated with the smaller exponent must be shifted to the right.
- This produces are alignment of two mantissas.
- The two mantissas are added or subtracted in segment 3.
- Finally result is normalized in segment 4.
- When a overflow occurs, the mantissa of the sum or difference is shifted right and the exponent is incremented by one.

➡ When an underflow occurs, the number of leading zeroes in the mantissa determines number of left shifts in the mantissa and the number that must be subtracted from the exponent.

7.7. INSTRUCTION PIPELINE DESIGN

A stream of instructions can be executed by pipeline in an overlapped manner. A typical instruction execution consists of a sequence of operations, including

(1) Instruction fetch
(2) Decode
(3) Operand fetch
(4) Execute
(5) Write back phases

We describe below instruction pipelines for CISC and RISC scalar processors.

Pipelined Instruction Processing

A typical instruction pipeline is depicted in Fig. 7.13. The fetch stage (F) fetches instructions from a

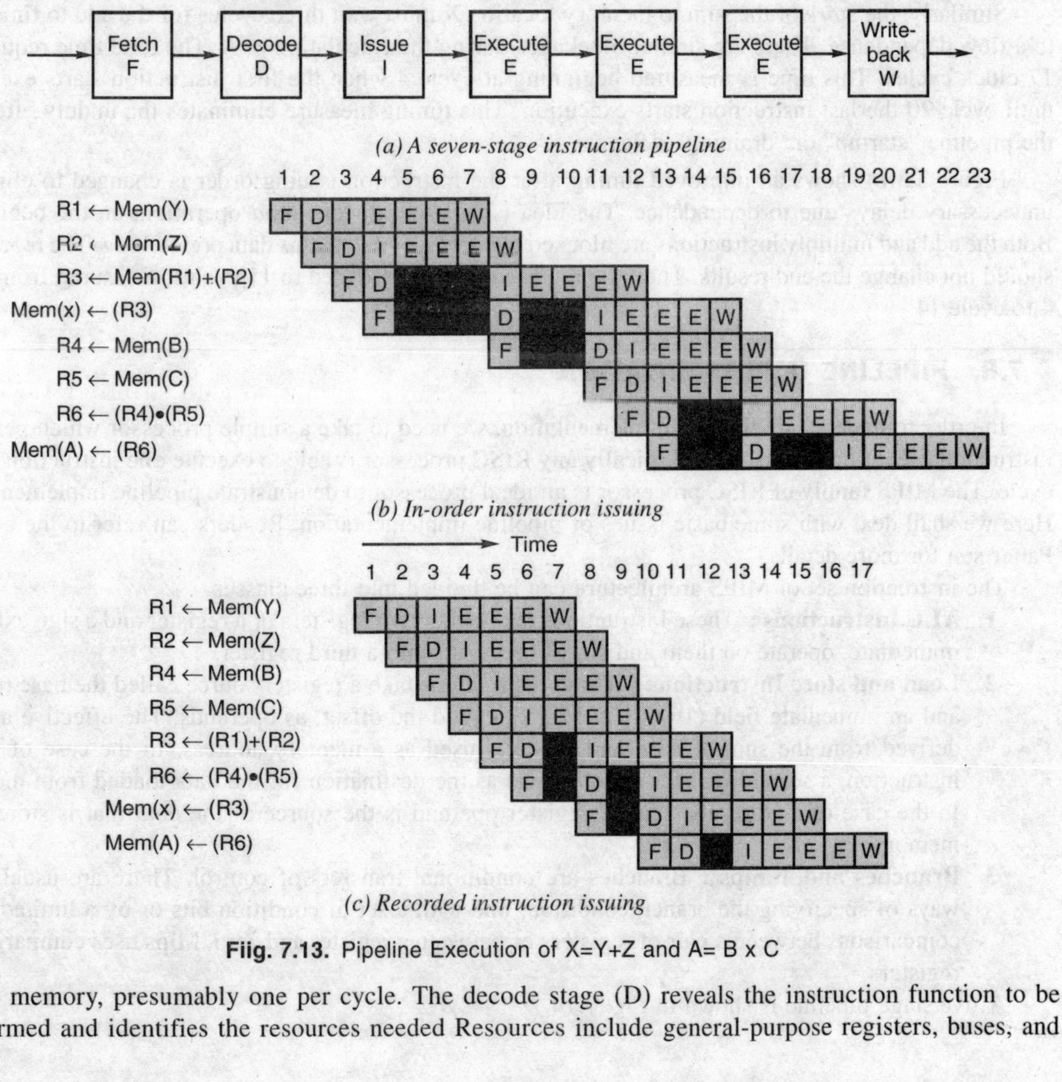

Fiig. 7.13. Pipeline Execution of X=Y+Z and A= B x C

cache memory, presumably one per cycle. The decode stage (D) reveals the instruction function to be performed and identifies the resources needed Resources include general-purpose registers, buses, and functional units. The issue stage (I) reserves resources. Pipeline control interlocks are maintained at this stage. The operands are also read from registers during the issue stage.

The instructions are executed in one or several execute stages (E). Three execute stages are shown in Fig. 7.13(*a*). The last write back stage (W) is used to write results into the registers. Memory load or store operations are treated as part of execution. Figure 7.13 shows the flow of machine instructions through a typical pipeline. These eight instructions are for pipelined execution of the high-level language statements X = Y + Z and A = B × C. Assume load and store instructions take four execution clock cycles, while floating-point add and multiply operations take three cycles.

The above timing assumptions represent typical values used in a CISC processor. In many RISC processors, fewer clock cycles are needed. On the other hand, Cray 1 requires 11 cycles for a load and a floating-point addition takes six. With in-order instruction issuing, if an instruction is blocked from issuing due to a data or resource dependence, all instructions following it are blocked.

Figure 7.13 (*b*) illustrates the issue of instructions following the original program order. The shaded boxes correspond to idle cycles when instruction issues are blocked due to resources latency or conflicts or due to data dependences. The first two *load* instructions issue on consecutive cycles. The *add* is dependent on both *loads* and must wait three cycles before the data (Y and Z) are loaded in.

Similarly, the *store* of the sum to memory location X must wait three cycles for the add to finish due to a flow dependence. There are similar blockages during the calculation of A. The total time required is 17 clock cycles. This time is measured beginning at cycle 4 when the first instruction starts execution until cycle 20 the last instruction starts execution. This timing measure eliminates the unduly effects of the pipeline "startup" or "draining" delays.

Figure 3.3(*c*) shows an improved timing after the instruction issuing order is changed to eliminate unnecessary delays due to dependence. The idea is to issue all four *load* operations in the beginning. Both the add and multiply instructions are blocked fewer cycles due to this data prefetching. The reordering should not change the end results. The time required is being reduced to 11 cycles, measured from cycle 4 to cycle 14.

7.8. PIPELINE IMPLEMENTATION

In order to understand pipeline implementation, we need to take a simple processor which performs instruction execution in one cycle. Typically any RISC processor is able to execute one instruction in one cycle. The MIPS family of RISC processor is an ideal processor to demonstrate pipeline implementation. Here we shall deal with some basic issues of pipeline implementation. Readers can refer to the book of Patter son for more details.

The instruction set of MIPS architecture can be divided into three classes.

1. **ALU Instructions :** These instructions take either two registers or a register and a sign extended immediate, operate on them and stores the result into a third register.
2. **Load and store Instructions :** These instructions take a register source called the base register and an immediate field (16 bits in MIPS), called the offset, as operands. The effective address derived from the sum of base and offset is used as a memory address. In the case of a load instruction, a second register operand acts as the destination for the data loaded from memory. In the case of a store, the second register operand is the source of the data that is stored into memory.
3. **Branches and Jumps :** Branches are conditional transfers of control. There are usually two ways of specifying the branch condition: one with a set of condition bits or by a limited set of comparisons between a pair of registers or between a register and zero. Mips uses comparison of registers.

A five stage pipeline is shown in Fig..7.14.

Pipeline Stages

➡ Instruction Fetch
➡ Instruction Decode
➡ Execute
➡ Memory access
➡ Write back

Time cycle	1	2	3	4	5	6	7	8	9
Instruction *i*	IF	ID	EX	MEM	WB				
Instruction *i* + 1		IF	ID	EX	MEM	WB			
Instruction *i* + 2			IF	ID	EX	MEM	WB		
Instruction *i* + 3				IF	ID	EX	MEM	WB	
Instruction *i* + 4					IF	ID	EX	MEM	WB

Fig. 7.14. Five stage pipeline.

In order to implement a hardware that performs operations as shown above it is necessary to take care of certain things,

1. We don't perform two different operations (or instructions) on the same data path resource (a register or a memory location for example) in the same clock cycle.
2. Instructions in the pipeline do not use the hardware resources at the same time.
3. Fast address calculation and transfer of control for branch instructions.

Based upon these simplified criterions one can start developing a pipeline implantation. However it is easier said than done as there are many issues that would require solution depending upon the instruction set of a processor. As in the case of an unpiplelined processor one can start the design by building the Data path of the processor. Many methods are used in the design of pipeline architecture and readers are requested to consult proper books for the same as the scope of this book is restricted in this regard.

Pipelined Control

Control for a pipelined processor is quite similar to the control for the single cycle CPU: it's just combinational logic. We can do this because as soon as we know what type of instruction we're executing, we know how the control signals need to be set for each stage of its execution.

The trick is to generate all the control signals for each instruction during decode, and to push the values of the control signals through the pipeline. This ensures that each instruction knows how its control signals need to be set for each stage of its execution.

Pipeline Timing Diagrams

A typical pipeline timing diagram (Fig. 7.15) looks something like this [f=fetch, d = decode, x = execute, m = memory, w = writeback] :

Instruction	cycle 1	cycle 2	cycle 3	cycle 4	cycle 5	cycle 6	cycle 7	cycle 8	cycle 9
Add \$1, \$2, \$3	f	d	x	m	w				
Sub \$4, \$10, \$20		f	d	x	m	w			
And \$6, \$16, \$11			f	d	x	m	w		
Lw \$8, 4(\$11)				f	d	x	m	w	
or \$10, \$18, \$11					f	d	x	m	w

Fig. 7.15. Pipeline timing diagram.

These timing diagrams carry a whole lot of information. They can be a bit confusing, and requires close attention. Each row of the diagram indicates when each instruction goes through each pipeline stage. So, if we read across the row marked "Sub \$4, \$10, \$20", we see that the subtract instruction is fetched in cycle 2, it decodes in cycle 3, it executes in cycle 4, etc.

Each column of the diagram shows which pipeline stages are in use at any instant in time, and it tells us which instructions are using those pipeline stages. As we read down the column marked "cycle 3", we know that in the third cycle, the add instruction is in the execute stage, the subtract instruction is in the decode stage, and the "and" instruction is in the fetch stage.

7.9. PIPELINE HAZARDS

Pipeline hazards are situations that prevent the next instruction in the instruction stream from executing during its designated clock cycle. The instruction is said to be stalled. When an instruction is stalled, all instructions later in the pipeline than the stalled instruction are also stalled. Instructions earlier than the stalled one can continue. No new instructions are fetched during the stall.

Types of Hazards

1. Structural hazards
2. Data hazards
3. Control hazards

7.9.1. Structural Hazards

Structural hazards occur when a certain resource (memory, functional unit) is requested by more than one instruction at the same time.

Instruction ADD R4,X fetches in the FO stage operand X from memory. The memory doesn't accept another access during that cycle.

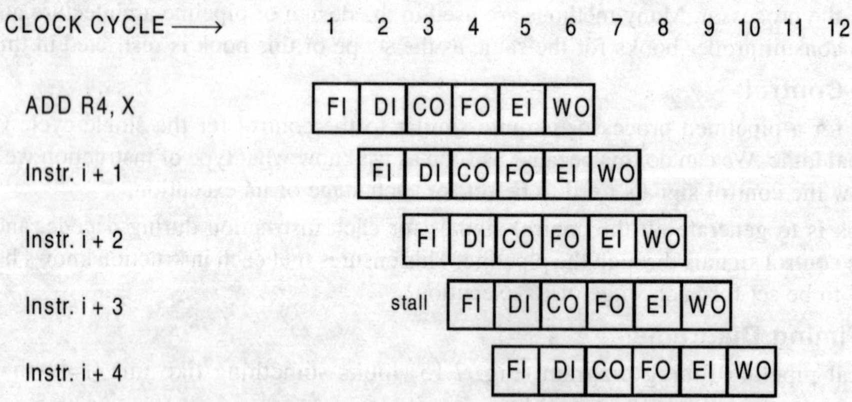

Certain resources are duplicated in order to avoid structural hazards. Functional units (ALU, FP unit) can be pipelined themselves in order to support several instructions at a time. A classical way to avoid hazards at memory access is by providing separate data and instruction caches.

7.9.2. Data Hazards

Data hazards occur when our processor tries to read data before the data has been written. Note that all data hazards can be resolved by waiting. We didn't have to worry about data hazards in our multiple cycle CPU - and we can turn our pipelined processor into a multiple cycle CPU by inserting four useless instructions ["no-ops"] in between every pair of adjacent instructions.

TIPS

We have two instructions, I1 and I2. In a pipeline the execution of I2 can start before I1 has terminated. If in a certain stage of the pipeline, I2 needs the result produced by I1, but this result has not yet been generated, we have a data hazard.

Let's look at the following code sequence:

```
add $1, $2, $3
sub $4, $1,$2
and $6, $6, $1
Iw $8, 4($1)
or $10, $8, $1
```

We have three data hazards: Read after Write (RAW), Write after Read (War) and Write after Write (WAW). The easiest way to identify them is to draw a pipeline timing diagram showing all the instructions moving through the pipeline normally:

Instruction	cycle 1	cycle 2	cycle 3	cycle 4	cycle 5	cycle 6	cycle 7	cycle 8	cycle 9
Add $1, $2, $3	f	d	x	m	w				
Sub $4, $10, $20		f	d	x	m	w			
And $6, $16, $11			f	d	x	m	w		
Lw $8, 4($11)				f	d	x	m	w	
or $10, $18, $11					f	d	x	m	w

Fig. 7.16. Finding data hazards.

Now we look for the trouble.

1. The add instruction will write $1 in cycle 5, but the sub instruction will read $1 in cycle 3.
2. The add instruction will write $ 1 in cycle 5, but the "and" instruction will read $1 in cycle 4
3. The 1w instruction will write $8 in cycle 8, but the "or" instruction will read $8 in cycle 6.

What can we do about these problems? The first two problems can be resolved by forwarding. Let's look at the first problem again. The add instruction won't write the value of $1 into the register file until cycle 5, but the add instruction actually computes the value of $1 in cycle 3. When it actually does the addition. The sub instruction doesn't really need the value of $1 until cycle 4, when it actually does the subtraction. Therefore, if we can forward the value computed by the add instruction in cycle 3 to the sub instruction in cycle 4. We can eliminate this hazard.

Similarly, the "and" instruction doesn't really need the value of $1 until cycle 5, when it actually does the "and" operation. So if we can forward the value computed by the add instruction in cycle 3 to the "and" instruction in cycle 5, we can eliminate this hazard.

The third problem is a bit trickier. The data read from memory by the Iw instruction won't be available until the end of cycle 7, at the earliest. But the "or" instruction needs this data at the beginning of cycle 7. There's no way to resolve this hazard by forwarding... the "or" instruction must wait one cycle for the data to be read from memory. This results in a pipeline stall.

7.9.3. Control Hazards

Branch instructions can cause a lot of trouble for pipelined processors. Like any other instruction, they can cause data hazards, since they require register values. They also introduce a new type of hazard:

the control hazard, we don't know which instructions to execute after the branch until we know whether the branch was taken or not. Consider the following piece of code [no branch delay slots for now]:

```
Iw $6, 0($10)
Iw $7,0($11)
add $1, $2, $3
beq$0, $1, label
sub $2, $3, $4
```

label :

```
Or $3, $4, $5
```

We can't push instructions into our pipeline until we know whether the branch is taken or not, because we don't know which instructions we're supposed to be executing. In the example above, until we know the outcome of the branch instruction, we don't know whether we should be fetching the subtract instruction, or the 'or' instruction.

TIPS

With conditional branch we have a penalty even if the branch has not been taken. This is because we have to wait until the branch condition is available.

As we discussed earlier, all hazards can be resolved by waiting. How long do we have to wait before we can start fetching instructions after a branch? In other words, how long does it take for us to figure out if the branch was taken or not?

We can reduce the branch delay by moving the branch-related data path logic (logic to determine taken/not taken and to calculate branch target address) into the decode stage. If we make this change, how long do we have to wait before we can start fetching instructions after a branch?

Assume we have moved branch logic into the decode stage. Draw a pipeline timing diagram for the code fragment above, assuming that the branch is taken. Don't forget about data hazards.

7.10. VECTOR PROCESSOR

Vector processor is a coprocessor specially designed to perform vector computations. A vector instruction involves a large array of operands. In other words, the same operation will be performed over a string of data. Vector processors are often used in a multilined supercomputer.

A vector processor can assume either a register-to-register architecture or a memory-to-memory architecture. The former uses shorter instructions and vector register files. The latter uses memory-based instructions which are longer in length, including memory address.

- A vector processor is a coprocessor designed to perform vector computations.
- A vector is a one-dimensional array of data items (each of the same data type).
- Vector processors are often used in multipipelined supercomputers.
- Architectural types include:
  - (*i*) register-to-register (with shorter instructions and register files)
  - (*ii*) memory-to-memory (longer instructions with memory addresses)

(*i*) **Register to Register Vector instruction**

- Assume V_i is a vector register of length n, s_i is a scalar register, M(1:n) is a memory array of length n, and "o" is a vector operation.

➡ Typical instructions include the following:

V_1 o $V_2 \rightarrow V_3$(element by element operation)

s_1 o $V_1 \rightarrow V_2$ (scaling of each element)

V_1 o $V_2 \rightarrow s_1$(binary reduction — i.e. sum of products)

$M(1{:}n) \rightarrow V_1$(load a vector register from memory)

$V_1 \rightarrow M(1{:}n)$(store a vector register into memory)

o $V_1 \rightarrow V_2$ (unary vector — i.e. negation)

o $V_1 \rightarrow s_1$(unary reduction — i.e. sum of vector)

(*ii*) Memory to Memory Vector instruction

Typical memory-to-memory vector instructions include these:

(*a*) $M_1(1{:}n)$ o $M_2(1{:}n) \rightarrow M_3(1{:}n)$ (binary vector)

(*b*) s_1 o $M_1(1{:}n) \rightarrow M_2(1{:}n)$ (scaling)

(*c*) o $M_1(1{:}n) \rightarrow M_2(1{:}n)$ (unary vector)

(*d*) $M_1(1{:}n)$ o $M_2(1{:}n) \rightarrow M(k)$ (binary reduction)

Pipelining in Vector Processor

Vector processors take advantage of unrolled-loop-level parallelism. The vector pipelines can be attached to any scalar processor, whether it is superscalar, super pipelined, or both.

Dedicated vector pipelines will eliminate some software overhead in looping control. Of course, the effectiveness of a vector processor relies on the capability of an optimizing compiler that vectorizes sequential code for vector pipelining.

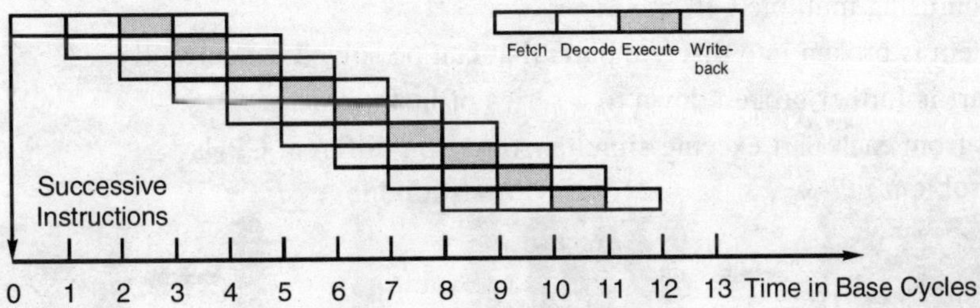

Scalar pipeline execution

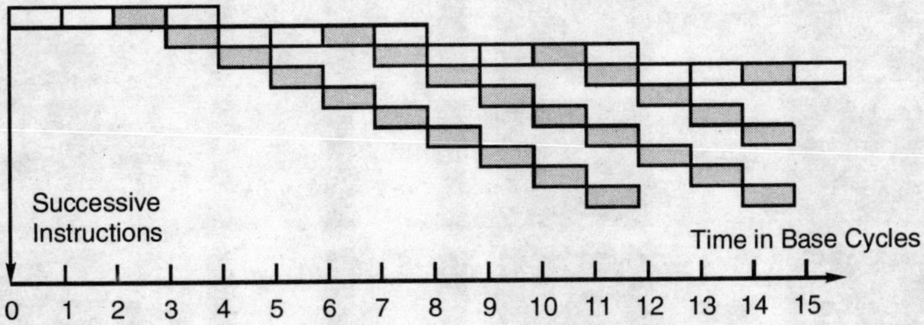

Vector pipeline execution

Pipelined execution in a base scalar processor and in a vector processor, respectively

Fig. 7.17

The pipelined execution in a vector processor is compared with that in a scalar processor in. Each scalar instruction executes only one operation over one data element. Each vector instruction executes a string of operations, one for each element in the vector.

7.11. INTRODUCTION TO PARALLEL COMPUTING

Traditionally, software has been written for ***serial*** computation:

- To be run on a single computer having a single Central Processing Unit (CPU);
- A problem is broken into a discrete series of instructions.
- Instructions are executed one after another.

Only one instruction may execute at any moment in time as shown in Fig. 7.18.

In the simplest sense, ***parallel computing*** is the simultaneous use of multiple compute resources to solve a computational problem.

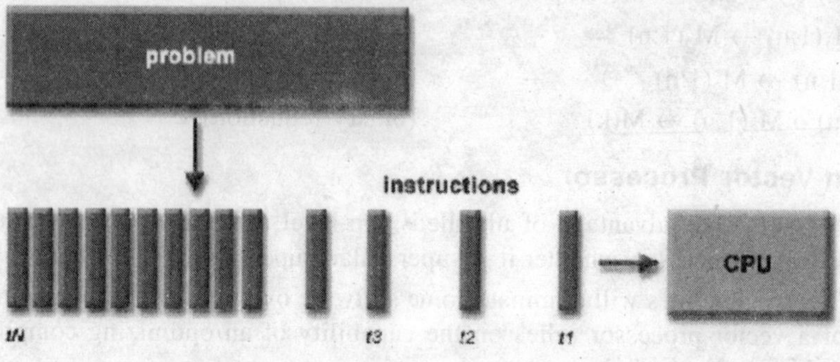

Fig. 7.18. Serial computation.

- To be run using multiple CPUs
- A problem is broken into discrete parts that can be solved concurrently
- Each part is further broken down to a series of instructions

Instructions from each part execute simultaneously on different CPUs

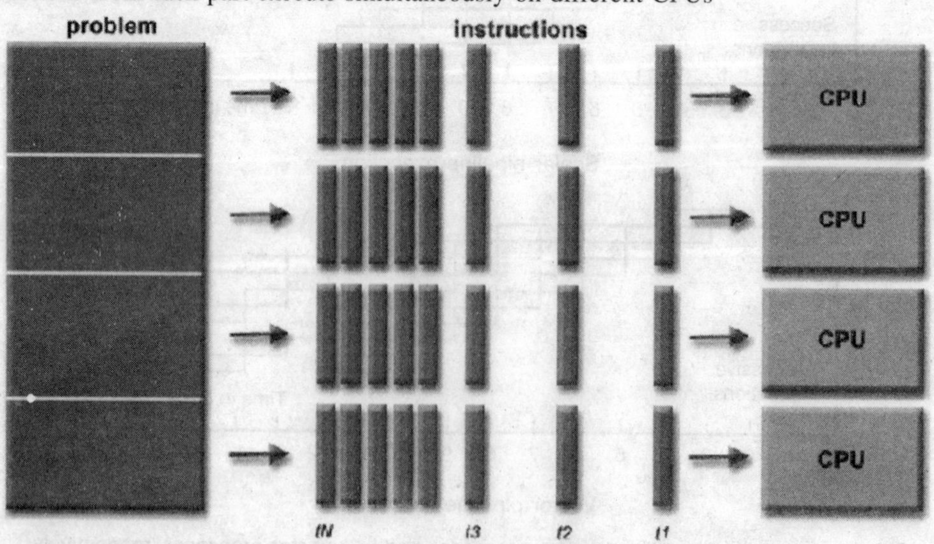

Fig. 7.19. Parallel computation.

The compute resources can include:

- A single computer with multiple processors;
- An arbitrary number of computers connected by a network;
- A combination of both.

> **Do You Know**
>
> *Parallel computing is an evolution of serial computing that attempts to emulate what has always been the state of affairs in the natural world: many complex, interrelated events happening at the same time, yet within a sequence.*

The computational problem usually demonstrates characteristics such as the ability to be:

- Broken apart into discrete pieces of work that can be solved simultaneously;
- Execute multiple program instructions at any moment in time;
- Solved in less time with multiple compute resources than with a single compute resource.

Some examples are as follows :

- Planetary orbits, galaxy formation
- Weather and ocean patterns
- Tectonic plate drift
- Rush hour traffic in LA
- Automobile assembly line
- Daily operations within a business
- Building a shopping mall
- Ordering a hamburger at the drive through.

We can say that there are **three reasons for using parallel computing** that are as

(*a*) The primary reasons for using parallel computing:
  - Save time - wall clock time
  - Solve larger problems
  - Provide concurrency (do multiple things at the same time)

(*b*) The second reasons might include:
  - Taking advantage of non-local resources - using available compute resources on a wide area network, or even the Internet when local compute resources are scarce.
  - Cost savings - using multiple "cheap" computing resources instead of paying for time on a supercomputer.
  - Overcoming memory constraints - single computers have very finite memory resources. For large problems, using the memories of multiple computers may overcome this obstacle.

(*c*) The third reason for using parallel computing is limits to serial computing . It means that both physical and practical reasons pose significant constraints to simply building ever faster serial computers:
  - *Transmission speeds :* The speed of a serial computer is directly dependent upon how fast data can move through hardware. Absolute limits are the speed of light (30 cm/nanosecond) and the transmission limit of copper wire (9 cm/nanosecond). Increasing speeds necessitate increasing proximity of processing elements.
  - *Limits to miniaturization* : The processor technology is allowing an increasing number of transistors to be placed on a chip. However, even with molecular or atomic-level components, a limit will be reached on how small components can be.
  - *Economic limitations* : It is increasingly expensive to make a single processor faster. Using a larger number of moderately fast commodity processors to achieve the same (or better) performance is less expensive.

So parallel computing is a form of computing in which many instructions are carried out simultaneously. Parallel computing operates on the principle that large problems can almost always be divided into smaller ones, which may be carried out concurrently ("in parallel"). Parallel computing has been used for many years, mainly in high performance computing, but interest in it has become greater in recent years due to physical constraints preventing frequency scaling. Parallel computing has recently become the dominant paradigm in computer architecture.

7.12. NEED FOR PARALLEL COMPUTING

There are various regions for using parallel computing. Some of them are given below :

(*i*) There are various hardware and software techniques that can be used in improving the degree of performance in uniprocessing systems.

(*ii*) To refine physical models such as those of complex molecules and crystals, to verify new theories and to solve newer problems of science we need computers which can perform several billions of arithmetic calculations per second. This requires parallel computers (or Supercomputers).

(*iii*) To model global weather forecasts, we need large set of non-linear partial differential equations. Hundreds of mega-flops speed and large memory size have been found essential to solve such real-time weather forecast equations.

(*iv*) To model interaction between atoms and understand micro behaviour of materials, quantum and structural chemists need high speed computing.

There are lots of problems become non-computatable when the size of the problem becomes too large to fit into computers power either in terms of its memory or speed. There applications require the speed of the order 3000 millions floating point operations per second (megaflops) which is unmatched with the available technology of single microprocessor. So, we require fast computers which can address these problems.

It should be noted that there may be a subset of problems which can still be computable on the present-day systems but they require a substantial amount of time which may be of the order of months. If we could wait for the output for so long then we don't need fast computer. But then again, if the application is of let's say weather forecasting, then the data received about currents, storms and wind motion need to be calculated and predicted so that it can be posted on news, then the conventional setup won't help.

7.13. PARALLEL PROCESSING

Parallel processing is defined as the process of information processing that emphasis the concurrent manipulation of data elements belonging to one or more problems solving a single problem.

7.13.1. Multiprocessor and Multicomputer

A multiprocessor is a single computer in which multiple processors exists. The complete system comprises of many (multiple) processors. The processors are interconnected through an Interconnection Network. They share common resources like memory and I/O devices. They have just one O.S. Even through many computer exist, it behaves as a single computer only. Whereas in a multicomputer system, multiple autonomous computers are interconnected by the network.

Shared memory MIMD computers are called, **multiprocessors:**

- Some multiprocessors have no shared memory which is central to the system and equally accessible to all processors. All the memory is distributed as local memory to the processors. However, each processor has access to the local memory of any other processor ⇒ a global physical address space is available.

This memory organization is called distributed shared memory.

- Communication between processors is through the shared memory. One processor can change the value in a location and the other processors can read the new value.
- From the programmers point of view communication is realised by shared variables; these are variables which can be accessed by each of the parallel activities (processes):
- With many fast processors memory contention can seriously degrade performance ⇒ multiprocessor architectures don't support a high number of processors.

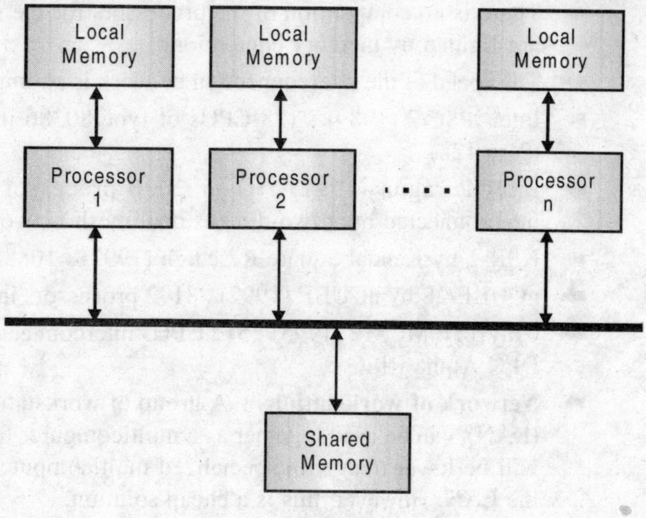

Fig. 7.20.

- IBM system/370 (1970s) : two IBM CPUs connected to shared memory.

  IBM system 370/XA (1981) : Multiple CPUs can be connected to shared memory.

  IBM System/390 (1990s) : Similar features like S370/XA, with improved performance. Possibility to connect several multiprocessor systems together through fast fibre-optic connection.
- Cray X-MP (mid 1980s) : From one to four vector processors connected to shared memory (cycle time : 8.5 ns).

  CRAY Y-MP (1988) : From one to 8 vector processors connected to shared memory; 3 times more powerful than CRAY X-MP (cycle time : 4 ns). C90 (early 1990s): further development of CRAY Y-MP; 16 vector processors.

  CRAY 3 (1993) : Maximum 16 vector processors (cycle time 2 ns).
- Butterfly multiprocessor system, by BBN Advanced Computers (1985/87) : Maximum 256 Motorola 68020 processors, interconnected by a sophisticated dynamic switching network; distributed shared memory organization BBN TC2000 (1990) : improved version of the Butterfly using Motorola 88100 RISC processor.

Multicomputers

MIMD computers with a distributed address space, so that each processor has its one private memory which is not visible to other processors, are called **multicomputers :**

- Communication between processors is only by passing messages over the interconnection network.
- From the programmers point of view this means that no shared variables are available (a variable can be accessed only by one single process). For communication between parallel activities (processes) the programmer uses channels and send/receive operations (see program in slide 10).

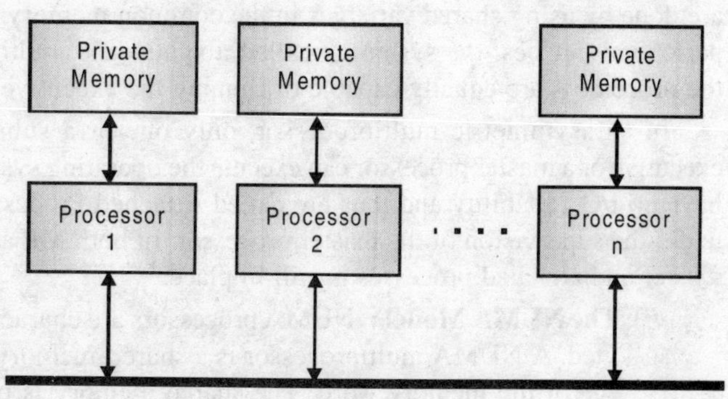

Fig. 7.21.

- There is no competition of the processors for the shared memory Þ the number of processors is not limited by memory contention.
- The speed of the interconnection network is an important parameter for the overall performance.
- Intel iPSC/2 (1989) : 128 CPUs of type 80386 inter-connected by a 7-dimensional hypercube ($2^7 = 128$).
- Intel Paragon (1991) : Over 2000 processors of type i860 (high performance RISC) interconnected by a two-dimensional mesh network.
- KSR-1 by Kendal Square Research (1992) : 1088 processors interconnected by a ring network.
- nCUBE/2S by nCUBE (1992) : 8192 processors interconnected by a 10-dimensional hypercube.
- Cray T3E MC512 (1995) : 512 CPUs interconnected by a tree-dimensional mesh; each CPU is a DEC Alpha RISC.
- **Network of workstations :** A group of workstations connected through a Local Area Network (LAN), can be used together as a multicomputer for parallel computation. Performances usually will be lower than with specialized multicomputers, because of the communication speed over the LAN. However, this is a cheap solution.

Now we will discuss three shared memory models that are as :

(*i*) The Uniform-Memory-Access (UMA) model.

(*ii*) The Nonuniform-Memory-Access (NUMA) model.

(*iii*) The Cache-Only-Memory-Architecture (COMA) model.

These models differ in how the memory and peripheral resources are shared or distributed.

(i) The UMA Model : In a UMA multiprocesosr model as shown in Fig. 7.22, the physical memory is uniformly shared by all the processors. All processors have equal access time to all memory words, and so the name **uniform memory access.** Each processor may use a private cache. Peripherals are also shared in the same manner. Multiprocessor (MP) systems are also called tightly coupled systems due to the high degree of resource sharing. The systems can be interconnected in any of the manner – a common bus, a crossbar switch or a multistage network. The UMA model is suitable for general-purpose and time sharing applications by multiple users. It can be used to speed up the execution of a single large program in time-critical applications. To coordinate parallel events, synchronization and communication among processors are done by using shared variables in the common memory. When all processors have equal access to all peripheral devices, the system is called a symmetric multiprocessor. In symmetric multiprocessors, all the processors are equally capable of running the executive programs—OS Kernel, I/O service routines.

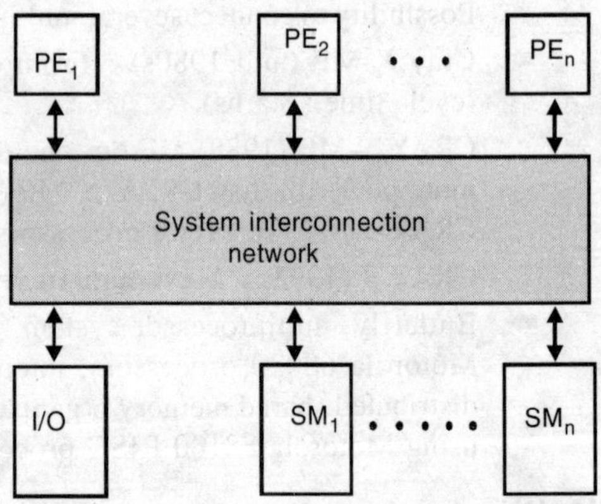

Fig. 7.22. The UMA multiprocessor model.

In an asymmetric multiprocessor, only one or a subset of processors are executive-capable. An executive or a master processor can execute the operating system and handle I/O. The remaining processors have no I/O capability and thus are called Attached Processors. Attached processor executes user codes under the supervision of the master processor. In both MP and AP configuration, memory sharing among master and attached processes is still in place.

(ii) The NUMA Model : NUMA processors are characterized by a shared address space and memory ...uted. A NUMA multiprocessor is a shared-memory system in which the access time varies with ... of the memory word. The shared memory is physically distributed to all processors called

local memories. The collection of all local memories forms a global address space accessible by all processors. It is faster to access a local memory with a local processor. The access of remote memory attached to other processors takes longer due to the added delay through the interconnection network. Besides distributed memories, globally shared memory can be added to a multiprocessor system. There are three memory access patterns :

(*a*) Local memory access (fastest)

(*b*) Global memory access.

(*c*) Remote memory access (slowest).

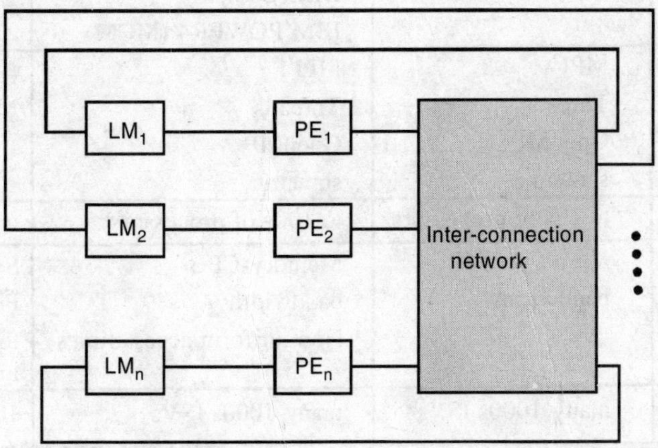

Fig. 7.23. Shared local memories model of NUMA multiprocesosrs System.

A hierarchically structured multiprocessor is modeled in Fig. 2.6. The processors are divided into several clusters. Each cluster is itself an UMA or NUMA multiprocessor. The clusters are connected to global shared memory modules. The entire system is considered a NUMA multiprocessor. All processors belonging to the same cluster are allowed to uniformly access the cluster shared-memory modules. All clusters have equal access to the global memory. The access time to the cluster memory is shorter than the global memory.

***(iii)* The COMA Model :** In COMA model, a multiprocessor using cache-only memory is assumed. The COMA model is a special case of a NUMA machine, in which the distributed main memories are converted to caches. There is no memory hierarchy at each processor node. All the caches forms a global address space. Remote cache access is assisted by the distributed cache directories denoted by D in Fig. 7.24. Depending on the interconnection network used, sometims hierarchical directories may be used to locate copies of cache blocks. Initial data placement is not critical because data will eventually migrate to where it will be used. Example, the KSR-1.

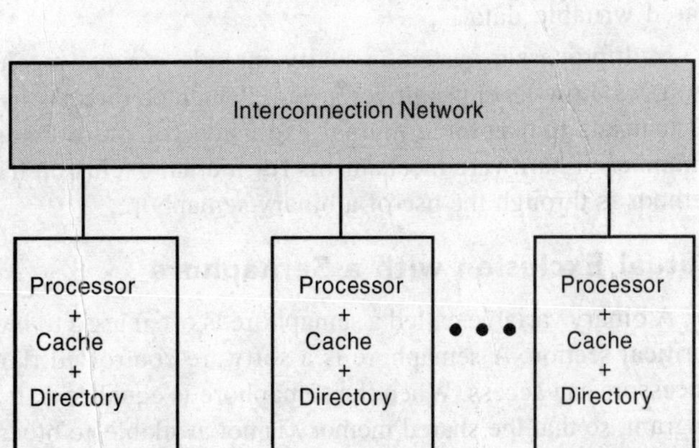

Fig. 7.24. The COMA Model of multiprocessor.

Comparison of Shared and Distributed Memory Architectures

The Table 7.1 shows the comparison of shared and distributed memory architectures.

Table 7.1.

Architecture	CC-UMA	CC-NUMA	Distributed
Examples	SMPs Sun Vexx DEC/Compaq SGI Challenge IBM POWER3	SGI Origin Sequent HP Exemplar DEC/Compaq IBM POWER4 (MCM)	Cray T3E Maspar IBM SP2
Communications	MPI Threads OpenMP shmem	MPI Threads OpenMP shmem	MPI
Scalability	to 10s of processors	to 100s of processors	to 1000s of processors
Draw Backs	Memory-CPU bandwidth	Memory-CPU bandwidth Non-uniform access times	System administration Programming is hard to develop and maintain
Software Availability	many 1000s ISVs	many 1000s ISVs	100s ISVs

7.14. INTERPROCESSOR COMMUNICATION AND SYNCHRONIZATION

The instruction set of a multiprocessor contains basic instructions that are used to implement communication and synchronization between cooperating processes. Communication refers to the exchange of data between different processes. For example, parameters passed to a procedure in a different processor constitute interprocessor communication. Synchronization refers to the special case where the data used to communicate between processors is control information. Synchronization is needed to enforce the correct sequence of processes and to ensure mutually exclusive access to shared writable data.

Multiprocessor systems usually include various mechanisms to deal with the synchronization of resources. Low-level primitives are implemented directly by the hardware. These primitives are the basic mechanisms that enforce mutual exclusion for more complex mechanisms implemented in software. A number of hardware mechanisms for mutual exclusion have been developed. One of the most popular methods is through the use of a binary semaphore.

Mutual Exclusion with a Semaphore

A binary variable called a semaphore is often used to indicate whether or not a processor is executing a critical section. A semaphore is a software-controlled flag that is stored in a memory location that all processors can access. When the semaphore is equal to 1, it means that a processor is executing a critical program, so that the shared memory is not available to other processors. When the semaphore is equal to 0, the shared memory is available to any requesting processor. Processors that share the same memory segment agree by convention not to use the memory segment unless the semaphore is equal to 0, indicating that memory is available. They also agree to set the semaphore to 1 when they are executing a critical section and to clear it to 0 when they are finished.

A semaphore can be initialized by means of a test and set instruction in conjunction with a hardware lock mechanism. A hardware lock is a processor- generated signal that serves to prevent other processors from using the system bus as long as the signal is active. The test-and-set instruction tests and sets a semaphore and activates the lock mechanism during the time that the instruction is being executed. This prevents other processors from changing the semaphore between the time that the processor is testing it and the time that it is setting it. Assume that the semaphore is a bit in the least significant position of a memory word whose address is symbolized by SEM. Let the mnemonic TSL designate the "test and set while locked" operation. The instruction

TSL SEM

will be executed in two memory cycles (the first to read and the second to write) without interference as follows:

$R \leftarrow M[SEM]$ TEST SEMAPHORE

$M[SEM] \leftarrow 1$ SET SEMAPHORE

The semaphore is tested by transferring its value to a processor register R and then it is set to 1. The value in R determines what to do next. If the processor finds that R = 1, it knows that the semaphore was originally set (the fact that it is set again does not change the semaphore value). That means another processor is executing a critical section, so the processor that checked the semaphore does not access the shared memory. R = 0 means that the common memory (or the shared resource that the semaphore represents) is available. The semaphore is set to 1 to prevent other processors from accessing memory. The processor can now execute the critical section. The last instruction in the program must clear location SEM to zero to release the share resource to other processors.

7.15. ARRAY PROCESSOR

An array processor is a processor that performs computations on large arrays on data. The term is used to refer to two different types of processors. An attached array processor is an auxiliary processor attached to a general-purpose computer. It is intended to improve the performance of the host computer in specific numerical computation tasks. An SIMD array processor is a processor that has a single-instruction multiple-data organization. It manipulates vector instructions by means of multiple functional units responding to a common instruction. Although both types of array processors manipulate vectors, their internal organization is different.

7.15.1. Attached Array Processor

An attached array processor is designed as a peripheral for a conventional host computer, and its purpose is to enhance the performance of the computer by providing vector processing for complex scientific applications. It achieves high performance by means of parallel processing with multiple functional units. It includes an arithmetic unit containing one or more pipelined floating-point adders and multipliers. The array processor can be programmed by the user to accommodate a variety of complex arithmetic problems.

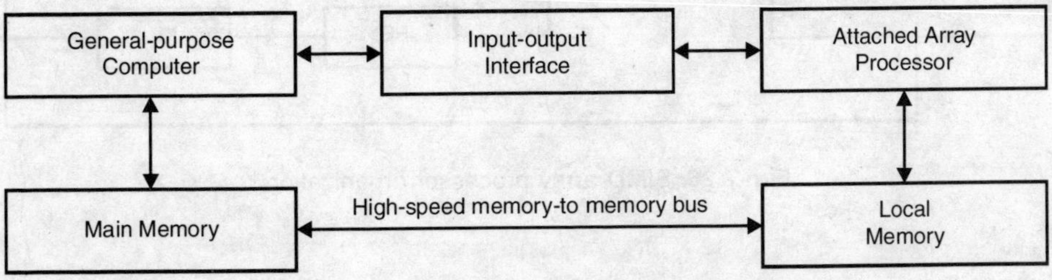

Fig. 7.25. Attached array processor with host computer.

Figure 7.25 shows the interconnection of an attached array processor to a host computer. The host computer is a general-purpose commercial computer and the attached processor is a back-end machine driven by the host computer. The array processor is connected through an input-output controller to the computer and the computer treats it like an external interface. The data for the attached processor are transferred from main memory to a local memory through a high-speed bus. The general-purpose computer without the attached processor serves the users that need conventional data processing. The system with the attached processor satisfies the needs for complex arithmetic applications.

7.15.2. SIMD Array Processor

An SIMD array processor is a computer with multiple processing units operating in parallel. The processing units are synchronized to perform the same operation under the control of a common control unit, thus providing a single instruction stream, multiple data stream (SIMD) organization. A general block diagram of an array processor is shown in Fig. 7.26. It contains a set of identical processing element (PEs), each having a local memory M. Each processor element includes an ALU, a floating-point arithmetic unit, and working registers. The master controlunits controls the operations in the processor elements. The main memory is used for storage of the program. The function of the master control unit is to decode the instructions and determine how the instruction is to be executed. Scalar and program control instructions are directly executed within the master control unit. Vector instructions are broadcast to all PEs simultaneously. Each PE uses operands stored in its local memory. Vector operands are distributed to the local memories prior to the parallel execution of the instruction.

Consider, for example, the vector addition $C = A + B$. The master control units first stores the ith components a_i and b_i of A and B in local M_i for $i = 1, 2, 3, ..., n$. It then broadcasts the floating-point add instruction $c_i = a_i + b_i$ to all PEs, causing the addition to take place simultaneously. The components of c_i are stored in fixed locations in each local memory. This produces the desired vector sum in one add cycle.

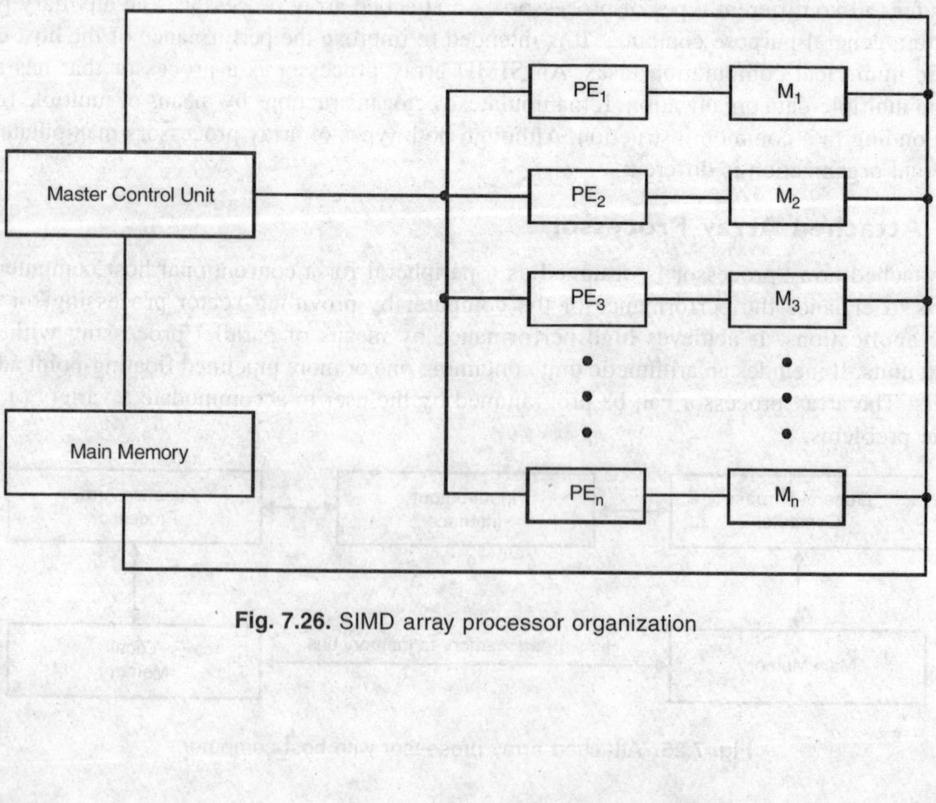

Fig. 7.26. SIMD array processor organization

Summary

- A linear Pipeline processor is a cascade of processing stages which are linearly connected to perform a fixed function over a stream of data flowing from one end to other. In modern computers, linear pipelines are applied for instruction execution, arithmetic computation, memory access operations.
- Data flow between adjacent stages in asynchronous pipeline is controlled by handshaking protocol. When stage S_1 is ready to transmit, it sends a ready signal to S_{i+1}.
- A dynamic pipeline can be reconfigured to perform variable functions at different times. The traditional linear pipelines are static pipelines because they are used to perform fixed functions. A dynamic pipeline allows feed forward and feedback connections in addition to the streamline connections. For this reason, we call such a structure as non-linear pipeline.
- The principles used in instruction pipelining can be used in order to improve the performance of computers in performing arithmetic operations such as add, subtract and multiply. In this case, these principles will be used to realize the arithmetic circuits inside the ALU.
- A stream of instructions can be executed by pipeline in an overlapped manner. A typical instruction execution consists of a sequence of operations, including

  (1) Instruction fetch

  (2) Decode

  (3) Operand fetch

  (4) Execute

  (5) Write back phases
- In order to understand pipeline implementation, we need to take a simple processor which performs instruction execution in one cycle. Typically any RISC processor is able to execute one instruction in one cycle.
- Pipeline hazards are situations that prevent the next instruction in the instruction stream from executing during its designated clock cycle. The instruction is said to be stalled.
- Vector processor is a coprocessor specially designed to perform vector computations. A vector instruction involves a large array of operands. In other words, the same operation will be performed over a string of data. Vector processors are often used in a multilined supercomputer.
- Parallel computing is the simultaneous use of multiple compute resources to solve a computational problem.
- Parallel processing is defined as the process of information processing that emphasis the concurrent manipulation of data elements belonging to one or more problems solving a single problem.
- All processors have equal access time to all memory words, and so the name uniform memory access.
- The shared memory is physically distributed to all processors called **local memories.**
- In COMA model, a multiprocessor using cache-only memory is assumed.
- Multiprocessor systems usually include various mechanisms to deal with the synchronization of resources. Low-level primitives are implemented directly by the hardware. These primitives are the basic mechanisms that enforce mutual exclusion for more complex mechanisms implemented in software. A number of hardware mechanisms for mutual exclusion have been developed. One of the most popular methods is through the use of a binary semaphore.

Review Questions

1. Define parallel computing ? Why parallel computing is required ?
2. Discuss various applications of parallel computing ?
3. Discuss various types of multiprocessor systems.
4. Write short note on multicomputers or NORMA.
5. Differentiate between UMA, NUMA, COMA and NORMA computers.
6. State and discuss various architectural classification schemes.
7. Discuss data dependences with the help of suitable example. Also define and draw the dependence graph.
8. Discuss the Berstein condition of parallelism.
9. Differentiate between hardware and software parallelism.
10. Write short notes on
    (*i*) Grain size and its type. (*ii*) Grain packing.
    (*iii*) Static multiprocessors scheduling with Node duplications.
11. Differentiate between control the data-flow and demand-driven machine models.
12. Discuss static and dynamic data flow architecture.
13. Discuss various parameters used for evaluation interconnection Networks.
14. Differentiate between static and dynamic interconnection networks ?
15. Write short note on
    (*i*) Dynamic Connection Networks. (*ii*) Cube-connected Cycles.
16. Analyze the data dependence among the following statements in a given program :
    I_1 : Load R1, 2014
    I_2 : Load R2, M(20)
    I_3 : Add R1, R2
    I_4 : Store M(2014), R1
    I_5 : Store M((R2)), 1024

❑❑❑

Appendices

Appendix A : RISC/CISC
Appendix B : PERFORMANCE EVALUATION

Appendix A

RISC/CISC

A.1. INTRODUCTION TO CISC

Aim of the computer architects is to design computers which are more cost effective than their predecessors. Cost effectiveness includes:

1. Cost of hardware to manufacture the machine
2. Cost of programming
3. Cost incurred in correcting both the initial hardware and subsequent programs on account of the new/old/compatible architecture.

If we review the history of computer families, we find that the changes were even more complex machine. A number of factors precipitated a return to RISC architectures:

- Designers discovered that compiler writers were not using all the complex instructions that they provided. Compiler writers discovered that they could often get better performance by using sequences of simpler, faster instructions than the complex, slow ones that were provided.
- Simpler instruction sets allowed simpler, faster logic. This became important as technology allowed much higher clock frequencies to be used.
- Compiler technology improved: optimising compilers started to analyse program and data flow so that they could re-order instructions or choose better sequences of instructions for increased throughput.
- The cost of designing a high-performance processor became a significant investment for any company. CISC processors had complex, expensive-to-design control logic. The regular logic of simpler processors was quicker and cheaper to design - and cheaper to verify.

A number of factors precipitated a return to RISC architectures:

Designers discovered that compiler writers were not using all the complex instructions that they provided. Compiler writers discovered that they could often get better performance by using sequences of simpler, faster instructions than the complex, slow ones that were provided.

Simpler instruction sets allowed simpler, faster logic. This becamc important as technology allowed much higher clock frequencies to be used.

Compiler technology improved: optimising compilers started to analyse program and data flow so that they could re-order instructions or choose better sequences of instructions for increased throughput.

The cost of designing a high-performance processor became a significant investment for any company. CISC processors had complex, expensive-to-design control logic. The regular logic of simpler processors was quicker and cheaper to design - and cheaper to verify.

A.2. CPU CATEGORIES

Some systems have machine languages that include single instructions that can perform multiple actions (so-called "Complex Instruction Set Computers," CISC) and others have machine languages that include only single-action instructions (so-called "Reduced Instruction Set Computers," RISC). A RISC machine operating at a higher number of instructions per second may still be slower to get work accomplished, because, on the average, it takes more instructions to accomplish any specific task on a RISC machine.

In general, at any given cost level it is possible to design the electronics so that RISC CPUs can execute so many more instructions per second that they will outperform CISC designs of comparable technology. This includes the effects of the greater ease with which compilers can be written for the simpler RISC architectures. Even though, in principle, a CISC design would require fewer instructions to accomplish the same task, in practice, reliable compilers for CISC machines rarely actually generate machine language code that takes full advantage of those more powerful instructions.

A.3. REDUCED-INSTRUCTION-SET COMPUTERS

A computer architecture that reduces chip complexity by using simpler instructions. RISC compilers have to generate software routines to perform complex instructions that were previously done in hardware by CISC computers. In RISC, the microcode layer and associated overhead is eliminated. RISC keeps instruction size constant, bans the indirect addressing mode and retains only those instructions that can be overlapped and made to execute in one machine cycle or less. The RISC chip is faster than its CISC counterpart and is designed and built more economically.

The RISC machine executes instructions faster because it doesn't have to go through a microcode conversion layer. The RISC compiler does more work than the CISC compiler. It has to generate routines using simpler instructions that would normally be performed by a single, complex instruction in the CISC computer.

Definitions

The technical term RISC has been swamped by the marketing term RISC to the point that it's lost almost all meaning as a technical term. Almost everything now is described as RISC, even when it isn't. A historical perspective can help illustrate the point in the development of computer architectures which "RISC" was invented to describe.

(1) Accumulator Architecture : The first computers were accumulator based, performing operations on data stored in a register with data loaded from memory locations. Addresses were coded in program memory, and could only be changed by modifying the program. Initially machines only had one (which all operations implicitly refered to), but later multiple accumulators were sometimes used.

(2) Index Registers : Index registers were used to hold memory addresses, while accumulators were still used for computation. Operations could specify an accumulator for the one operand, and an index register to fetch the second operand from memory.

(3) Memory to Memory Architectures : With a little modification, index and accumulator registers could be used interchangably - they were now general purpose, meaning that a register could hold either data or an address. Some designs took this further, allowing the value in memory to be an indirect address itself - sometimes in an unlimited chain of references.

(4) Stack Architectures : General registers can be made even more flexible allowing them to hold multiple values (either data or addresses) in a useful order - in other words, stacks. At the same time, the design can be simplified back to a single stack (as an accumulator architecture) with no loss of flexibility, providing *swap* and *duplicate* operations are provided.

Because of the complexity of the VAX architecture, a "back to basics" movement started in early 80s. Mechanisms for simple instructions but efficient executions were sought.

We shall define a RISC as a computer that has most of the following properties

- Instructions are conceptually simple
- Instructions are of a uniform length
- Instructions use one (or very few) instruction formats
- The instruction set is orthogonal
- Instructions use one or very few addressing modes
- The architecture is a load-and-store architecture
- The architecture supports two or a few more datatypes

Let us see what are the reasons for increased complexity which resulted in complex instruction set computers (CISC) and what exactly we mean by this.

➡ **Speed of memory Vs. the speed of CPU :** The CPU, in general, were about 10 times as fast as core memory. Thus, more and more instructions were put into hardware. Since then many higher level instructions have been added to machines in an attempt to improve performance. This trend began towards the imbalance in speed.

The advances in semiconductor memory have made several changes to the assumptions about the relative difference in speed between CPU and the main memory. Semiconductor memories are both fast and relatively inexpensive. Cache memories have further reduced the difference between the CPU and the memory Speed.

➡ **Code density :** With early computers memory was very expensive. It was, therefore, cost effective to have very compact programs. It was assumed that more instructions will result in small programs. While code compaction is important, however, the cost of 10 percent more memory is often far cheaper than the cost of squeezing code by 10 percent out of the CPU architecture innovations.

➡ **Support for high level language** : With the increasing use of more and more high level languages, manufacturers have become eager to provide more powerful instructions to support them. It is argued that richer instruction set would alleviate the software crisis and would simplify the compilers. However, RISC designers did not agree to it. They conducted research to find out what constructs are most frequently used in average programs.

There research shows that integer constants appeared almost as frequently as arrays or structures, and more than 80 percent of the scalars were local variables to subroutines and more than 90 percent of the arrays or structure were global variables.

In addition, it was found that procedure calls and returns are an important aspect of High level language programs. These are also the most time consuming operations in complied high level language programs. Thus, it will be profitable to consider ways of implementing these operations efficiently. Two aspects are significant in this respect : the number of parameters and variables that procedure deals with and the depth of nesting.

Further, it has been statistically found that 98% of dynamically called procedures pass lower than six arguments, and 92% of them use fewer than six scalar variables.

Thus, in the 1980s, a new philosophy evolved called RISC. These machines would have fewer instructions-(reduced set) and the instructions would be simple and would generally execute in one cycle, hence, the reduced instruction set computers (RISCs). Optimising compilers are used in these machines to compile the programs into simple machine instructions.

A.3.1. Register to Register Architectures

Although conceptually powerful, memory to memory operations are limited in speed because memory access is relatively slow, while registers are fast. The obvious improvement is to increase the number of registers, and restrict or simplify memory access (limited to load and store operations) to simplify the overall design. The first major system to follow this idea was the CDC 6600, but the idea was also explored in the IBM 801 project. However, it wasn't until the Berkeley RISC design named this type of design a "Reduced Instruction Set Computer". The term CISC was also invented at this time solely to give a name to the prior generation (memory-memory designs).

A diagram can show the approximate lineage of processor architectures:

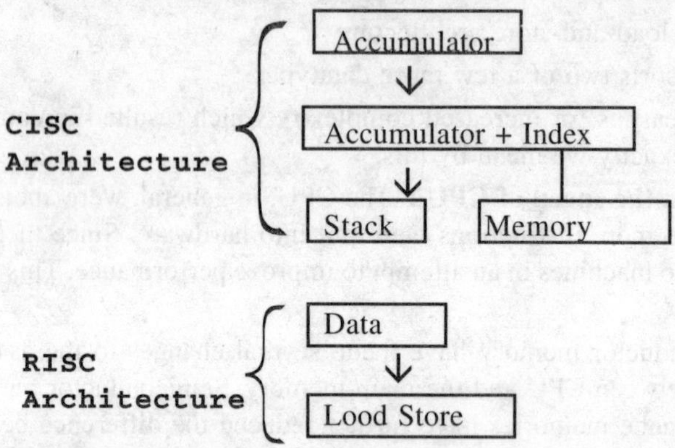

A.3.2. Characteristics of RISC Architecture

We shall first list out the various characteristics of RISC architecture

- Hardware Instructions should be kept simple unless there is a very good reason to do otherwise.
- Microcode, a programming way of implementing control unit, is not magic. Moving software into microcode does not make it better, it just makes it hard to change.
- Simple decoding and pipelined execution are more important than program size. Imagine a model in which the total work per instruction is broken into pieces, and different pieces for each instruction execute in parallel. At the peak rate a new instruction is started every cycle. Pipelined execution gives a peak performance of one instruction every step, so the longest piece determines the performance rate of the pipelined machine, so ideally each piece should take the same amount of time.
- The large number of registers should be utilized optimally.

A.3.3. Design of RISC Machines

As a first approximation, we can think of the Mic-2 as being RISC-like, but in reality the situation is more complicated than that. In this section we will discuss the technical design of RISC machines in detail. However, let us first present the basic RISC design philosophy in a nutshell. Designing a RISC machine has five steps.

- Analyze the applications to find the key operations.
- Design a data path that is optimal for the key operations.
- Design instructions that perform the key operations using the data path.
- Add new instructions only if they do not slow down the machine.
- Repeat this process for other resources.

The first step says the designer should find out what the intended programs actually do. For traditional algorithmic languages some statistics. However, for applications in COBOL, Smaltalk, or Lisp, one would have to start all over collecting statistical information about real programs in these languages. Similarly, when designing dedicated machines for banking, robotics, or computer-aided design applications, yet other information would be required.

The heart of any computer is the data path, which contains the registers, the ALU, and the buses connecting them. This circuit should be optimized for the language or application in question. The time required to fetch the operands from their registers, run them through the ALU, and store the result back into a register, called the DATA PATH CYCLE TIME, should be made as short as possible.

The next step is to design machine instructions that make good use of the data path. Only a few instructions and addressing modes are typically needed. Additional instruction should only be added if they will be frequently used and do not reduce the performance of the most important ones. RISC Golden Rule number **1** states: **"Sacrifice everything to reduce the data path cycle time."**

Whenever an attractive new feature rears its ugly head, it should be looked at in this light: how does it affect the data path cycle time? If it increases the cycle time, it is probably not worth having.

Finally the same process should be repeated for other resources within the CPU, such as cache memory, memory management, floating-point coprocessors, and so on. As Antonie de St. Expurey so aptly put it: "Perfection is achieved, not when there is nothing left to add, but when there is nothing left to take away."

A.4. CHARACTERISTICS OF RISC ARCHITECTURE

In brief, we can now say that the main characteristics of RISC architecture are as follows :

- One instruction per cycle
- Register-to-register operands
- Simple address modes
- Simple instruction formats

Machine cycle is defined to be the time taken to fetch two operands from registers, perform an ALU operation and store the result in a register. Thus, RISC machine instructions should be no more complicated, than, and execute as fast as,microinstructions on CISC machines. With simple, one cycle instructions, there is little or no need for microcode, the machine instructions can be hardwired. Such instructions should execute faster than comparable machine instructions on other machines, since it is not necessary to access a microprogram control store during the execution.

A second characteristic is that most operations should be register-to-register, with only simple LOAD and STORE operations accessing memory. This design feature simplifies the instruction set and, therefore,the control unit. For example, a RISC instruction set may include only one or two ADD instructions (e.g. integer add and add with carry); the VAX-11 has 25 add instructions. Another benefit is that such an architecture encourages the optimization of register use, so that frequently used operands remain in high speed memory.

The third characteristic is the use of simple addressing modes. Almost all instructions use simple register addressing mode. Several additional addressing modes such as displacement and PC relative may be included. Other more complex modes can be synthesized in software from the simple ones. Again, this design feature simplifies the instruction set and the control unit.

A final characteristic is the use of simple instruction formats. Generally, only one or a few formats are used. Instruction length is fixed and aligned on word boundaries. Fields location especially the opcode, are fixed. This design feature has a number of benefits. With fixed fields, opcode decoding and register operand accessing can occur simultaneously. Simplified formats simplify the control unit. Instruction fetching is optimized since word length units are fetched. This also means that single instruction does not cross page boundaries.

Taken together these characteristics can be accessed to determine the potential of RISC architecture. These potentials fall into two categories : those related to performance and those related to VLSI implementation.

With respect to performance, firstly more optimizing compilers can be developed. With primitive instructions there are more opportunities for moving functions out of loops, reorganizing code for efficiency, maximizing register utilization, and so forth. It is even possible to compute part of complex instruction during the compile time.

A second point, already noted, is that most instruction generated by a compiler are relatively simple anyway. It would seem reasonable that a control unit built specifically for those instructions are using little or no microcode could execute them faster than a comparable CISC. A third point relates to the instruction pipelining. Instruction pipelining technique can be applied more effectively with a reduced instruction set.

The second area of potential benefit, which is more clear cut related to VLSI implementation. When VLSI is used, the design and the implementation of the CPU are fundamentally changed. Traditional CPUs like VAX-11 consist of one or more printed circuit boards containing standardized MSI and SSI packages. With the advent of LSI and VLSI, it is possible to put an entire CPU on a single chip. For a single chip CPU, there are two motivations for following a RISC strategy. First, there is the issue of performance. On chip delays are of much shorter durations than inter-chip delays. Thus, it makes sense to have the rare functions built on a separate chip. RISC chips are mainly designed with this consideration in mind. Whereas a typical microprocessor dedicates about half of its area to the microcode control store, the RISC chip devotes only about 6% of its area to the control unit.

A second VLSI related issue is design and implementation time. A VLSI processor is difficult to develop. Instead of relying on available SSI and MSI parts, the designer must perform circuit design, layout, and modelling at the device level. With a reduced instruction set architecture, this process is far easier as evidenced by table A.1. If, in addition, the performance of RISC chip is equivalent to omparable CISC microprocessors, then the advantage of the RISC approach becomes evident.

Table A.1 : Design and layout efforts for some microprocessors

CPU	Transistors (thousands)	Design (person-months)	Layout (person-months)
RISC 1	44	15	12
RISC 11	41	18	12
M68000	6A	100	70
Z8000	18	60	70
inter iAPX-432	110	170	90

A.5. RISC IMPLEMENTATION TECHNIQUES

Major techniques used to implement RISC computers:

- RISC s use pipelining to speed up instruction decoding and execution (thus RISC s have separate functional units)
- RISC s usually do not allow program self-modification
- RISC s use Harvard architectures (separate instruction stream and bus stream)
- RISC s use large register set to reduce CPU-to-memory traffic

- RISC s often use delayed branches to avoid branch penalty, i.e. put off the instruction execution for the branch, fill in other instructions (if possible)
- RISC s use delayed loads to avoid waiting, i.e. insert an instruction that is independent of the load right after the load instruction and always executes it. (very similar idea to the use of delayed branches)
- Pre-fetch branch-target to reduce the branch delay
- Use special cache to decrease memory-to-CPU delay
- Use optimizing compilers that can fill load-delay and branch-delay slots
- allocate registers to reduce result-register dependencies
- move load instruction as early as possible
- move instructions that evaluate branch addresses and branch condition test as early as possible to reduce branch penalty

A.6. DIFFERENCE BETWEEN RISC AND CISC

At that time, RISC and "load-store" were often synonymous, but RISC usually referred to a list of features such as:

- All instructions are a single, fixed length.
- Operations are simple enough to execute in a single cycle, using an execution pipeline.
- Operations are performed on registers only, the only memory operations are load and store.
- No microcode is needed, because of the simpler design.
- A large number of registers are available.
- Register windows are used to speed subroutine calls.

Register windows turned out to not be a useful enough idea to catch on, so have been mostly forgotten. RISC is more commonly used to refer to a design philosophy than a list of features (actually implementation techniques), especially since most of them have been applied to CISC designs (pipelines as far back as the Zilog Z8000, register windows in the Hitachi H16 and H32, and microcode eliminated in most modern designs). Basically, RISC asks whether hardware (for complex instructions or memory-to-memory operations) is necessary, or whether it can be replaced by software (simpler instructions or load/store architecture). Higher instruction bandwidth is usually offset by a simpler chip that can run at a higher clock speed, and more available optimisations for the compiler.

By contrast, the CISC philosophy has been that if added hardware can result in an overall increase in speed, it's good - the ultimate goal of mapping every high level language statement on to a single CPU instruction. The disadvantage is that it's harder to increase the clock speed of a complex chip. The Power PC is a good example of this idea applied to a load-store architecture.

RISC machines can differ from their CISC counterparts in eight critical ways, as listed in Table A.1 which illustrates the differences between than.

Table A.2. Difference between RISC and CISC Architecture

RISC	CISC
1. Simple instructions taking 1 cycle	1. Complex instructions taking multiple cycles
2. Only LOADS/STORES reference memory	2. Any instruction may reference memory
3. Highly pipelined	3. Not pipelined or less pipelined
4. Instruction executed by the hardware	4. Instructions interpreted by the microprogram
5. Fixed format instructions	5. Variable format instructions
6. Few instructions and modes	6. Many instructions and modes
7. Complexity is in the compiler	7. Complexity is in the microprogram
8. Multiple register sets	8. Single register set

A.7. ONE INSTRUCTION PER DATA PATH CYCLE

In a sense the name Reduced Instruction Set Computer is something of a misnomer. While it is true that most RISC machines have relatively few instructions, the most important single characteristic that distinguishes them from CISC machines is that RISC instructions are completed in a single data path cycle. A data path cycle consists of fetching two operands from the register scratch pad, putting them on the internal buses, running them through the ALU, and finally storing the result back in the scratch pad. All of this activity takes one clock cycle. In this respect, a RISC instruction is very much like a microinstruction. In contrast, an interpreted instruction, such as the Mac-1 instruction, typically takes 10 clock cycles.

A consequence of the principle that every RISC instruction must take one cycle is that any operation that cannot be completed in one cycle cannot be included in the instruction set. Thus, many RISC machines lack instructions for multiplication or division. In practice, most multiplications are by small constants known at compile time, so they can be simulated by sequences of shifts and adds. The remaining multiplications and all divisions are handled by library procedures. Floating-point instructions are executed by a coprocessor.

A.8. LOAD/STORE ARCHITECTURE

Given our desire to have every instruction take one clock cycle, it is clear that instructions that reference memory are going to be a problem. Instruction that fetch their operands from registers and store their results in registers can be handled in one cycle, but instructions that load from or store into memory take too long. Increasing the clock cycle by a factor of two or three to accomodate loads and stores violates Golden Rule Number 1 of RISC design.

The solution is twofold. First, ordinary instructions may only have register operands. Thus, unlike on the 80386, 68030, and other CISCs, they do not have addressing modes for direct addressing, indexed addressing, or register or memory indirect addressing. Only register addressing is permitted.

However, some instructions must be reference memory, so special LOAD and STORE instructions are added to the architecture. Only these instructions may be reference memory. Fig. 8-5 shows a typical set of LOAD/STORE instructions for a 32-bit RISC machine. The two signed LOAD instructions load 8 or 16 bits into the low-order part of the specified 32-bit register and fill all the remaining bits with sign bit (bit 7 or 15) to produce a 32-bit signed integer. The unsigned LOAD instructions do not propagate the sign bit. For full word LOADs and for STORES, the issue does not arise.

It should be emphasized that the only instructions capable of addressing memory are those shown in Fig. A.1. The instructions format used for ADD, MOVE, AND, and so on, typically contains 5-bit operand fields for addressing 32 registers, but no fields for memory addresses.

Table A.3. Assembly Language definition for RISC I

Instruction	Operands		Comments
Arithmetic-logical instructions			
ADD	Rs,S2,Rd	Rd-Rs + S2	integer add
ADDC	Rs,S2,Rd	Rd-Rs + S2 + carry	add with carry
SUB	Rs,S2,Rd	Rd-Rs-S2	integer subtract
SUBC	Rs,S2,Rd	Rd-RS-S2-carry	subtract with carry
SUBR	Rs,S2,Rd	Rd-Rs-S2-Rs	integer subtract
SUBCR	Rs,S2,Rd	Rd-S2-Rs-carry	subtract with carry
AND	Rs,S2,Rd	Rd-Rs&S2	logical AND

	Instruction	Operands	Comments
OR	Rs,S2,Rd	Rd-Rs i S2	logical OR
XOR	Rs,S2,Rd	Rd-Rs xor S2	logical xor
SLL	Rs,S2,Rd	Rd-Rs	shifted by S2 shift left
SRL	Rs,S2,Rd	Rd-Rs	shifted by S2 shift right logical
SRA	Rs,S2,Rd	Rd-Rs	shifted by S2 shift right arithmetic
		Memory Access Instructions	
LDL	(Rx)S2,Rd	Rd -M[Rx+S2]	load long
LDSU	(Rx)S2,Rd	Rd -M[Rx+S2]	load short unsigned
LDSS	(Rx)S2,Rd	Rd -M[Rx+S2]	load short signed
LDBU	(Rx)S2,Rd	Rd -M[Rx+S2]	load byte unsigned
LDBS	(Rx)S2,Rd	Rd -M[Rx+S2]	load byte signed
STL	Rm,(Rx)S2	M [Rx + S2] -Rm	store long
STS	Rm,(Rx)S2	M [Rx + S2] -Rm	store short
STB	Rm,(Rx)S2	M [Rx + S2] -Rm	store byte
		Branch Instruction	
JMP	COND,S2(Rx)	PC	conditional jump
JMPR	COND,Y	PC-PC+Y	conditional relative
	Instruction	**Operands**	**Comments**
CALL	Rd,S2(Rx)	Rd -PC, next	call
		PC-Rx+S2;CWP	and change window
		-CWP-1<t/d	
CALLR	Rd, Y	Rd-PC,next	call relative
		PC ,-PC + Y;	and change window
		CWP- CWP-1 <t/d	
RET	Rm,S2	PC-Rm+S2;	return and change window
		CWP.-CWp+l<t/d	
CALLINT	Rd Rd—last	PC; next	disable interrupts
		CWPO-CWP-1	
RETINT	Rm,S2	PC,-Rm + S2;	enable interrupts
		next CWP-CWP + 1	
		Miscellaneous Instruction	
LDHI	Rd, Y	Rd<31:13.-Y	load immediate high
		;Rd<12:0-.,—0	
GTLPC	Rd	Rd -last PC	to restart delayed jump
GETPSW	Rd	Rd -PSW	load status word
PUTPSW	Rm	PSW-RM	set status word
		Load signed byte	Store byte
		Load unsigned byte	
		Load signed halfword	Store halfword
		Load unsigned halfword	
		Load word	Store word

Fig. A.1. LOAD and STORE instructions for a typical 32-bit RISC machine.

A.8.1. Pipelining

Of course, forbidding ordinary instructions from accessing memory does not solve the problem of how to make the LOADs and STORE operate in one cycle. The solution lies in a bit of a trick. We are going to relay our goals ever so slightly. Instead of demanding that every instruction be executed in one cycle, we will merely insist that we will be able to start one instruction in each cycle, without regard to when it finishes. If in n cycles we manage to start n instructions, on the average we have achieved one instruction per cycle, which is good enough.

In order to achieve this modified goal, all RISC machines are pipelined. The CPU contains several independent units that work in parallel. One of them fetches the instructions, and other ones decode and execute them. At any instant, several instructions are in various stages of processing.

In Fig. A.2 we show a simple pipeline with three stages, for fetching instructions, executing instructions, and performing memory references. Ordinary instructions are in the machine for two cycles, for fetching and executing. LOADs and STOREs require a third cycle for the memory reference. In cycle 1, instruction 1 is fetched. In cycle 2, instruction 2 is fetched and instruction 1 is executed. In cycle 3, instruction 3, (marked L for LOAD) is fetched and instruction 2 is executed. In cycle 4, the LOAD is started, but it cannot be completed in one cycle.

	Cycle									
	1	2	3	4	5	6	7	8	9	10
Instruction fetch	1	2	L	4	5	6	S	8	9	10
Instruction execution		1	2	L	(4)	5	6	S	(8)	9
Memory reference					L				S	

Fig. A.2. A pipelined RISC machine with delayed LOAD, L, and STORE, S.

In cycle 5 something interesting happens (marked by the circle). Instruction 4 is executed, even though the LOAD started in the previous cycle has not yet completed. As long as instruction 4 does not try to use the register in the process of being LOADED, everything is fine and the machine continues execution at full speed. It is up to the compiler to make sure that the instruction following a LOAD does not use the item being fetched from memory. If the compiler cannot find any instruction to put after the LOAD, it can always insert a NO-OP instruction and waste one cycle. The analogous situation after a STORE (see cycle 9) does not cause a problem unless instruction 8 cannot operate properly until the STORE has completed.

The concept of having a LOAD instruction, not take effect immediately, is known as a delayed load. In our example the result is not available for one cycle but longer delays are also possible, depending on the ratio of the processor speed to the memory speed. With some RISC machines it is up to the compiler to fill the slots following LOADS with something useful. The MIPS machine uses a somewhat different approach, however. Here the compiler generates ordinary sequential code, and a program called the reorganizer shuffles the assembly language instructions around, trying to fill up the slots behind the LOADS. This method not only keeps the compiler simpler, but allows the same reorganizer to be used with all compilers, so that the problem only has to be dealt with once.

In a sense, we have seen the concept of the delayed load before. In the Mic-2, the BEGRD microinstruction starts a memory operation, the result of which is not available for two cycles. In the Mic-r, the slot following the BEGRD must be filled with a CONRD, which is essentially a NO-OP. We could easily change the design to allow other instructions following the BEGRD, provided that they did not touch the register in the process of being loaded.

Even though it is not supposed to, what happens if a program tries to use a newly loaded register before it has arrived? RISC machines differ on this point. Most of them have a hardware interlock, effectively inserting automatic NO-OPs until the register has been loaded.

On the other hand, there are also RISC machines, like the MIPS, that simply use the incorrect value in the register. If the reorganizer cannot fill the slot, it must insert an explicit NO-OP, increasing the program length. The advantage of this approach is a simpler (and potentially faster) CPU, since the hardware does not have to check if the register being accessed is currently valid or not.

LOADs and STOREs are not the only instructions that can be delayed. Conditional jumps wreak havoc with any pipelined machine. The RISC solution is the same as with the LOADs and STOREs, namely delayed jumps. The instruction following a JUMP is always executed, no matter whether the JUMP is ultimately taken or not. In a RISC machine there is no time to run them. It is up to the compiler or reorganizer to find a useful instruction to put after each Jump instruction. Failing that, a NO-OP is used.

A.9. RISC ARCHITECTURE INSTRUCTIONS

The RISC 1 architecture has 31 instructions, most of which do simple ALU and shi operations on registers. As shown in Table A.4, they have been grouped into four categories :

1. Arithmetic-logical
2. Memory Access
3. Branch
4. Miscellaneous

Data, address and registers are 32 bits. The execution time of a RISC 1 cycle is given by the time it takes to read and add two registers, and then store the result back into a register. Register 0 which always contains zero, allows us1to synthesize a variety of operations and addressing modes.

Load and Store instructions move data between registers and memory. Rather than, lenghten the general cycle to complete a memory access, these instructions use two CPU cycles. There are eight variations of memory access instructions to accomodate sign extended or zero-extended eight bit, 16 bit and 32-bit data. Although there appears to be only the index-plus-displacement addressing mode in data transfer instructions, absolute and register-indirect addressing can be synthesized using register 0 (see table A.4).

Table A.4. Synthesizing RISC addressing modes Addressing VAX RISC Equivalent

REGISTER	Rx	Rx
IMMEDIATE	#literal	S2 (13,bit literal)
INDEXED	Rx + DISPL	Rx + S2 (13-bit displacement)
ABSOLUTE	@#address	RO + S2(RO = 0)
REGISTER INDIRECT	(Rx)	Rx + RO

Branch instructions include call, return, conditional and unconditional jump.

As a concrete example consider the architecture of a general RISC processor in Fig. A.2. The central part of the Fig. A.2 is a cyclical 3-stage pipeline. It starts at the bank of the general registers whose outputs are latched in one clock cycle in A, B and D registers. The A and B registers drive an arithmetic logic unit to an output register, F. The F register in turn, stores back into the general registers.

The interface in the data cache is through two paths, one for data to cache and for the cache address. The cache address is developed at the output of ALU and is clocked into the D-cache address registers. Data from the cache passes from the general registers to the D- register, and from there through a shift/logic unit to the D-cache data out register. Data from the cache to the processor is latched into the D-cache data in register, and from there it is stored into the general registers.

At the left of the Fig. A.3 is the program counter, a 1-cycle increment path, and a separate adder for updating the program counter from the instruction stream. The interface with the instruction cache is similar to the interface with the data cache, except for that there is no provision for writing into the instruction stream.

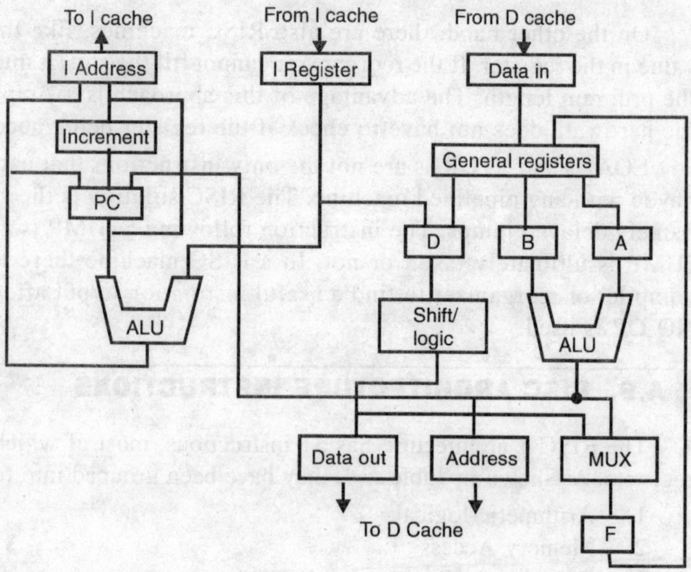

Fig. A.3. The structure of a pipelined RISC processor.

The pipeline in the RISC architectures is different from the pipeline of the conventional architectures because the pipeline here does not contain a separate stage for address generation. Most instructions are register-to-register, and the instruction cycle has the following two phases:

1. I : Instruction Fetch
2. E : Execute. Performs an ALU operation with register input and output.

For LOAD and STORE operations:

1. I : Instruction Fetch
2. E : Execute, calculate memory address
3. D : Memory, memory to register or register to memory, operation

The pipeline does not contain a separate stage for address generation. Address generation is done in the ALU for LOADs and STORES. For STORES, the address generation in ALU is done concurrently with the flow of data to cache through the D register path. But the arithmetic and logic instructions use the ALU for execution, and thus these instructions do not generate both address and perform arithmetic manipulation.

The advantage for removing address generation is that the pipeline is shorter, and the penalty is less when interlocks or conditional execution causes the pipeline to empty. The disadvantage is that the arithmetic instructions cannot also load data from memory, this architecture requires more LOADs than the one that has an address generation stage.

The machine has three read and two write ports to the general register array to keep the pipelined filled at all times. All five ports can be active concurrently without any conflict. With the potential for performing two writes and three reads concurrently on any cycle, the processor can execute a sequence of register instructions at the rate of one per cycle, provided the register instructions are non-conflicting. That is, the input register for any instruction is not altered by its immediately preceding instructions.

The instruction decode pipeline is independent of the execution pipeline except that the conditional branches have to be interlocked to the results of data registers.

A.10. USE OF LARGE REGISTER FILE IN RISC

The register storage is the faster storage device, faster than even the main memory and the cache. Thus, a strategy is needed that will allow the most frequently accessed operands to be kept in registers and to minimize register memory operations.

Two basic approaches are possible, one based on the software and the other based on the hardware. The software approach is to rely on the compiler to maximize register usage. The compiler will attempt to allocate the registers to those variables that will be used most in the given time period. This approach requires the use of sophisticated program-analysis algorithms. The hardware approach is simply to use more registers so that more variables can be held in registers for longer periods of time. RISC follows the hardware approach.

Since most operand references are to local scalars, the obvious approach is to store these in registers, with perhaps a few registers reserved for global variables. The problem is that the definition of local changes with each procedure call and return, operations that occur frequently. On every call, local variables must be saved from the registers into memory, so that the registers can be reused by the called program. Furthermore, the parameters must be passed. On return, the variables of the parent program must be restored and the results must be passed back to the parent program.

RISC takes care of these with the help of register windows. Multiple small sets of registers are used, each assigned to different procedure,. A procedure call automatically switches the CPU to use a different fixed size window of registers, rather than saving registers in memory. Windows for adjacent procedures are overlapped to allow parameter passing.

A.11. RISC PIPELINING

We have already discussed in the previous units, instruction pipelining is often used to enhance performance. Let us consider this in context of RISC architecture. In RISC architectures most of the operations are register-to-register, and the instructions are executed in two phases:

1. **I :** Instruction Fetch
2. **E:** Instruction Execute

All the memory access in RISC is done through LOAD and STORE operations. Three phases are required for these:

1. **I:** Instruction Fetch
2. **E:** Effective address calculation of the operand
3. **D:** Memory to register or register to memory operation.

In operations all these were being done sequentially, as depicted in Fig. A.4, clearly we see what a wasteful situation it will be. We now see, how to execute these instructions in pipelined fashion, and later how to optimize the pipelining.

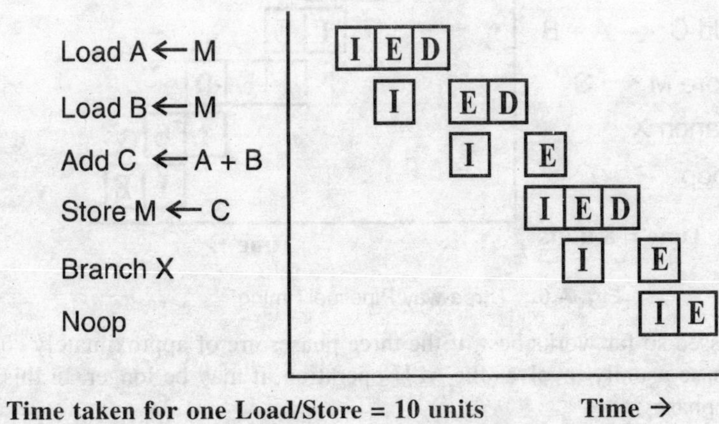

Fig. A.4. Timing of sequential execution.

A.11.1. Pipelining with Regular Instructions

Consider Fig. A.4 again, let us now try to perform the same operation in pipeline. Fig. A.5 shows a two way pipelining scheme, in which I and E phases of two different instructions are performed simultaneously. This scheme can yield upto twice the execution rate of the sequential scheme.

Two problems prevent the pipelining from achieving the maximum speed, firstly, because of a single port memory, the access is limited to one device only at a time. This requires the insertion of wait state in some memory access phases for some instructions. Second, branch instruction interrupts the sequential flow of execution, To accommodate this with minimum circuitry, a NOOP instruction can be inserted into the instruction stream by the compiler or the assembler.

Pipelining can be improved further by permitting two memory accesses per phase. This yields the sequence shown in Fig. A.6.

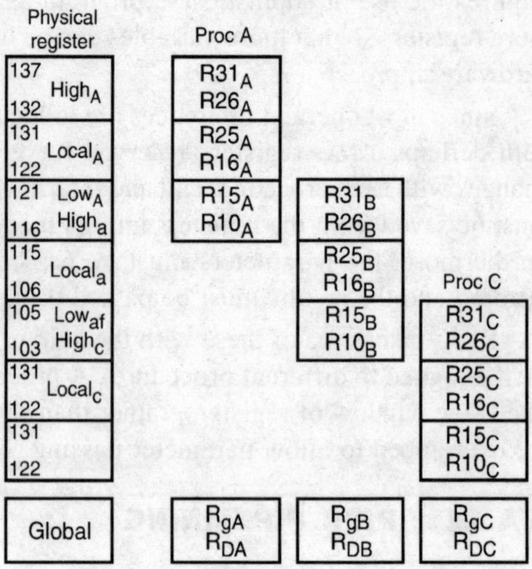

Fig. A.5. Two Way Pipelined Timing

Now, upto three instructions can be overlapped, and the improvement is as much as a factor of three. Again the branch instructions hamper the speed of the pipeline preventing it to reach at its maximum. also, if there are any data dependencies between the pipelines, those also cause the speed to slow down. Data dependencies are like, if an instruction needs an operand that is altered by the previous instructions, a delay is required to wait for the previous instruction to complete its execution. Again this can be accomplished by NOOP.

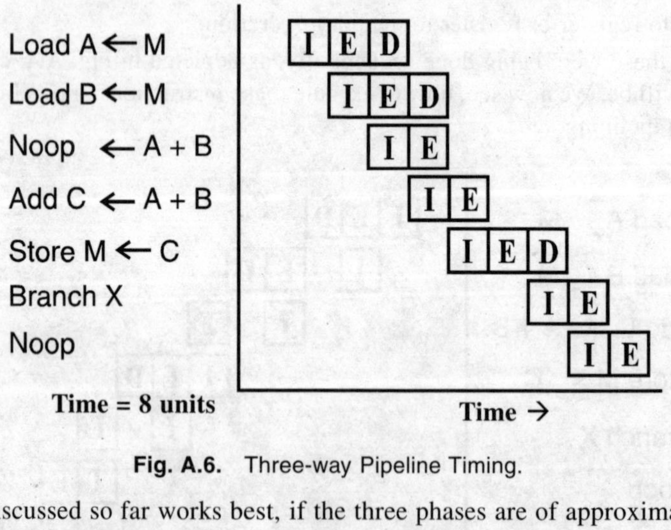

Fig. A.6. Three-way Pipeline Timing.

The pipelining discussed so far works best, if the three phases are of approximately equal duration. Because the execution phase usually involves the ALU operation, it may be longer. In this case we can divide the operation into phases:

1. **E1 :** Register file read
2. **E2 :** ALU operation and register write

Because of the simplicity and the regularity of the instruction set, the design of the phasing into three or four phases is equally accomplished. Fig. A.7 shows the result with a four way pipeline.

Upto four instructions at a time can be executed simultaneously, and the maximum potential speed is a factor of four. NOOPs can be inserted for conflict in memory accesses and branch operations.

A.12. INTERPRETATION IN RISC MACHINES MICROCODE

The instructions generated by a compiler for a RISC machine are directly executed by the hardware. They are not interpreted by microcode. Eliminating this level of interpretation is the secret of the RISC machine's speed.

How these instructions are executed is not difficult, to understand, the "microinstructions" come from the output of the instruction fetch unit of the pipeline, that is, from main memory. They are not microinstructions at all, but user program instructions.

The OP decode box consists of hardwired logic (or a PLA) that takes the opcode field as input and generates all the control signals needed to drive the data path. As long as the instruction fetch unit is capable of providing one instruction every cycle, one data path operation can be done on each cycle.

RISC defenders often say that it is better to implement complex instructions in microcode, instead of user code. This statement leads us to RISC Golder Rule number 2: **"Microcode is Not Magic".**

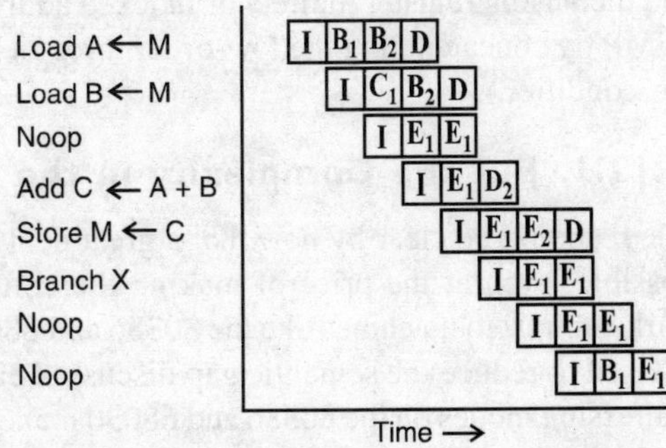

Time = 10 units, however, unit size will be smaller

Fig. A.7. Four-way Pipelined Timing

If a complex instruction (say, decimal addition), takes 10 microinstructions at 100 nsec each, it will take 1 usec to execute on a CISC machine. On a RISC machine, the same instruction will take 10 RISC instructions instead of 10 microinstructions. If the RISC instructions also take 100 nsec, it will still take 1 usec.

The only advantage to the CISC implementation is saving a little memory. However, if the operation occurs relatively often, it can be made into a library procedure at a minor cost in speed. Remember that most programs do simple things, so it is worthwhile accepting a small penalty on unusual instruction in return for getting the register moves and simple additions to go fast.

A.12.1. Fixed-Format Instructions

The individual bits in each instruction are used as inputs to the OP decode box and other parts of the hardware. Having variable-length instruction, ranging from 1 to 17 bytes (as on the 80386), just would not do here. That only works when there is a microprogram inside, sequentially pulling the bytes out of the instruction fetch queue and analyzing them one at a time in software.

A.12.2. Reduced Instruction Set

Finally, we come to the point for which RISC machines are named. Actually, there is no insurmountable objection to having many instructions, provided that each instruction executes in one cycle. The only real problem is that the complexity of the OP decode box grows exponentially with the number of instructions, and thus consumes increasingly large amounts of scarce chip area.

The situation with addressing modes is different. For reasons of speed and complexity, it is undesirable to have more than a minimal number of addressing modes. Fig. 6.9 shows the basic instruction format used in the RISC I, as an example. It specifies an opcode, two 5-bit registers, a mode bit (immediate/not immediate), and an offset. Using this one format, and the fact that register 0 is hardware to the constant 0, we can produce most of the useful addressing modes.

For ordinary instructions, such as ADD, the operands depend on the I bit. If it is cleared, one operand is taken from the source register, the second is taken from the register specified by the low-order 5 bits of the OFFSET field, and the result is stored in the destination register. This combination gives register addressing. If the I bit is set, the second operant is a 13-bit constant, giving immediate addressing.

For LOAD and STORE instructions, the OFFSET is added to the source register to form the effective memory address (indexed addressing). If OFFSET is 0, this mode reduces to register indirect addressing. If register 0 is specified, we have direct addressing of the bottom 8K of memory, which is useful for accessing global variables. Other modes can be constructed at run time by building an address in a register and then using register indirect or indexed addressing. The RISC II design added a PC-relative conditional JUMP by concatenating the low-order three fields to form a 19-bit signed offset (the DEST field specifies the condition).

A.12.3. Put the Complexity in the Compiler

It should be clear by now that a great deal of effort has been made to keep the hardware as simple as possible, even at the price of making the compiler considerably more complicated. This strategy is in stark contrast to machines like the 80386 and 68030, with their highly complex addressing modes, allegedly devised to reduce the semantic gap discussed earlier. In practice, the existence of some of the more exotic addressing modes on the 80386 and 68030 make both the compiler and the microprogram very complicated.

Still, CISC machines do not have nasty features like delayed loads, stores, and jumps. These certainly add considerably to the complexity of the compiler or reorganizer. In addition, the fact that ordinary instructions cannot use memory operands means that it is essential for RISC compilers to go to great lengths to optimize register usage. The penalty for suboptimal use is much greater than for CISC compilers, which gives the compiler writer an additional burden.

Finally, the reduced instruction set itself also causes trouble, Slome instructions that CISC compiler writers take for granted-MULTIPLY, for example-have to be synthesized in a variety of ways, depending on the operands.

A.13. IBM SYSTEM 360/370/390: THE MAINFRAME(1964)

The IBM System/360 is a sort of geologic feature in the computer world, and isn't at all a microprocessor, but was certainly influential (and enough people asked for it to be included in this list). It was designed to be an "all around" (as in, 360 degrees) system usable for any computing task, and as a result created many of the standards for the computing industry, such as 8-bit bytes and byte addressable memory, 32-bit words, segmented and paged memory (see the Intel 80386), packed decimal and the EBCDIC character set (the latter isn't really a standard, as most systems use ASCII, except for the fact that immense amounts of data are stored on IBM System/360s in EBCDIC format).

The S/360 has sixteen 32 bit general purpose registers(occasionally paired up as 64 bit registers), four 64 bit floating point registers (or two 128 bit registers), and a Program Status Word like that in the DEC VAX , except that in the S/360 the PSW includes the program counter [24 bits in the S/360, 31 bits in the S/370 XA (eXtended Architecture, pre 1983) and later versions]. The S/370 (pre 1977) also includes sixteen control registers used by the operating system.

A two stage pipeline was first introduced in the IBM 3033 (1977). Instructions are fetched from the cache into three 32 bit buffers. The Instruction Pre-Processing Function (IPPF) then decodes them, generates operand addresses and stores them in operand address registers, and places source operands in operand buffers. Decoded instructions were placed into a 4 entry queue until the execution unit was ready.

In some models (such as 360/91) when a conditional branch occurs, the most likely next instruction is loaded into the IPPF buffer, but the previous next instruction is not discarded, so either can be executed without penalty. Two speculative branches can be buffered this way. Some had a "loop mode" like the Motorola 68010.

Addressing was originally 24 bit, but was extended to 31 bits (the high bit indicated whether to use 24 or 32 bits) with the XA architecture (This caused problems with software which stored type information in the unused 8 bits of a 32 bit word. The same thing happened when the Motorola 68000 was expanded from 24 to 32 bit addressing). The S/360 used completely position independent (register+offset and register+index) addressing modes. Virtual memory was added in the S/370, and used a segment and paging method - the first 8 bits of an address indicated an entry in a segment table which is added to the next 4 or 8 bits to get the page table index which contains the upper (12 or 20) bits of the physical memory address, and the rest of the address provides the lower 12 bits (the Intel 80386 uses a similar method, while the Motorola 68030 uses fixed length logical/physical pages instead of variable length segments).

Like the DEC VAX, the S/370 has been implemented as a microprocessor. The Micro/370 discarded all but 102 instructions (some supervisor instructions differed), with a coprocessor providing support for 60 others, while the rest are emulated (as in the MicroVAX). The Micro/370 had a 68000 compatible bus, but was otherwise completely unique (some legends claim it was a 68000 with modified microcode plus a modified 8087 as the coprocessor, others say IBM started with the 68000 design and completely replaced most of the core, keeping the bus interface, ALU, and other reusable parts, which is more likely).

More recently, with increased microprocessor complexity, a complete S/390 superscalar microprocessor with 64K L1 cache (at up to 350MHz, a higher clock rate than the 200MHz Intel's Pentium Pro available at the time) has been designed.

A.14. VAX: THE PENULTIMATE CISC (1978)

The VAX architecture wasn't designed as a microprocessor, though single chip versions were implemented (around 1984). However, it and its predecessor, the PDP-11, helped inspire design of the Motorola 68000, Zilog Z8000, and particularly the National Semiconductor 32xxx series CPUs. It was considered the most advanced CISC design, and the closest so far to the ultimate CISC goal. This is one reason that the VAX 11/780 is used as the speed benchmark for 1 MIPS (Million Instructions Per Second), though actual execution was apparently closer to 0.5 MIPS.

The VAX was a 32 bit architecture, with a 32 bit address range (split into 1G sections for process space, process specific system space, system space, and unused/reserved for future use). Each process has its own 1G process and 1G process system address space, with memory allocated in pages.

It features sixteen user visible 32 bit registers. Registers 12 to 15 are special - AP (Argument Pointer), FP (Frame Pointer), SP and PC (user, supervisor, executive, and kernal modes have separate SPs in R14, like the 68000 user and supervisor modes). All these registers can be used for data, addressing and indexing. A 64 bit PSL (Program Status Longword) keeps track of interrupt levels, program status, condition codes, and access mode [kernal (hardware management), executive (files/records), supervisor (interpreters), user (programs/data)].

The VAX 11 featured an 8 byte instruction prefetch buffer, like the 8086, while the VAX 8600 has a full 6 stage pipeline. Instructions mimic high level language constructs, and provide dense code. For example, the CALL instruction, which not only handles the argument list itself, but enforces a standard procedure call for all compilers. However, the complex instructions aren't always the fastest way of doing things. For example, the INDEX instruction was 45% to 60% faster when replaced by simpler VAX instructions. This was one inspiration for the RISC philosophy.

Further inspiration came from the MicroVAX (VAX 78032) implementation, since in order to reduce the architecture to a single (integer) chip, only 175 of the 304 instructions (and 6 of 14 native data types) were implemented (through microcode), while the rest were emulated - this subset included 98% of instructions in a typical program. The optional FPU implemented 70 instructions and 3 VAX data types, which was another 1.7% of VAX instructions. All remaining VAX instructions were only used 0.2% of the time, and this allowed MicroVAX designs to eventually exceed the speed of full VAX implementations, before being replaced by the Alpha architecture.

A.15. RISC HISTORY: CDC 6600 (1965)

Most RISC concepts can be traced back to the Control Data Corporation CDC 6600 'Supercomputer' designed by Seymour Cray (1964), which emphasized a small (64 op codes) load/store and register-register instruction as a means to greater performance.

The CDC 6600 was a 60-bit machine ('bytes' were 6 bits each), with an 18-bit address range. It had eight 18 bit A and 18 bit B (address) and eight 60 bit X (data) registers, with useful side effects - loading an address into A1, A2, A3, A4 or A5 caused a load from memory at that address into registers X1, X2, X3, X4 or X5. Similarly, A6 and A7 registers had a store effect on X6 and X7 registers - loading an address into A0 had no side effects.

As an example, to add two arrays into a third, the starting addresses of the source could be loaded into A2 and A3 causing data to load into X2 and X3, the values could be added to X6, and the destination address loaded into A6, causing the result to be stored in memory. Incrementing A2, A3, and A6 (after adding) would step through the array. Side effects such as this are decidedly anti-RISC, but very nifty. This vector-oriented philosophy is more directly expressed in later Cray computers.

Only one instruction could be issued per cycle, but multiple independent functional units in the CDC 6600 meant instruction execution in different units could overlap (a scoreboard register prevented instructions from issuing to a unit if the operands weren't available). The units weren't pipelined until the CDC 7600 (1969), at which point instructions could be issued without waiting for operands (they would wait for them in the functional unit if necessary).

Compared to the variable instruction lengths of other machines, instructions were only 15 or 30 bits, packed within 30 bit half-words (a 30 bit instruction could not occupy the upper 15 bit "parcel" of one half-word and the lower 15 bits of the next, so the compiler would insert NOPs to align instructions) to simplify decoding (a RISC-like feature). Branches were 60-bit-word aligned. Like the DEC Alpha, there were no byte or character operations, until later versions added a CMU (Compare and Move Unit) for character, string and block operations.

A.16. RISC FORMALISED: IBM 801

The first system to formalise these principles was the IBM 801 project (1975), meant for a simple network switching controller and named after the building it was developed in. Like the VAX, it was not a microprocessor (ECL implementation), but strongly influenced microprocessor designs. The design goal was to speed up frequently used instructions while discarding complex instructions that slowed the overall implementation. Like the CDC 6600, memory access was limited to load/store operations (which were delayed, locking the register until complete, so most execution could continue). Branches were delayed, and instructions used a three operand format common to load-store processors. Execution was pipelined, allowing 1 instruction per cycle.

The 801 had thirty two 32 bit registers, but no floating point unit/registers, and no separate user/supervisor mode, since it was an experimental system - security was enforced by the compiler. It implemented Harvard architecture with separate data and instruction caches, and had flexible addressing modes.

IBM tried to commercialise the 801 design starting in 1977 (before RISC workstations first became popular) with the ROMP CPU (Research OPD (Office Products Division) Mini Processor), 1986, first chips as early as 1981 used in the PC/RT workstation, but it wasn't successful. Originally designed for wordprocessor systems, changes to reduce cost included eliminating the caches and Harvard architecture (but adding 40 bit virtual memory), reducing registers to sixteen, variable length (16/32 bit) instructions (to increase instruction density), and floating point support via an adaptor to an NS32081 FPU (later, a 68881 or 68882 were available). This allowed a small CPU, only 45,000 transistors, but an average instruction took around 3 cycles.

The 801 Itself Morphed into an I/O Processor for the IBM 3090 Mainframes. This wasn't the only innovative design developed by IBM which never saw daylight. Slightly earlier (around 1971) the Advanced Computer System pioneered superscalar (seven issue) design, speculative execution, delayed condition codes, multithreading, imprecise traps and instruction streamed interrupts, and load/store buffers, plus compiler optimisation to support these features. It was expensive and incompatible with the System/360, so was not pursued, but many ideas did find its way into the expensive high end mainframes.

A.17. RISC REFINED: BERKELEY RISC, STANFORD MIPS

Some time after the 801, around 1981, projects at Berkeley (RISC I and II) and Stanford University (MIPS) further developed these concepts. The term RISC came from Berkeley's project, which was the basis for the fast Pyramid minicomputers and SPARC processor. Because of this, features are similar, including a windowed register file (10 global and 22 windowed, vs 8 and 24 for SPARC) with R0 wired to 0. Branches are delayed, and like ARM, all instructions have a bit to specify if condition codes should be set, and execute in a 3 stage pipeline. In addition, next and current PC are visible to the user, and last PC is visible in supervisor mode.

The Berkeley project also produced an instruction cache with some innovative features, such as instruction line prefetch that identified jump instructions, frequently used instructions compacted in memory and expanded upon cache load, multiple cache chips support, and bits to map out defective cache lines.

The Stanford MIPS project was the basis for the MIPS R2000, and like the case with Berkeley project, there are close similarities. MIPS stood for Microprocessor without Interlocked Pipeline Stages, using the compiler to eliminate register conflicts. Like the R2000, the MIPS had no condition code register, and a special HI/LO multiply and divide register pair.

Unlike the R2000, the MIPS had only 16 registers, and two delay slots for LOAD/STORE and branch instructions. The PC and last three PC values were tracked for exception handling. In addition, instructions were 'packed' (like the Berkeley RISC), in that many instructions specified two operations that were dispatched in consecutive cycles (not decoded by the cache). In this way, it was a 2 operation VLIW, but executed sequentially. User assembly language was translated to 'packed' format by the assembler.

Being experimental, there was no support for floating point operations.

A.18. SOAR (SMALLTALK ON A RISC) MODIFIED THE RISC II DESIGN TO SUPPORT SMALLTALK

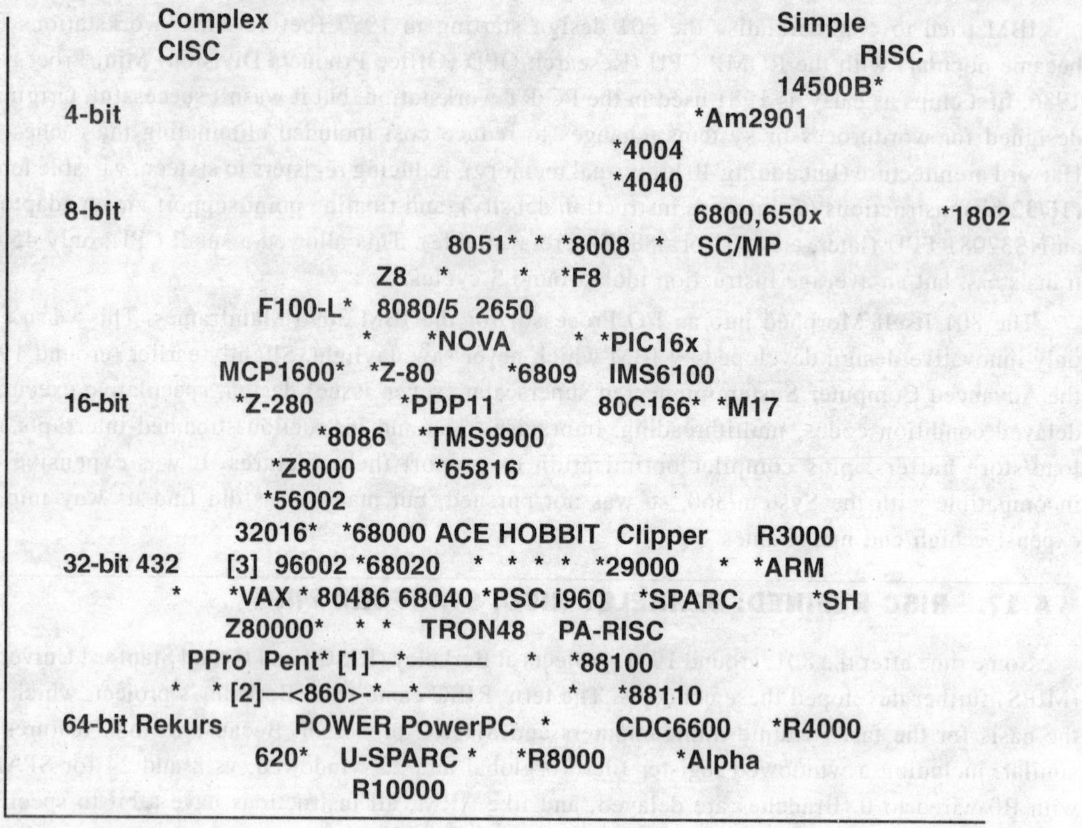

Fig. A.8.

Since this is only a 2-dimensional graph, and I want to get a lot more across than that allows, design features 'pull' a CPU along the RISC/CISC axis, and the complexity of the design (given the number of bits and other considerations) also tug it - thus the much of the POWER's RISC-ness is offset by its inherently complex (though effective) design.

A.19. MULTIPLE REGISTER SETS

By not having any microcode, a substantial amount of chip area is freed up on RISC chips for other purposes. Some, but not all, RISC machines use this newfound chip area for implementing a large number of CPU registers, to reduce the number of LOADs and STOREs. The subject of how these registers are organized is so important that we will devote the entire next section to the subject of registers.

A.20. OPEN ISSUES

Despite all these points of commonality, RISC designers do not agree on everything. A number of design issues are by no means settled. For one thing, a key question is how much of the hardware should be visible to the compiler writer. The problem of interlocks following a LOAD as discussed above is a typical one. One camp says that the machine simply cannot do a LOAD in one cycle, the compiler writer should know about this property and learn to live with it. The other camp says it should be hidden by automatic interlocking.

If interlocking is visible, what about the cache? Can the compiler writer make the assumption that a variable just used will still be in the cache, and thus can be accessed more quickly than one in memory? Can he skip the delay slot following a LOAD and use the newly-loaded register in the very next instruction.

Byte ordering (little endian versus big endian) is still a hot topic. The MIPS chip finesses the problem by being configurable either way. Is this a good idea? It clearly costs some hardware and chip area that could be used for other things.

The RISC I has a bit in each instruction that tells whether the condition codes should be set. While setting the condition codes is frequently on arithmetic instructions, there is no consensus about whether MOVEs should set them. Adding a bit gives the compiler writer the choice on every MOVE. On the other hand, the MIPS chip does not have condition codes at all. It can test and jump in one instruction. Which way is better is a hotly debated issue. There are numerous other design issues that have not been settled.

A.21. REGISTER USAGE

The goal of every RISC machine is to execute one instruction per cycle on the average. Since LOADs and STOREs typically require two cycles, this average can only be achieved if the compiler or reorganizer is successful in filling 100 percent of the delay slots behind each of them (not to mention the delay slots following JUMPs). It stands to reason that the fewer LOADs and STOREs there are, the fewer will be wasted due to the compiler's inability to fill them with something useful.

For this reason, compilers for RISC machines make heavy use of registers in order to reduce the memory traffic (i.e., the number of LOADs and STOREs). Not only do RISC machines have substantially more registers than machines like the 80386 and 68030 (500 registers is not unusual), but more importantly, they are organized in a different way. Building a machine like the PDP-11 or 680x0, except with 512 registers instead of 8 or 16, would be counterproductive. Not only would the 16-bit instructions all become 32 bits to accomodate the 9-bit register fields, but the time to save all the registers on procedure calls would become prohibitively expensive.

In sec. 6.2 and 6.3, we discussed several design principles for RISC machines, the first of which says to analyze the applications to see how they spend their time. People have done precisely that with respect to memory traffic. The most important conclusion is that a large percentage of the total memory traffic is related to procedure calls. Parameters must be passed, registers must be saved, and thereturn address must be stacked on the call and unstacked on thereturn. All of these actions generate memory traffic.

The designers of the RISC I (Patterson and Sequin, 1982) thought of an ingenious way to eliminate nearly all of this traffic. Their method, now known as overlapping register windows, has been adopted by other RISC machines as well. We will describe it in general terms below. The exact details vary slightly among implementations.

When overlapping register windows are used, the CPU contains a large number of registers, but at any instant only a subset of them, usually 32, is visible. These 32 registers are shown in Fig.A.9, where they are designated R_O through R_{31}. On most RISC machines the registers are 32 bits wide, but 64 bits or other widths are also possible.

The registers are divided into four distinct groups as illustrated in Fig.A.9. In the RISC I, these groups consisted of 10, 6, 10, and 6 registers, respectively. However, more recent designs have made them all the same size (8 registers per group). For implicitly, in our discussion we will make this latter assumption.

```
<————————32-Bits————————>
RO
.
. Global variables
.
R7
R8
.
.Incoming parameters
.
R15
R16
.
.  Local variables
.
R23
R24
.
.Outgoing parameters
.
R31
```

Fig. A.9. The 32 registers visible to a program at any instant of time.

The first group holds global variables and pointers. These are not specific to any one procedure, but are used by many procedures throughout the program. It is up to the compiler to decide what to put in each register. In some RISC machines, R_0 is hardwired to the constant 0. Reads from it yield a 0 and writes to it have no effect.

The second group of registers, R8 through R_{15}, holds the incoming parameters. Most procedures have parameters, supplied by the calling procedure. In traditional computers, these parameters are pushed onto the stack one at a time, just before the CALL instruction. When overlapping register windows are used, the procedure parameters are put into R8, R9 and so on up to R15 instead of being pushed onto the stack. Not putting them on the stack eliminates STOREs when they are being passed and eliminates LOADs when they are accessed by the called procedure. If there are more than eight parameters, or if a parameter is more than one word, those parameters that cannot be passed in registers are passed on the stack in the conventional way. It is a matter of experience that for most procedures eight parameter is probably overkill.

Registers R16 through R_{23} are available for local variables. For those procedures that need more than eight words of local variables, the excess are kept on the stack, just as with excess procedure parameters. Still, if 80 or 90 parameters or for their local variables, we have made a substantial dent in the memory traffic and hence greatly reduced the numbers of LOADs and STOREs and thus unfilled delay slots.

The final group of registers, R_{24} through R_{31}, is used to pass outgoing parameters to called procedures. Instead of pushing these parameters onto the stack, they are put here. As usual, if we run out of registers, the excess parameters go on the stack.

By itself this scheme is cute, but not terribly interesting. The real power comes from the presence of a large amount of on-chip storage, in this case in the form of a register file, say, 520 registers for this example. In Fig. A.9, with the addition of a variable CWP (Current Window Pointer) that points to register R8. When the current procedure, A, calls another procedure, B, the CWP pointer is incremented by 16 so that it points to 24.

When procedure B starts up, the situation is as shown in Fig. A.10(*b*). References to R0 through R7 still use the eight global registers, as before, but R8 through R32 now refer to a different set of 24 registers. In particular, R8 (B' s first input parameter) how contains A's first output parameters. Thus A can pass its parameters to B without using memory, and B can access them without using memory. The reason that A and B can communicate this way is that eight of the registers overlap, is shown. When B calls C, CWP is advanced again, and the same relative situation holds.

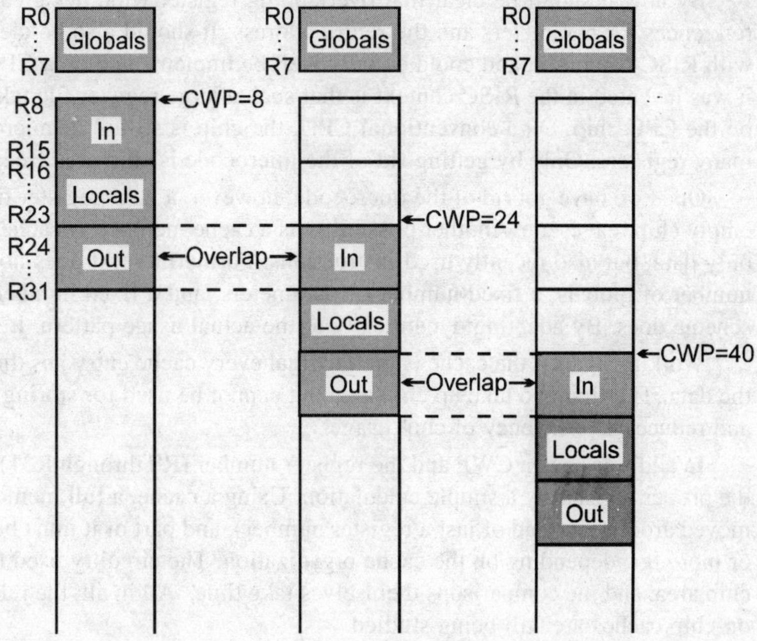

Fig. A.10. Overlapping register windows.

In effect, the register file is being used as a kind of stack with CWP being the stack pointer. Whenever a machine instruction references R0 through R7, the global registers are used, but R8 always means the register pointed to by CWP. Only the 24 registers starting at CWP (plus the eight globals) are accessible at any instant. CWP is advanced upon procedure calls and set back upon returns, but otherwise cannot be touched by ordniary user programs. (The preparing system can save and restore it, however, when switching processes).

To make the scheme of Fig.A.10 work, the compiler must obey certain conventions, but these are not difficult to understand. When passing parameters to a called procedure, the caller must put the first one in R24, the second one in R25, and so on, with the seventh one in R30 and the eighth one, if any, as the first one on the real memory stack. R31 is reserved for thereturn address. Similarly, when the called procedure accesses its parameters, it must use R8 to access the first one, R9 to access the second one, and so on, with the seventh one coming from R14 and the eighth one, if any, from the real memory stack. The return address can be found in R15. As long as all calling procedures and all called procedures obey the rules, there is no ambiguity and the scheme works fine.

When a parameter is longer than one word, it is up to the compiler to handle it, but a typical solution is to put the parameter itself onto the memory stack, and pass a pointer to it in the approproate register. For long parameters after the seventh one, there is no need to pass a pointer in memory to a parameter in memory (even a one-word parameter); it is sufficient just to pass the parameter itself.

If the program gets nested deeply enough in calls (for example, due to recursion), the entire regtister the eventually will fill up. An additional call will have no place to put its registers. The usual solution is to design the hearware so that the attempt to advance CWP causes a trap. The trap handler then saves the first register window, starting at R8 in Fig.A.10(a), by copying it to memory. If the calling continues, another trap occurs on the next call, and more registers are saved to memory. In this manner, the register file is used as a kind of circular buffer. Some administration is needed to keep track of what is in the registers and what is in memory but that is not difficult to do. In practice, programs that are so deeply recursive that they overflow the register file are relatively rare.

By now it shoult be clear that overlapping register windows greatly reduce the number of memory references for parameters and the return address. It should also be clear that this scheme has little to do with RISC machines, and could equally well be implemented on a CISC chip. The reason, however, that it was invented in the RISC context is that such a large register file takes up a substantial amount of area on the CPU chip. On a conventional CPU, the chip is so full of microcode that there is no room for so many registers. Only by getting rid of the microcode is chip area freed up for the large register file.

Once we have got rid of the microcode, however, a large register file is not the only candidate for the empty chip real estate. Another possibility is a cache memory. A cache has the advantage that it holds not only data, but also recently used instructions. Furthermore, it does not enforce a rigid split with a fixed number of globals, a fixed number of parameters, and a fixed number of locals, as the register window scheme does. By adapting dynamically to the actual usage pattern, it is potentially more efficient.

Working against the cache is the fact that every cache entry has three parts, the valid bit, the tag, and the data. The first two take up chip area but cannot be used for storing data. They are a form of overhead and reduce the efficiency of chip usage.

In addition, given CWP and the register number (R0 through R31), it is easy for the hardware to find the proper register by a simple calculation. Using a cache, a full memory address must be calculated and moved around (instead of just a register number), and part of it must be extracted and compared with one or more tgs, depending on the cache organization. The circuitry used for these comparisons also uses up chip area, and the comparisons themselves take time. All in all, the relative merits of register files versus on-chip caches are still being studied.

A.22. REGISTER ALLOCATION

The overlapping register window architecture presented above works best when all, or nearly all, of the local variables are in registers. For small procedures, there is often no problem, but for larger ones there may be more variables than there are registers available for them. In this case some must be spilled into memory, as described above.

To minimize this problem, and in keeping with the RISC design philosophy of pushing as much work as possible onto the compiler, most RISC compilers make a great effort to optimize register usage to reduce the number of variables that must be kept in memory. A common approach is to time-share individual registers over several variables that are used during disjoint intervals within a procedure. The basic data is best illustrated by an example, such as the program of Fig. A.11

At the left of Fig. A.11, is a (somewhat arbitrary) 20-line Pascal function that takes one parameter, n, computes n2, and then counts the number of 0 bits and 1 bits in the binary representation of n2. If there are more 0s than 1s, it returns true otherwise it returns false. High-order 0s (i.e., 0s to the left of the most significant bit of n2) are not counted. It operates by successively dividing the

```
function MostlyZero(n:integer):boolean;
var num, ones, zeros, half : integer;
res : boolean;
begin
num : = N * N;
ones : = O;
zeros : = o:
while num > 0 do
begin
half : = num dlv 2;
If 2*half = num
then zeros : = zeros + 1
else ones : = ones + 1
num : = half
end;
If zeros > ones
then res : = true
else res : = false;
MostlZero : = res
end ;
```

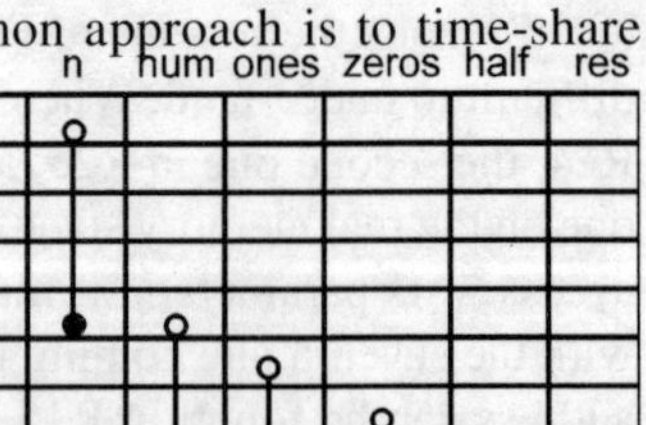

Fig. A.11. An example program and the state of its variables. The empty circles show when each variable becomes alive and the filled-in ones show when they become dead.

number being tested in half to see if it is odd or even, thus telling it the value of the low-order bit. The test for odd-even is done on lines 10 and 11 (remember that dividing an odd number in half, and then doubling the quotient, does not give the original number back, whereas doing this with an even number does).

The function uses six variables: n, njm, ones, zeros, falf, and res (it could obviously have been written with fewer, but that is not the point here). The first of these, n, is a parameter; the rest are locals. At each point during the execution of the function, each variable is in one of two states.

A variable is said to be live if the value it contains is needed. For example, if the compiler were to generate code to store random garbage in the variable zeros on line 12, the function would fail to work properly. However, the same garbling of zeros on line 5 would not matter, because zeros will be assigned to shortly.

More formally, a variable V is live at a point P if there is a control flow path from the procedure's entry point through a statement that assigns a value to V and then on through P to another statement U which uses V without an intervening assignment to V between P and U. In Fig. A.12 (a), for example, we see that V is live between P and U because its value matters. In Fig. A.12 (b), on the other hand, V is not live between P and Q because it is going to get a new value at Q anyway. A variable that is not live is said to be dead.

The concept of liveness is important to register allocation because two or more variables can use the same register if they are never live simultaneously. Going back to Fig A.11 we see the live periods for each variable marked in the six columns at the right-hand side of the Fig.. In particular, note that n, half, and res are live in completely disjoint portions of the function. An optimizing compiler can assign all three variables to the same register. Because n is a input parameter, this register will be in the range R_8 to R_{15} (see Fig. A.9), but there is no problem with using an input register for local variables once the input parameter is dead. If there are unused input registers, they can be used for locals, too, of course.

A number of algorithms are known for maximizing the packing of variables into registers (Chaitin et al., 1981; Chow and Hennessy, 1984). The most popular of these use what is called graph colouring, in which each procedure is represented as a directed graph, with the statements as the nodes and the control paths between them as the arcs. Each variable is assigned a colour in such a way that no two variables that are simultaneously live have the same colour. The goal of the algorithm is to find the minimum number of colours required for the graph, the so-called chromatic number. If the chromatic number is less than or equal, to the number of registers available, all variables can be kept in registers, and no LOADs or STOREs are needed.

A.23. RISC VERSUS CISC

While RISC machines have increasingly many supporters, they also have no shortage or critics (e.g., see Colwell et al., 1985a, 1985b; Flynn et al., 1987; Patterson and Hennessy, 1985). Given the fundamental premise of the RISC camp-complex microprogrammed CPUs of the type that have dominated the computer industry for decades are a dead end and should be abandoned. It is not surprising that some people who are closely associated with CISC machines have taken a close and critical look at RISC technology and found problems that the RISC proponents have tried to sweep under the rug. In this section we will examine four key questions that go to the heart of the RISC-CISC controversy and present the arguments offered by both sides, to let the reader make an informed judgment.

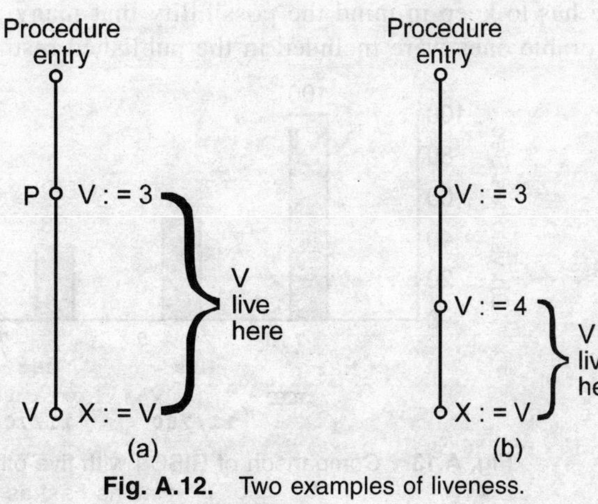

Fig. A.12. Two examples of liveness.

A.23.1. Suitability with High-Level Languages?

No sane person programs in assembly language any more, so everyone wants to know whether RISC architectures are better than CISC architectures when it comes to running programs written in Ads, C, C++, FORTRAN, Modula 2, Pascal, and other high-level languages. The question is simple but the answer is not. To start with, what does "better" mean?

The obvious answer is: "faster." One could simply run a collection of important high-level language programs on RISCs and CISCs to see which is faster. Unfortunately, benchmarking, as this process is called, opens a real can of worms (Serlin, 1986). The first problem is: which language should the benchmark programs be written in? FORTRAN programs tend to be large and unstructured, with few procedure calls and many GOTOs . Since RISC machines are good at procedure calls and bad at jumps, would such a comparison be fair to the RISC side?

The next question is: what kind of programs should be included in the benchmarks? Small highly recursive programs like the towers of Hanoi and Ackermann's function really fly on RISC machines, but are they representative of real work loads?

What about I/O? Experimental RISC machines typically have only sedimentary I/O, whereas many real programs are I/O limited. Is it fair to exclude I/O from the comparison?

Should floating-point programs be included, and if so, how much weight should they be given ? Without special hardware assistance, RISC machines are not good at floating-point calculations. Many cannot even do integer multiplication and division, after all. And floating point comes in three varieties: single (32-bit), double (64-bit), and extended (128-bit) precision. Which should be used?

Another major factor that makes benchmarking tricky is factoring out the influence of the compiler. If a RISC machine outperforms a VAX or an 80386, who is to say that the RISC architecture is better? May be it simply has an optimizing compiler that does a great job on the graph coloring problem or otherwise makes much better use of the registers than the VAX or 80386 compiler.

In a similar vein, real programs use other resources than just the CPU. How can one tell if the result of running complex test programs reflects differences in the machine architecture, or perhaps differences in the performance of the underlying operating system, page size, or paging algorithm? Suppose a RISC machine beats a VAX when both of them run UNIX but suppose the same VAX then outperforms the RISC machine when running VMS. Which machine is really faster, and how much of the performance depends on the architecture, as opposed to the different compilers, operating systems, and the specific test programs chosen?

One final point regarding published comparisons is the old adage: "Fig.s don't lie, but liars Figs." If the person or organization making the study (e.g., a computer vendor) has an axe to grind, one has to keep in mind the possibility that many programs were benchmarked and only the most favorable ones were included in the published results.

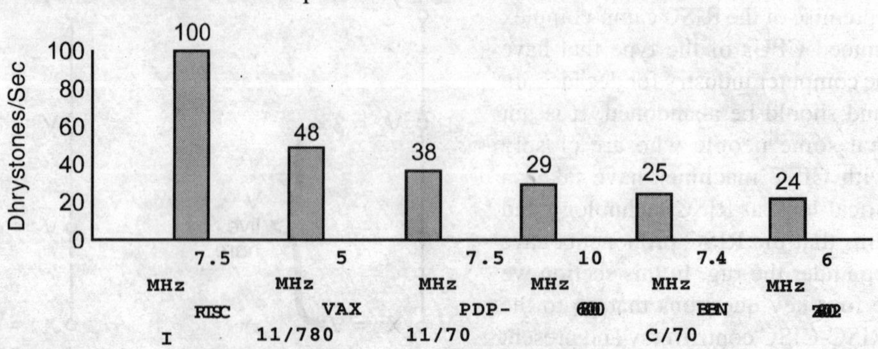

Fig. A.13. Comparison of RISC I with five other computers. None of the others is Even Half as Fast as the RISC I.

All this, said, in Fig. A.13 we present some comparisons of the Berkeley RISC I with popular CISC machines on a collection of UNIX benchmarks devised by the Department of Defence for comparing architectures (Patterson and Piepho, 1982). While one can quibble with many issues relating to benchmarking in general, it seems safe to say that if two professors and a class of graduate students at Berkeley can design and implement a RISC chip that outperforms a number of microprocessors and minicomputers designed by practicing professional engineers, the RISC idea undoubtedly has some merit.

Given all the difficulties with running real programs, a commonly used approach is to create synthetic benchmark programs that have the same instruction mix as some measured sample of real programs. These artificial benchmarks typically do not I/O at all, all they really test is the compiler and CPU performance. In the 1970s, Curnow and Wichman (1975) of the National Physical Laboratory in England developed a synthetic benchmark that measured floating-point performance in whetstones per seconds. (A whetstone is defined as a statistically "average" "floating-point" instruction).

A successor to the whetstone, naturally called the dhrystone (Weicker, 1984), is now widely used to compare RISC and CISC machines. Unlike the whetstone, the dhrystone measures only integer computations. Although the results are still affected by factors such as compiler quality, register allocation, and cache performance, the number of dhrystones per second that a computer can execute gives at least a rough measure of its performance.

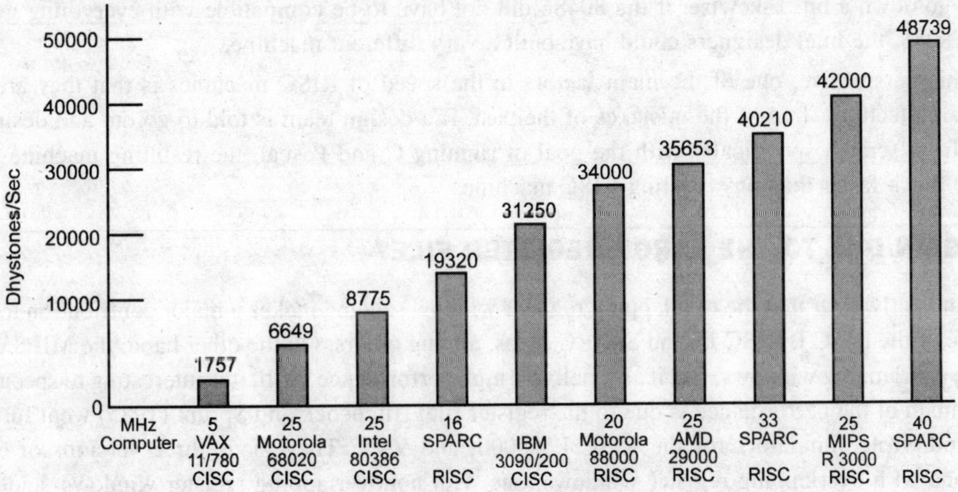

Fig. A.14. Number of dhrystones/sec for a variety of RISC and CISC computers. As a rough approximation, 2000 dhrystones is 1 MIPS.

Fig. A.14. shows the number to dhrystones for several RISC and CISC machines. As usual, all benchmarks should be taken with a grain (or better yet, an imperial gallon) of salt. Still, when four RISC chips from four different vendors all outperform IBM's top-of-the-line 3090/200 mainframe, it becomes hard to deny that RISC machines are very fast.

Perhaps the most difficult fact or remove when comparing two machines is the technology. Is an ECL RISC machine faster than a CMOS CISC machine? Probably ECL transistors switch much faster than CMOS transistors. The RISC and CISC machines being compared invariably use different kinds of semiconductors, different clock rates, different buses, and different memory chips (not to mention different caches, different memory management units, different numbers of wait states, and different pipelining techniques). Under these conditions it is hard to compare the architectures independent of the implementations.

One measure that is sometimes used to factor out the implementation is program size. However, with memory getting cheaper every year, size is increasingly less important, and besides, it does not correlate well with speed. Another metric thought to be implemented independent is memory traffic. The idea here is that one can write a simulator for any computer and then have the simulator count the total number of memory references, both for code and data. If two simulated computers run the same benchmark program and one of them executes it in 1,000,000 memory references while the other needs only 800,000 all things being equal, the second computer is better.

However, even this seemingly technology-independent measure is not as foolproof as it first appears. The 80386 and many other computers prefetch instructions before they are needed. Every time a jump is taken, the contents of the prefetch buffer must be discarded. The memory cycles needed to prefetch these discarded instructions do not show up in the simulation runs, but can often add 25 percent extra to the bus load.

Although the performance Fig.s favour the RISC machines, in all fairness it should be pointed out that the CISC machines being benchmarked are greatly burdened by having to be compatible with ancient machines. When a group at IBM is told to build the fastest computer it can, the desingers are all excited at the challenge. Several milliseconds later, when they are also told that this new wonder must run every program anybody, anywhere in the world has ever written for any IBM mainframe since 1964, their spirits may go down a bit. Likewise, if the 80486 did not have to be compatible with everything going back to the 8088, the Intel designers could have built a very different machine.

From this perspective, one of the main factors in the speed of RISC machines is that they are all brand new architectures, free of the mistakes of the past. If a design team is told to go out and design a new CISC from scratch specifically with the goal of running C and Pascal, the resulting machine will certainly be much faster than any existing CISC machine.

A.24. GAIN DUE TO THE LARGE REGISTER FILE?

It is abundantly clear that the overlapping register window scheme makes a major contribution to the performance of the RISC I, RISC II, and SPARC chips, among others. On the other hand, the MIPS chip does not use overlapping windows, and it also delivers high performance. Still, it is interesting to speculate about how much of the performance is due to the register file. Hitchcock and Sprunt (1985) went further than that, and wrote simulators for the RISC I, 68000, and VAX. They made three versions of each simulator, one with overlapping register windows, one with nonoverlapping register windows, and one with a single set of registers. They then ran a standard set of benchmarks on all nine machines by hand-modifying the compiled assembly code to fit the register model. In this way the effect of the register file can be clearly separated from the large (VAX), MEDIUM, (68000) or small (RISC I) instruction set. The performance metric used was the total volume of CPU-memory traffic.

On the whole, the results are as one might suspect. The modified VAX with overlapping register windows was up to twice as fast as a normal one, and the 68000 with overlapping windows was up to four times as fast. On the other hand, removing the window scheme from the RISC I slowed it down by as much as a factor of nine. The only counterintuitive result is that on extremely recursive windows (a full window has to be saved on almost every call and restored on every return). Their conclusion is that overlapping register windows are a good idea, but that this says little about RISC versus CISC architectures.

To this, the RISC supporters reply that a large register window is only possible due to the simplicity of the RISC chips, and especially the lack of microcode. At any point in time, it is possible to put n transistors on a chip. Some of them are needed for the basic control unit, but the rest are available for other use. They can be used for storing a large microprogram or for storing a large register file, but these two uses are clearly in competition for the same transistors.

Thus the reply to Hitchcock and Sprunt is that the comparison of an overlapping tegister VAX with an overlapping register RISC machine is not fair. The former would take up much more chip area than the latter. Put in other words, if a RISC designer were given the amount of chip area an overlapping register VAX would need, he could either include a much larger register file, an on-chip memory cache, or other features to improve performance.

A.25. SUPERIORITY OF RISC MACHINES

The initial benchmark programs presented in the RISC literature tended to be very small. The longest test reported by Patterson and Piepho (1982), for example, ran less than 7 seconds. From these data it is clear that if your application consists of running tiny recursive functions like the towers of Hanoi all day, RISC is the way to go. But what if you want to run fat COBOL programs? The results are less clear.

The initial attempts to run Smalltalk and Lisp programs on the RISC I were disappointing. This led Patterson to design the SOAR (Smalltalk On A Risc) and SPUR (Symbolic Processing Using Risc) chips (Unger et al., 1984).

Even though these chips were more successful, the issue arises of whether RISC technology is suitable for computers that have to run multiple applications, and especially languages appreciably different from C. Some evidence comes from Coutant et al. (1986). They report on building a COBOL compiler for a RISC machine. Since no decimal arithmetic or other COBOL-type instructions were available, they wrote a library of procedures to do those things that a CISC machine normally does in microcode.

Furthermore, they were able to take advantage of compile-time information to enhance performance. For example, when moving a string, the compiler can call one procedure when the source and destination fields are the same size, and a different one when they are not. All in all, the resulting performance met the project's goals.

In addition to raw performance and language issues, other factors also are important to vendors and users. RISC chips are much simpler than CISC chips and use fewer transistors (even taking into account the large register file). This fact makes them easier to design and less likely to be subject to recall due to microcode bugs. It also reduces the time between design and shipment, a crucial issue in a rapidly moving and competitive industry.

Because RISC designs use relatively few transistors, they are well suited to very high-speed chips based on GaAs (Gallium Arsenide) instead of silicon. These fast chips have many fewer transistors than conventional silicon chips, so there is often not enough room for complex microprograms.

On the other hand, CISCV machines lend themselves better to a complete family of machines spanning a wide range in price and performance, such as the 360 series. Computer vendors like the family idea, because it makes it possible to provide personal computers, minicomputers, and mainframes that all run the same architecture. The smaller models can be microprogrammed and achieve their functionality in software (microcode) while the larger ones can be hardwired for speed. RISC architectures do not lend themselves as well to this concept.

Speculative Execution and Branch Prediction

As discussed in the section on the execution process for x86 translating procedures, some processors have the ability to execute multiple instructions at once. In some cases, not all of the results of the execution will be used, because changes in the program flow may mean that the instruction should never have been encountered in the first place. In particular, this occurs in the vicinity of branches, where a condition is tested and the program path is altered depending on the results.

Branches are very common in x86 code and represent a real problem for pipelining, because they mean that you can't always rely that instructions will go in a linear sequence. Pipelining means starting the next instruction before the first one has completed. When we come to a conditional test instruction (an "if.. then" instruction), we don't know until the condition test has been executed which instruction is next, so what are we supposed to put into the pipeline next?

A less sophisticated processor will stall the pipeline until the results are known, hurting performance. More advanced processors will speculatively execute the next instruction anyway, with the hope that it will be able to use the results if the branch goes the way it thinks it will. Still more advanced processors combine this with branch prediction, where the processor can actually predict with fairly good accuracy which way the branch will go based on past history.

Take the following code as an example:

```
IF A = B THEN
C = C + 1
ELSE
C = C - 1
END IF
```

Until the first instruction has been completely executed, we don't know if the next one will be the addition instruction or the subtraction instruction. A processor that speculatively executes may start both the addition and subtraction going at the same time, and then simply discard whichever one it turns out not to need. Or, it may make use of branch prediction to start only one of them, the one it feels is more likely to be the result of the "if" statement.

Branch prediction improves the handling of branches by making use of a special small cache called the branch target buffer or BTB. Whenever the processor executes a branch, it stores information about it in this area. When the processor next encounters the same branch, it is able to make an informed "guess" about which way the branch is likely to result. This helps keep the pipeline flowing and improves performance. (It's pretty cool too.) The larger the BTB is, the more branches it can maintain information for.

A.26. OUT-OF-ORDER EXECUTION

Processors that use multiple execution units can in theory complete the processing of instructions in the wrong order; instruction 2 can be executed before instruction 1 has completed. This flexibility improves performance since it allows execution with less "waiting" time. However, care must be taken to make sure that the results of execution are "reassembled" into the correct order to ensure that the program runs correctly. This is normally done by the retirement unit on translating processors.

Register Renaming

Register renaming is a technique used to allow multiple execution paths without conflicts between different execution units trying to use the same registers. Instead of just one set of registers being used, multiple sets are put into the processor. This allows different execution units to work simultaneously without unnecessary pipeline stalls.

Write Buffers

Write buffers are used to hold the results of execution until they can be written back to registers or memory locations. More write buffers allow more execution to occur without stalling the pipelines.

Processor Performance

Considering how important it is, the value of a processor is measured using surprisingly few parameters. Since most processors work in a similar fashion and have similar external characteristics (*i.e.*, in terms of how a user would see them) the most important question becomes: "how fast does it run?" This is a very fair question to focus on, because while the internal details of how the processor works are interesting, what really counts to the computer user is the performance level of the processor in question.

A.27. SOME EXAMPLE PROCESSORS

In the following sections, a selection of modern processors are described. They provide examples of the manner in which the devices described in previous sections have been implemented in production processors.

- **PowerPC 601**

The Power architecture was designed by IBM for a new family of RISC processors which started with the RS/6000 machines. The original RS/6000's used a five-chip set to form the CPU.

The architecture was adopted by Apple and Motorola and the three companies commenced work on single chip implementations of the architecture. The first of these was the PowerPC 601, which powered the first of the PowerMacs.

- **Instruction Unit**

The instruction unit fetches a cache line (32-bytes or 8 instructions) from the cache in a single cycle. The processor is a 3-way superscalar, so it can potentially issue as many as 3 instructions from the bottom of the 8-word instruction queue in one cycle. (The upper four words of the instruction queue are used as a buffer to reduce the frequency of cache accesses.)

Instructions may be issued out-of-order but the processor ensures that they have the same effect as if they were issued and completed strictly in program order.

- **Functional Units**

The Integer Unit (IU), Floating Point Unit (FPU) and Branch Processing Unit (BPU) can take an instruction each in one cycle.

The IU contains a general purpose register file of 32 32-bit registers (GPR).

The FPU contains a floating point register file of 32 double-precision (64-bit) registers.

- **Cache**

The cache is a unified, 8-way set associative physically addressed cache containing 128 sets of 8 32-byte lines for a total of 32K bytes.

- **MMU**

The processor emits 32-bit addresses - 4 bits of which select one of 16 segment registers containing 24 bit segment addresses. These 24 bits are added to the lower 28 bits of the address to form a 52-bit virtual address. This is translated by the MMU to a 32-bit physical address. There are separate instruction (ITLB) and data (UTLB) TLBs.

Block Diagram

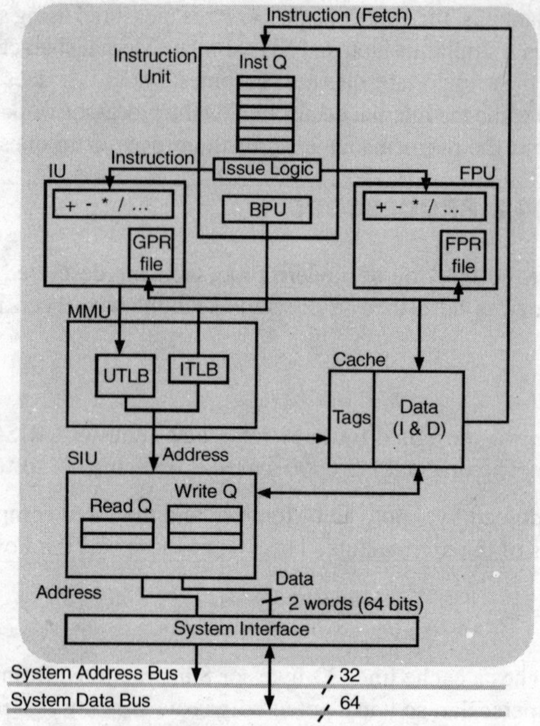

Fig. A.15

A.28. BRANCH TARGET BUFFERS

- **Branching Costs**

Branching is always expensive in a heavily pipelined processor. Even unconditional branches require interruption of current instruction fetching operations and restarting the instruction stream from a new memory location - any instructions which may have been fetched from memory and are waiting in instruction buffers or cache lines must be discarded. Conditional branches can be much more disruptive

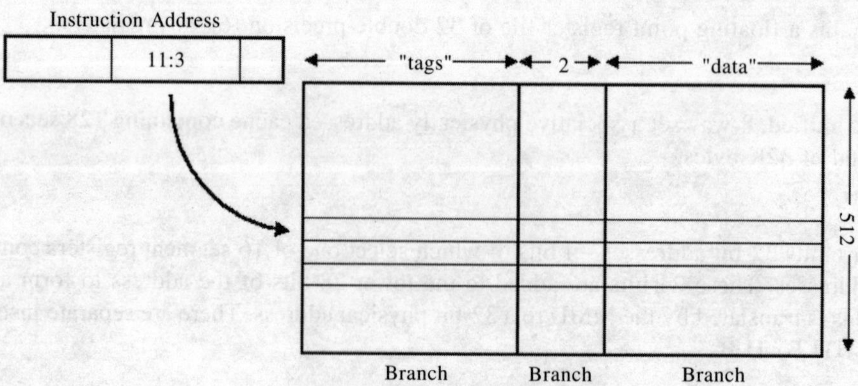

Fig. A.16

as they must often wait for operands to be generated or status bits to be set before the direction of the branch (taken or not taken) can be determined: an efficient processor may have fetched and partially executed many instructions beyond the branch before it is known whether to take it or not. Thus considerable effort has been devoted to reducing the cost of branches.

- **Branch Prediction**

Predicting that a branch will be taken turns out to be a good choice most of the time. The majority of branches are those at the end of loops which branch back to the start of the loop. If the loop iterates n times, then the branch will be taken (n-1)/n of the time.

- **Branch Target Buffers**

To further increase the prediction accuracy, many systems now incorporate branch target buffers.

- **Commercial Processor Data**

In an almost futile effort to keep up to date with recent commercial releases, some statistics for various commercially available processors are collected here. An estimate of release date is included also, giving some idea of the rate of change.

A.29. PAGING AND VIRTUAL MEMORY

On most modern computers, each program can potentially address a vast amount of memory, four gigabytes on a typical 32 bit machine. Few computers actually have that much memory, and even the ones that do need to share it among multiple programs. Paging hardware divides a program's address space into fixed size pages, typically 2K or 4K bytes in size, and divides the physical memory of the computer into page frames of the same size. The hardware includes page tables with an entry for each page in the address space, as shown in Fig. A.17.

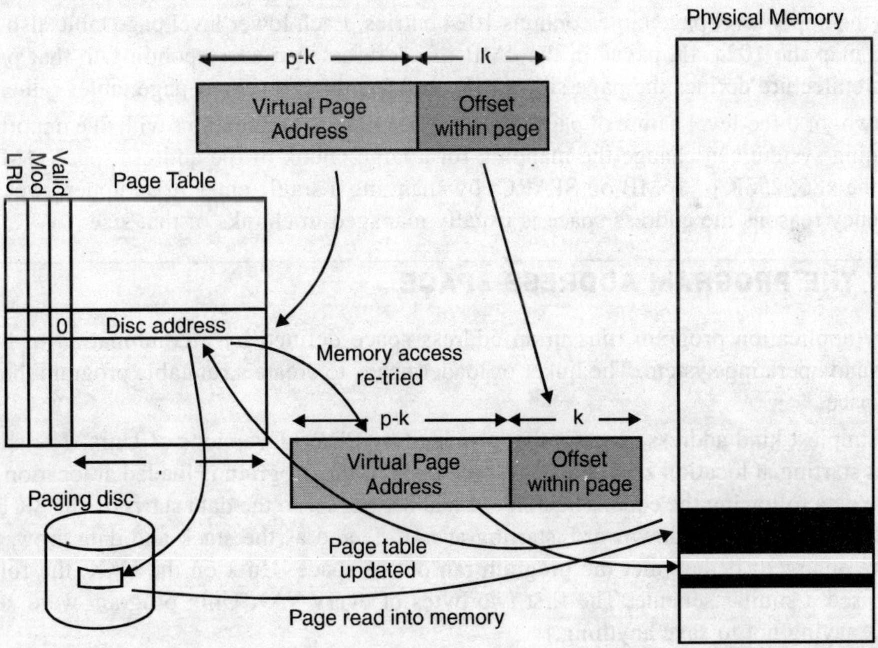

Fig. A.17. Page mapping. Picture of pages mapped through a big page table to real page frames.

A page table entry can contain the real memory page frame for the page, or flag bits to mark the page "not present." When an application program attempts to use a page that is not present, hardware generates a page fault which is handled by the operating system. The operating system can load a copy of the contents page from disk into a free page frame, then let the application continue. By moving pages back and forth between main memory and disk as needed, the operating system can provide virtual memory which appears to the application to be far larger than the real memory in use.

Virtual memory comes at a cost, though. Individual instructions execute in a fraction of a microsecond, but a page fault and consequent page in or page out (transfer from disk to main memory or vice versa) takes several milliseconds since it requires a disk transfer. The more page faults a program generates, the slower it runs, with the worst case being thrashing, all page faults with no useful work getting done. The fewer pages a program needs, the fewer page faults it will generate. If the linker can pack related routines into a single page or a small group of pages, paging performance improves.

If pages can be marked as read-only, performance also improves. Such pages don't need to be paged out since they can be reloaded from wherever they came from originally. If identical pages logically appear in multiple address spaces, which often happens when multiple copies of the same program are running, a single physical page suffices for all of the address spaces.

An x86 with 32 bit addressing and 4K pages would need a page table with 2^{20} entries to map an entire address space. Since each page table entry is usually four bytes, this would make the page tables an impractical 4 megabytes long. As a result, paged architectures page the page tables, with upper level page tables that point to the lower level page tables that point to the actual page frames corresponding to virtual addresses.

On the 370, each entry in the upper level page table (called the segment table) maps 1MB of address space, so the segment table in 31 bit address mode may contain up to 2048 entries. Each entry in the segment table may be empty, in which case the entire segment is not present, or may point to a lower level page table that maps the pages in that segment. Each lower level page table has up to 256 entries, one for each 4K chunk of address space in the segment. The x86 divides up its page tables similarly, although the boundaries are different. Each upper level page table (called a page directory) maps 4MB of address space, so the upper level page table contains 1024 entries. Each lower level page table also contains 1024 entries to map the 1024 4K pages in the 4MB of address space corresponding to that page table. The SPARC architecture defines the page size as 4K, and has three levels of page tables rather than two.

The two- or three-level nature of page tables are invisible to applications with one important exception: the operating system can change the mapping for a large chunk of the address space (1MB on the 370, 4MB on the x86, 256K or 16MB on SPARC) by changing a single entry in an upper level page table, so for efficiency reasons the address space is usually managed in chunks of that size.

A.30. THE PROGRAM ADDRESS SPACE

Every application program runs in an address space defined by a combination of the computer's hardware and operating system. The linker or loader needs to create a runnable program that matches that address space.

The simplest kind address space is that provided by PDP-11 versions of Unix. The address space is 64K bytes starting at location zero. The read-only code of the program is loaded at location zero, with the read-write data following the code. The PDP-11 had 8K pages, so the data starts on the 8K boundary after the code. The stack grows downward, starting at 64K-1, and as the stack and data grow, the respective areas were enlarged; if they met the program ran out of space. Unix on the VAX, the follow-on to the PDP-11, used a similar scheme. The first two bytes of every VAX Unix program were zero (a register save mask saying not to save anything.)

As a result, a null all-zero pointer was always valid, and if a C program used a null value as a string pointer, the zero byte at location zero was treated as a null string. As a result, a generation of Unix programs in the 1980s contained hard-to-find bugs involving null pointers, and for many years, Unix ports to other architectures provided a zero byte at location zero, because it was easier than finding and fixing all the null pointer bugs.

Modern Unix systems typically use an only slightly more complex scheme. Programs are still loaded at location zero, or more often one page after location zero, in order to make zero a deliberately invalid address and catch any lurking null pointer bugs. Data starts on the page after the end of the code, and the stack grows downward from a fixed high address.

Unix systems put each application program in a separate address space, and the operating system in an address space logically separate from the applications. (Sometimes the operating system and the running application share a single address space, where addresses with the high bit set are for the system and those with the high bit clear are for the application, but the application has no access to the system space.) Other systems put multiple programs in the same address space, making the linker and particularly the loader's job more complex, because a program's actual load address isn't known until the program is about to be run.

MS-DOS systems use no hardware protection, so the system and running applications share the same address space. When the system runs a program, it finds the largest chunk of free memory, which can be anywhere in the address space, loads the program into it, and starts it. IBM mainframe operating systems do roughly the same thing. In both cases, either the program loader or in some cases the program itself has to adjust to the location where the program is loaded.

MS Windows has an unusual loading scheme. Each program is linked to load at a standard starting address, but the executable program file contains relocation information. When Windows load the program, it places the program at that starting address if possible, but may load it somewhere else if the preferred address isn't available.

A.31. MAPPED FILES

Virtual memory systems move data back and forth between real memory and disk, paging data to disk when it doesn't fit in real memory. Originally, paging all went to "anonymous" disk space separate from the named files in the file system. Soon after the invention of paging, though, designers noticed that it was possible to unify the paging system and the file system by using the paging system to read and write named disk files. When a program maps a file to a part of the program's address space, the operating system marks all of the pages in that part of the address space not present, and uses the file as the paging disk for that part of the address space, as in Fig. A.18. The program can read the file merely by referencing that part of the address space, at which point the paging system loads the necessary pages from disk.

There are three different approaches to handling writes to mapped files. The simplest is to map a file read-only (RO), so that any attempt to store into the mapped region fail, usually causing the program to abort. The second is to map the file read-write (RW), so that changes to the memory copy of the file are paged back to the disk by the time the file is unmapped. The third is to map the file copy-on-write (COW, not the most felicitous acronym). This maps the page read-only until the program attempts to store into the page. At that time, the operating system makes a copy of the page which is then treated as a private page not mapped from a file. From the program's point of view, mapping a file COW is very similar to allocating a fresh area of anonymous memory and reading the file's contents into that area, since changes in the program makes are visible to that program but not to any other program that might have mapped the same file.

A.32. SHARED LIBRARIES AND PROGRAMS

In nearly every system that handles multiple programs simultaneously, each program has a separate set of page tables, giving each program a logically separate address space. This makes a system considerably more robust, since buggy or malicious programs can't damage or spy on each other, but it potentially could cause performance problems. If a single program or single program library is in use in more than one address space, the system can save a great deal of memory if all of the address spaces share a single physical copy of the program or library. This is relatively straight forward for the operating system to implement — just map the executable file into each program's address space. Unrelocated code and read only data are mapped RO, writable data are mapped COW. The operating system can use the same physical page frames for RO and unwritten COW data in all the processes that map the file. (If the code has to be relocated at load time, the relocation process changes the code pages and they have to be treated as COW, not RO.)

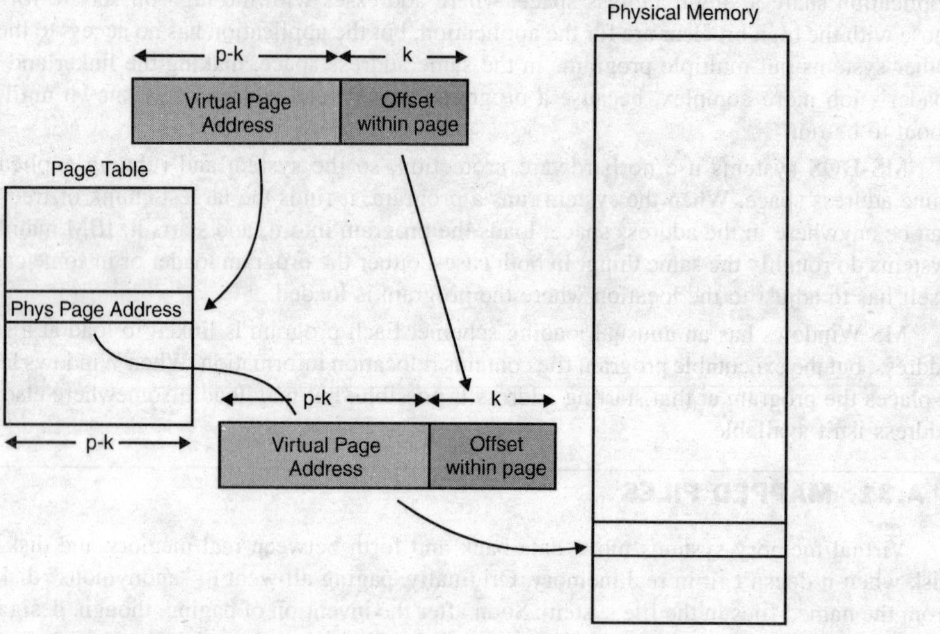

Fig. A.18. Mapping a file Program points to set of page frames that map to disk file or local RAM.

Considerable linker support is needed to make this sharing work. In the executable program, the linker needs to group all of the executable code into one part of the file that can be mapped RO, and the data into another part that can be mapped COW. Each section has to start on a page boundary, both logically in the address space and physically in the file. When several different programs use a shared library, the linker needs to mark the each program so that when each starts, the library is mapped into the program's address space.

A.33. POSITION-INDEPENDENT CODE

When a program is in use in several different address spaces, the operating system can usually load the program at the same place in each of the address spaces in which it appears. This makes the linker's job much easier, since it can bind all of the addresses in the program to fixed locations, and no relocation need be done at the time the program is loaded.

Shared libraries complicate this situation considerably. In some simple shared library designs, each library is assigned a globally unique memory address either at system boot time or at the time the libraries are created. This puts the each library at a fixed address, but at the cost of creating a serious bottleneck to shared library administration, since the global list of library memory addresses has to be maintained by the system manager, not by individual users. Furthermore, if a new version of a library appears that is larger than the previous version and doesn't fit into the address space assigned, the entire set of shared libraries and, potentially, all of the programs that reference them, may need to be relinked.

The alternative is to permit different programs to map a library to different places in the address space. This eases library administration, but the compiler, and linker, and program loader need to cooperate so that the library will work regardless of where in the address space the library appears.

One simple approach is to include standard relocation information with the library, and when the library is mapped into each address space, the loader can fix up any relocatable addresses in the program to reflect the loaded addresses. Unfortunately, the process of fixing up involves writing into the library's code and data, which means that the pages will no longer be shared, if they're mapped COW, or the program will crash if the pages are mapped RO.

To avoid this problem, shared libraries use Position Independent Code (PIC), code which will work regardless of wherein memory it is loaded. All the code in shared libraries is usually PIC, so the code can be mapped read-only. Data pages still usually contain pointers which need relocation, but since data pages are mapped COW anyway, there's little sharing lost.

For the most part, PIC is pretty easy to create. All three of the architectures we discussed in this chapter use relative branching, so that branch instructions within the routines need no relocation. References to local data on the stack use based addressing relative to a base register, which doesn't need any relocation, either. The only challenges are calls to routines not in the shared library, and references to global data. Direct data addressing and the MIPS high/low register loading trick won't work, because they both require run-time relocation. Fortunately, there are a variety of tricks one can use to let PIC code handle inter-library calls and global data.

A.34. INTEL 386 SEGMENTATION

The final topic in this chapter is the notorious Intel architecture segmentation system. The x86 series is the only segmented architecture still in common use, other than some legacy ex-Burroughs Unisys mainframes, but since it's so popular, we have to deal with it.

The original 8086 was intended as a follow-on to Intel's quite popular 8-bit 8080 and 8085 microprocessors. The 8080 has a 16 bit address space, and the 8086 designers were torn between keeping the 16 bit address space, which made translation of 8085 easier and permitted more compact code, and providing a larger address space to give "headroom" for future applications in larger programs. They compromised, by providing multiple 16 bit address spaces. Each 16 bit address space was known as a segment.

A running x86 program has four active segments defined by the four segment registers. The CS register defines the code segment, from which instructions are fetched. The DS register defines the data segment, from which most data are loaded and stored. The SS register defines the stack segment, used for the operands of push and pop instructions, the program address values pushed and popped by call and return instructions, and any data reference made using the EBP or ESP as a base register. The ES register defines the extra segment, used by a few string manipulation instructions. The 386 and later chips define two more segment registers FS and GS. Any data reference can be directed into a specific segment by using a segment override. For example, the instruction MOV EAX,CS:TEMP fetches a data value from the location TEMP in code segment rather than the data segment. The FS and GS segments are only used via segment overrides.

The segment values need not all be different. Most programs set the DS and SS values the same, so that pointers to stack variables and global variables can be used interchangeably. Some small programs set all four segment registers the same, providing a single address space known as tiny model.

On the 8086 and 186, the architecture defined a fixed mapping from segment numbers to memory addresses by shifting the segment number four bits to the left. Segment number 0x123 would start at memory location 0x1230 for example. This simple addressing is known as real mode. Programmers often refer informally to paragraphs, 16-byte units of memory that a segment number can address.

The 286 added a protected mode, in which the operating system can map segments to arbitrary places in real memory and can mark segments as not present, providing segment based virtual memory. Each segment can be marked executable, readable, or read/write, providing segment-level protection. The 386 extended protected mode to 32 bit addressing, so that each segment can be up to 4GB in size rather than only 64K.

With 16 bit addressing, all but the smallest programs have to handle segmented addresses. Changing the contents of a segment register is quite slow, 9 clock cycles on a 486 compared to 1 cycle to change the contents of a general purpose register. As a result, programs and programmers to go great lengths to pack code and data into as few segments as possible to avoid having to change the contents of the segment registers. Linkers aid this process by providing "groups" that can collect related code or data into a single segment. Code and data pointers can be either near, with an offset value but no segment number, or far, with both segment and offset. Compilers can generate code for various memory models which determine whether code and data addresses are near or far by default. Small model code makes all pointers near and has one code and one data segment. Medium model code has multiple code segments (one per program source file) using far calls, but a single default data segment. Large model code has multiple code and data segments and all pointers are far by default. Writing efficient segmented code is very tricky, and has been well documented elsewhere.

Segmented addressing places significant demands on the linker. Every address in a program has both a segment and an offset. Object files consist of multiple chunks of code which the linker packs into segments. Executable programs to be run in real mode have to mark all of the segment numbers that occur in the program so they can be relocated to the actual segments where the program is loaded. Executable programs to be run in protected mode further have to mark what data is to be loaded into what segment and the protection (code, read-only data, read-write data) for each segment.

Although the 386 supports all of the 16 bit segmentation features of the 286, as well as 32 bit versions of all of the segmentation features, most 32 bit programs don't use segmentation at all. Paging, also added in the 386, provides most of the practical benefits of segmentation without the performance cost and the extra complications of writing segment manipulation code. Most 386 operating systems run applications in the tiny model, creating a single code segment and a single data segment each 4GB long and mapping them both to the full 32 bit paged address space. Tiny model is a misnomer, since even though the program is only using a single segment, that segment can be the full size of the address space.

The 386 makes it possible to use both 16 bit and 32 bit segments in the same program and a few operating systems, notably Windows 95, take advantage of that ability. Windows 95 runs a lot of legacy Windows 3.1 code in 16 bit segments in a shared address space, while each new 32 bit program runs in its own tiny model address space, with the 16-bit programs' address space mapped in to permit calls back and forth.

B

Appendix

Performance Evaluation

B.1. ISSUES OF SYSTEM PERFORMANCE

We all want better performance. Let us take the case of speed performance. What do we mean by that?

- Elapse time to reach 60 KMph? (11.6 sec., 8 sec. 10 sec. etc)
- What is the Top speed?
- Kilometers per litre?
- Reliability?
- Comfortable?

Different users can come to different conclusion. This is the same for a computer.

B.2. PROCESSOR PERFORMANCE FACTORS

Since the performance of a processor is (in most cases) based on how many instructions it can execute in a given time, it has become common to use the words "performance" and "speed" somewhat interchangeably. Unfortunately, the word "speed" has too many meanings when it comes to processors, and as a result frequently one person will use it to mean one thing and another person to mean quite another.

There are two major factors that determine the performance level of a processor:

Clock Speed: The processor's clock speed is a measure of how fast it is running in raw terms, meaning, how many clock cycles it has to work within a given period of time. Using the analogy of a bicycle, this would be equivalent to how fast you are pedaling.

Architecture: The internal design and architecture of the processor measures how efficiently the processor does its work and how much it does with each clock cycle. In the bicycle analogy this would be a measure of how hard you are pedaling.

Since clock speed is easy to see and understand—it's just a number—and architecture is both complex and difficult to understand, it's not surprising that the former receives much more attention than the latter. This is unfortunate, because looking at just the clock speed of a processor is very deceptive, because it tells only one part of the picture. In fact, this is more so today than ever before due to the much greater variety in processor designs and technologies. The "P" rating scheme was invented by Intel competitors AMD and Cyrix to provide what they feel is an assessment of the value of their processors that is more fair than using just clock speed.

One great example showing the importance of architecture is AMD's K5 series of processors. Some of these processors actually have very different performance ratings despite running at the same clock speed. For example, the K5-PR100 and the K5-PR133 both run at the same clock speed, 100 MHz, but the PR133 has approximately 33% better performance, due to changes that were made to the processor's internal architecture.

Clock speed can be used to compare processors only if they are identical internally. (Obviously, the K5-PR100 and K5-PR133 aren't the same internally if they have the same clock speed and different performance. I think AMD should have given them more clearly different designations to avoid this confusion.) This means you can only use clock speed to compare the performance of otherwise identical processors. A Pentium 200 is in fact 20% faster than a Pentium 166. But it isn't 20% faster than a Pentium with MMX 166, because of the latter's architectural improvements, such as a larger internal cache.

B.3. INDICATORS OF PERFORMANCE MEASUREMENT

(1) MIPS (million instructions per second)

This measures the absolute speed of a CPU. This is an indication how fast a CPU can run. Since different CPU have different instructions set, this is often meaningless. (for example: which instruction out of all possible instructions? If multiple instructions are used, what are they? How are they determined?)

Sometimes, a VAX unit π_6 performance (VUP) is used as a basic unit.

Sometimes MIPS is stated as "Meaningless indicator of performance statistic": the primary difficulty is that different types of computers have different sets of usable machine language instructions, and different relative speeds for the different instructions.

A secondary difficulty with the use of MIPS to predict performance is that in many situations the speed of the CPU is not the limiting factor on system performance. We go into more detail in other sections, but other possible bottlenecks include the following:

- The delay before data starts coming back from disk drives,
- The rate at which data is returned from disk drives once it starts coming back,
- The delay in retrieving information from system RAM,
- The rate at which data can be sent through network connections,

The speed with which the screen display can be updated.

MIPS is useful only when the two machines being compared are of very similar design: same machine language, same types of input/output circuitry, same types of peripheral devices, etc., and are expected to perform very similar work.

(2) Clock Speed

If two systems have the same type of CPU (both 80486, both 68040, etc.) then comparing their clock speeds provides a good first approximation to the relative speeds of the systems for CPU-bound tasks. Several factors can complicate the comparison.

CPU-external vs CPU-internal: the internal speed of the CPU may be a multiple of the external clock signal. For example, so-called 486 DX2 chips operate with an internal speed twice that of the external clock. (You can readily guess which number will appear in advertisements.)

CPU vs primary bus: the CPU may operate at a speed that is a multiple of the speed used by the primary system bus, the interconnection between the CPU, RAM, and I/O interface circuitry.

CPU vs memory: a common high-end CPU speed is 300 MHz. That is a clock that takes 3.333 nanoseconds per cycle. The fastest common RAM, likely to be installed on such a system, takes perhaps 10 nanoseconds to store or to retrieve information. Therefore many CPU cycles can be wasted in idle "wait states" until a memory read or write is completed. See the discussion of memory cache, in the next section.

CPU vs I/O bus: in systems that have a separate I/O bus, its speed may be different from that of the CPU.

Performance Measures

One commonly quoted measure of performance is peak performance, which is the maximum rate at which some operation can be executed. For example, the number of single-cycle operations, perfectly scheduled to keep all pipelines operating at 100% utilization, that can be executed in a tight loop, entirely in cache.

MIPS – million instructions per second

BIPS, BOPS, GIPS, GOPS–billions (giga–109)of instructions/ operations per second

TOPS – Trillions or tera operation per second. Often teraops.

FLOPS – floating point operations per second (MFLOPS, GFLOPS, TFLOPS)

LIPS – logical inferences per second

CPS or COPS – Connections per second

Beyond tera (1012) are peta (1015), exa (1018), yota (1021) and zeta(1024).

Absolute peak performance (instructions per second) is based on the clock rate of the processor (Rclock), and on the minimum number of clock Cycles Per Instruction (CPI) attainable. The clock rate depends on the technology used to implement the processor. The minimum CPI depends on the instruction set (usually just a subset of instructions have the minimum CPI) and the speed of the innermost cache, which is usually on the same chip as the CPU in modern processors. Note that CPI may be less than one if multiple function units are employed in parallel.

peakabs = Rclock/CPImin

Thus, absolute peak performance ignores the fact that there is a mix of CPI values that depend on the instruction set, the cache behavior, and the proportions in which instructions are executed.

Note, however, that as Rclock increases, peak performance increases linearly. This is why clock rate is the most often quoted performance figure. It is only a bit more meaningful than peak performance — can you think of a situation where a clock rate quotation may be useful?

When comparing two machines with the same architecture, but different clock rates, the ratio of the clock rates can indicate the difference in performance. When is this comparison invalid? (when memory, I/O, communication, do not increase in speed by the same proportion)

What people really want to know is how fast a machine runs a given program. But that is a much more complex performance measurement. What factors determine this performance?

- Algorithm and data set (size, length of execution, pattern of access — affects memory system)
- Compiler (efficiency of code generated, ability to optimize access patterns)
- Instruction set (how many instructions does it take to encode the program?)
- Available operations (floating point support?)
- Number and length of pipelines

- Schedule of instructions, pipeline efficiency
- Branch prediction accuracy, branch frequency, branch penalty
- Operating system (timeout period and cost, other processes, etc.)
- CPI
- Clock rate
- Memory system performance (cache hit rate, miss penalty)
- I/O system performance and overhead

The only real way to accurately measure the performance of a system is to code a program of interest and execute it. Benchmarks have been created to serve as standard codes that can be tried on different machines for comparison purposes. Some popular ones include SPECmark, Perfect Club, Linpack, Livermore loops. There are also specialized benchmarks, such as those that test performance of PCs on popular packages, graphics benchmarks, transaction processing benchmarks, network benchmarks, etc.

Machine	Clock Rate	Peak Performance	CPU Time
VAX 11/780	5 MHz	1 MIPS	12 seconds
IBM RS/6000	25 MHz	18 MIPS	1 second

Throughput is the number of transactions per second for a program, where a transaction is some meaningful unit of work for the application. For example, queries per second for a database, or gates per second for a VLSI layout tool.

B.3.1. Calculating Performance

Let's assess the performance of our simple processor. Assume that the whole system is driven by a clock at f MHz. This means that each clock cycle takes

$t = 1/f$ microseconds

Thus a processor with a clock running at 100MHz is operating with 10ns clock cycles. Generally, a processor will execute one step every cycle, thus, for a memory load instruction, our simple processor needs:

Step	Operation	Time (cycles) Notes
1	PC to bus	1
2	Memory response	*tac*
3	Decode and register access	1
4	ALU operation and latch result to MAR	1
5	Memory response	*tac*
6	Increment PC -	Overlap with step 3
	Total	3 + 2**tac*

If the memory response time is, say, 100ns, then our simple processor needs 3x10+2*100 = 230ns to execute a load instruction. For the add instruction, we make a similar table:

Step	Operation	Time (cycles) Notes
1	PC to bus	1
2	Memory response	*tac*
3	Decode and register access	1
4	ALU operation and latch result to destination register	1
5	Increment PC -	Overlap with step 3
	Total	3 + *tac*

So an add instruction requires $3 \times 10 + 100 = 130$ ns to execute. A store operation will also need more than 200 ns, so instructions will require, on average, about 150 ns.

B.3.2. Performance Measures

One commonly used performance measure is MIPS or millions of instructions per second. Our simple processor will achieve:

$$1/(150\text{x}10\text{-}9) = \sim 6.6 \times 10^6 \text{ instructions per second}$$
$$= \sim 6.6 \text{ MIPS}$$

As you will know from reading the popular literature, 100MHz is a very common figure for processors in 1998 (leading edge commercial processors have clocks which are more than 5 times faster) and a MIPS rating of 6.6 is very ordinary. In fact to be competitive, a 100MHz processor should be achieving of the order of 100MIPS - or one instruction for each machine cycle. One of the main aims of this course is to examine how this is achieved.

Bottlenecks

From the simplistic analysis presented above, it will be obvious that access to main memory is a major limiting factor in the performance of a processor. Management of the memory hierarchy to achieve maximum performance is one of the major challenges for a computer architect. Unfortunately, the hardware maxim "Smaller is faster" conflicts with programmers' and users' desires for more and more capabilities and more elaborate user interfaces in their programs - resulting in programs that require megabytes of main memory to run. This has led the memory manufacturers to concentrate on density (improving the number of bits stored in a single package) rather than speed. They have been remarkably successful in this: the growth in capacity of the standard DRAM chips which form the bulk of any computer's semiconductor memory has matched the increase in speed of processors. However the increase in DRAM access speeds has been much more modest - even if we consider recent developments in synchronous RAM and FRAM. Another reason for the manufacturer's concentration on density is that a small increase in DRAM access time has a negligible effect on the effective access time which needs to include overheads for bus protocols. (The 100ns figure used above assumes 60ns of DRAM access time and a - very optimistic - allowance of 40ns for bus overhead.)

Cache memories are the most significant device used to reduce memory overheads and they will be examined in some detail later. However, a host of other techniques such as pipelining, pre-fetching, branch prediction, etc are all used to alleviate the impact of memory fetch times on performance.

B.4. ENHANCING SYSTEM PERFORMANCE USING CACHE MEMORY

System performance enhancements typically address one or more bottlenecks. Cache memory addresses bottlenecks in the RAM subsystem and bottlenecks in the data paths between RAM and the CPU. A memory cache is a small, very high-speed memory that is connected to the CPU through a very high-speed bus. The cache stores recently accessed data, keeping track of the data and of the address from which it came. Since it is common for software to contain loops, which re-access the same memory locations after small numbers of intervening instructions, a modest size cache can often provide requested information (instructions or data) promptly without having to go out on the system bus to fetch it (slowly) from the RAM.

The original IBM PC's CPU operated at 11.8 MHz (208 ns/cycle), with RAM that responded in 150 ns, so that the CPU never had to wait an extra cycle. A modern CPU may operate at 300 MHz (3.33 ns/cycle), faster by a factor of at least 60. That modern CPU will, however, be connected to RAM that responds in 10 ns, faster by just a factor of 15. Thus, the modern CPU will routinely have to wait until the third or fourth cycle before the information is available from RAM. There is every reason to expect that these trends will continue: CPUs will get faster by leaps and bounds, and RAM will get faster too, but much more gradually. Therefore, we can see that memory cache technology is already (and will continue to be) vital to achieving optimum performance from personal computers.

Modern CPU chips usually have some cache designed into the chip itself; this is known as "primary cache," or "level 1 cache." Some CPU circuit boards include cache memory; this is known as "secondary cache," or "level 2 cache." The expensive part of the cache is the very high speed RAM. Often the CPU will be designed to include the circuitry that keeps track of the data, the cache controller, but with an empty socket for the cache RAM to be added as an optional upgrade.

An even more expensive technique is to install the L2 cache on a dedicated bus, operating a much higher speed than the system bus. In the most extreme case, such a "lookaside cache" will operate at the same speed as the CPU itself, so that there are no wait states for data retrieved from the L2 cache. The expense occurs because the L2 cache RAM needs to be very fast and because the CPU has to be constructed with interface circuitry to drive both the lookaside cache bus and the regular system bus. This technique is being used with Digital Alpha AXP and with PowerPC G3 systems.

B.5. PERFORMANCE ENHANCING ARCHITECTURAL FEATURES

It is important to remember that all of the architectural features of a processor matter very little taken by themselves. They are a means to an end, namely better performance. For most users, what really matters is how the design of the processor translates to faster execution, much the way most car drivers don't know what the displacement of their engine is, or the intricacies of fuel injection. Still, those who want to know how the processor works "under the hood", will find the descriptions of these features useful.

B.5.1. Superscalar Architecture

Superscalar architecture refers to the use of multiple execution units, to allow the processing of more than one instruction at a time. This can be thought of as a form of "internal multiprocessing", since there really are multiple parallel processors inside the CPU. Most modern processors are superscalar; some have more parallel execution units than others.

B.5.2. Superpipelining

Superpipelining simply refers to pipelining that uses a longer pipeline (with more stages) than "regular" pipelining. In theory, a design with more stages, each doing less work, can be scaled to higher clock frequency. However, this depends a lot on other design characteristics, and it isn't true by default that a processor claiming superpipelining is "better" than one that does not.

B.6. PROCESSOR V/S SYSTEM PERFORMANCE

It is important to remember that the processor is not the only component in the system that determines overall system performance. It is an important one, but it isn't the only one. Many hardware companies like to overstate the value of the processor's performance (sometimes focusing just on the clock speed). For example, you'll see claims that a system running with a Pentium 150 is "50% faster" than one with a Pentium 100. Or a retailer will try to sell you a 120 MHz Pentium OverDrive for your Pentium 60 system with claims that it will "double performance". These claims are, in almost every case, totally untrue. In fact, they are usually nowhere close to being true.

The reason is that speeding up the processor only improves system performance for those aspects of system use that depend on the processor. In most systems, the processor is already fast enough, but it is other parts of the system—the memory, system buses, hard disk and video card especially—that are the "bottlenecks" to system performance. Since most processors are already much faster than the devices that support them, they spend a great deal of time waiting around for data that they can use. Putting a still faster processor in place of the current one, will not yield a very large performance increase if this is the case, because the faster processor will just spend more time waiting.

One of the most important factors that influences overall system performance is memory bus speed. The availability of Pentium PCs with both 60 and 66 MHz system bus speeds has emphasized this. Since the processors on these machines are so fast, they end up waiting a great deal on data from the memory bus. As a result, a slower processor running on a 66 MHz system bus can provide comparable performance to a faster one on a 60 MHz bus.

The classical example here is the Pentium 133, which provides performance virtually identical to the Pentium 150 in most cases. In fact, the Pentium 150 scores below the Pentium 133 in some benchmarks. Another way to look at this: setting a Pentium 100 to run with a memory bus speed of 50 MHz and a clock multiplier of 2 will result in a significant decrease in system (not processor) speed compared to its normal setting of 66 MHz and clock multiplier of 1.5, even though in both cases the CPU is running at 100 MHz.

The faster processors get, the more this phenomenon is observed. This is sort of a law of diminishing returns in processor speed. When you get to very high clock speeds, the improvement in overall system performance is minimal even if the processor's performance increases a great deal. The classic example is the Pentium 200, which by most benchmarks provides less than 10% system performance improvement over the Pentium 166, despite having 20% higher benchmarks when looking just at the processor.

B.7. INTEGER AND FLOATING-POINT PERFORMANCE

Most performance measures look at the processor's ability to process integer data: whole numbers, text and the like. This is perfectly valid, because most processing is done on this sort of data. However, intensive floating point applications, such as spreadsheets and some graphics programs and games, make heavy use of the processor's floating point unit (FPU).

Processors that have similar performance on integer tasks can have very different performance on floating point operations. In particular, a number of competitors to Intel chips have come onto the market that have comparable integer performance, but far inferior floating point performance. For most users this is not an issue since heavy floating-point applications are unusual, but it is something to be aware of.

B.8. THE "P-RATING"

When Intel competitors AMD and Cyrix developed their chips, they had the advantage of using a newer design that enabled them to make architectural changes to improve performance. The result is that many of their processors have greater performance than a Pentium processor running at the same clock speed. Since the market has a tendency to over-value raw clock speed instead of looking at bottom line performance, the "P rating" scheme was invented to provide what AMD and Cyrix consider to be more reasonable comparisons to the Pentium line than using clock speed.

The "P rating" scheme is quite simple really; it uses standardized system-oriented benchmarks to compare processors against Pentiums. (System-based benchmarks reflect "real world" usage better than processor-oriented benchmarks, and the theory is that as long as the systems are identical, this can be used to compare just the processors.) A standard system configuration is set up with a Pentium processor, a good-quality benchmark such as Winstone 96 is run on it, and the score recorded. The Pentium is then replaced with different speed Pentiums and their scores recorded. Finally, the Pentium is replaced with the chips whose "P ratings" we are trying to determine.

The scores are tallied up, and the AMD or Cyrix chip is assigned a P rating equal to the speed of the fastest Pentium whose benchmark the AMD or Cyrix chip exceeded. This yields a reasonable comparison between the processor families, although not one that is definitely foolproof. Since the rating uses system-based benchmarks, it is dependent on other components in the system. While of course all the processors are tested with the same motherboard, memory and chipset, some processors might be better tuned to work with certain support hardware compared to others, especially when comparing chips with different internal architecture (which these tests almost always do).

Warning: The "P rating" is based on system benchmarks that concentrate mostly on integer performance, not floating point.

This is important to remember, because many Intel-compatible processors have far inferior floating point units compared to the Intel processors that they claim to be equivalent to with their P rating, and will not perform based on their "P rating" in floating-point-intensive applications.

B.9. WHAT IS A COMPUTER BENCHMARK

A computer **benchmark** is a program which measures one or more performance attributes of a computer system, and provides a relative score to allow comparison between different computer systems.

Computer benchmarks exist to measure specific system resources, such as CPU or memory performance. Others focus on a more general level of hardware performance, such as video performance (which is affected by many factors, including the video card in use, the type of bus, the speed of the bus, the CPU type/speed, etc), while others provide an overall score for the entire computer system.

B.10. BENCHMARKS

Whetstone Benchmark

The Whetstone benchmark is a collection of codes originally written in Algol 60 that test floating point math library performance. These routines are designed to be either scalar or unvectorizable array operations, although modern compilers are now able to optimize some of them. The benchmark reports Whetstones/second - an intentionally meaningless value that is only useful in comparison to other executions of the benchmark. Unfortunately, Whetstone has been recoded in other languages over the years, in both single and double precision, and "improved" from time to time (sometimes in ways that favored particular machines) so that it is almost impossible to compare one Whetstone rating to another. However, if it indicates anything at all about a machine, it is probably most related to scalar floating point performance.

Dhrystone Benchmark

Another popular benchmark is the (the name is taken from a play on Whetstone and the fact that Dhrystone doesn't employ floating point — i.e. if it doesn't float, it must be dry) synthetic benchmark, which involves integer and string processing. The benchmark has been implemented in Pascal Ada and C. It reports Dhrystones Instructions per Second (DIPS) — another intentionally meaningless number. However, it has been reported that 1657 Dhrystones/second equals one VAX-11/780 MIPS. Unfortunately, when DEC reported that the VAX-11/780 is a 1 MIPS machine, they exaggerated by about a factor of 2 (the VAX-11/780 was actually a 0.47 MIPS machine). DEC later tried to correct this error by renaming VAX MIPS as VAX Units of Performance (VUPS).

The Dhrystone benchmark does not execute any real task (which is why it is termed synthetic). Nor is its balance of processing particularly representative of many applications. It heavily favors string processing, has only a shallow subroutine call tree, and its subroutines are smaller than would be found in programs doing real work. Dhrystone is also small enough that it can be completely cache-resident during execution on many machines,

Linpack Benchmark

One of the more famous floating-point benchmarks is Linpack, created by Jack Dongarra, which get its name from a linear algebra package that it uses to solve a dense system of linear equations with Gaussian elimination. The benchmark keeps track of execution time and then divides this into the number of floating point operations that it performs to get a MegaFLOPS rating. The version that is most often reported is based on a 100 x 100 matrix using double precision floating point. Valid results must be reported from a standard FORTRAN compiled execution — hand tuning is forbidden if the result is to be considered official. Unfortunately, this version of the benchmark is usually cache resident in any machine with more than about 300 KB of cache.

The core of the 100 x 100 benchmark is a routine called daxpy that is often referred to in journal articles about compiler optimization. This is because the importance of this benchmark for marketing purposes has become so great that everyone seeks to have their compiler give the absolute best performance on it. Thus, compilers today often have special code just to recognize daxpy and other parts of Linpack — technically, this is not hand optimization because it is part of the compiler, but it skews the meaning of the reported performance nonetheless. Some compilers even have special optimization switches that are only used for compiling Linpack.

The daxpy routine is based on a loop of the following form:

```
do 30 i = 1,n
   dy(i) = dy(i) + da*dx(i)
30 continue
```

It can be seen that this loop provides an excellent opportunity to exploit a multiply-add operation as found in most vector-processing machines and even some microprocessors and digital signal processors. Thus, one may occasionally hear a reference to MACs — Multiply-ACcumulate operations per second, which are often derived from a daxpy-like loop benchmark.

Another version of Linpack that is often reported involves a 1000 x 1000 matrix. While this may seem to be more realistic than the 100 x 100 version, the caveat is that implementations of the larger version are allowed to be hand-optimized. Thus, the performance for this non-cache-resident benchmark can be surprisingly better than for its smaller cousin.

One final note about Linpack is that it is an officially administered benchmark. In order to have permission to quote it, a vendor must report their results to Oak Ridge National Laboratory, where it is administered. The results are available on-line through an auto-mailer. Simply send the message "send performance from benchmark" to netlib@ornl.gov and the latest report will be sent as a reply.

SPECxx

The Standard Performance Evaluation Corporation (SPEC) is a nonprofit corporation set up by a group of vendors who decided they needed more meaningful benchmark than the simple synthetic ones in use at the time that they could quote in their market literature. The original SPEC89 benchmark was replaced by a 1992 version, and that was in turn replaced by a 1995 version, so one must take care in identifying the version used when comparing machines of different periods. One way to tell the difference is that SPEC89 reports a single figure called a SPECmark, while the newer benchmarks report separate integer and floating point figures (SPECint92, SPECfp92, SPECint95, SPECfp95).

In the SPEC92 version, most of the individual codes are written in FORTRAN although 8 of the 20 are in C. Of the 14 floating point routines, 9 are double precision (64 bit) while the others are single precision. The floating point routines account for about 58,000 lines of code all together, while the integer codes are about 124,000 lines long. The SPEC rating is obtained by dividing the wall clock time for a run of a routine by a reference time (the time for a specified reference machine — a VAX-11/780 for SPEC92, a SPARCStation 10/40 for SPEC95) and then combining the ratios (separately for the integer and floating point suites) with a geometric mean.

The rules for running the SPEC92 benchmarks are very flexible. The only restriction is that a compiler flag cannot contain the name of one of the benchmark routines. As a result, vendors have taken ever opportunity to stretch the benchmark to its limits. The SPEC ratings are sometimes called a "guaranteed never to exceed" performance indicator. Some of the tricks used include disabling the network and display, using only an external terminal to interface to the machine, running in single user mode, using different compile flag settings and even different compilers for each routine, using third-party optimizers, modifying the OS kernel, and formatting the disk in a nonstandard manner.

A variation on the SPEC benchmark was introduced in 1994 to address some of these problems and is called the SPECBase. Vendors must report the SPECBase figure to SPEC but can continue to quote SPECPeak as the existing benchmark was renamed. The SPECBase restricts the run to use only 4 compiler flags and they must be set the same for all of the routines. The same compiler must be used for all routines. No assertion flags are allowed that give the compiler information regarding the properties of the program that cannot be identified from the source code. For SPECBase92, run-time profile feedback was not allowed but SPECBase95 relaxes this restriction.

Each of the SPEC routines is referred to by a reference number and a name. The SPEC92int routines are all written in C.

The SPEC95 benchmark reuses many of the existing codes, eliminates a few and adds a few new codes. For all of the routines new data sets were developed that force longer runs. In one case, a routine was modified slightly to use larger arrays. All of the SPECint routines are in C and all of the floating point routines are in Fortran 77. The SPECint95 routines are:

099.go an artificial intelligence program playing the game of Go.

124.m88ksim is a simulation of a Motorola 88K RISC microprocessor running a test program.

The SPEC CPU 2000 benchmarks retired some of the SPEC95 benchmarks and added new ones. These are meant to be more challenging. However, one analysis has shown that even though the benchmarks increase memory usage, they are less challenging in terms of branch prediction. In particular, none of the SPEC 2000 benchmarks contains the density of difficult-to-predict branches as the 099.go benchmark from SPEC 95. Good branch prediction is critical to modern processors with multiple pipelines, and thus SPEC 2000 does not provide applications that are as effective in helping to guide architects in advancing the state of the art in branch prediction. Branching is, however, becoming more of an issue with increased use of object-oriented programming.

The SPEC 2000 CPU integer benchmarks are still mainly written in C, although one is now in C++.

The SPEC CPU 2000 floating point benchmarks are now written in C, Fortran 77, and Fortran 90. It should be noted that, from a research perspective, this diversity of languages makes it more difficult to evaluate compiler optimizations on the full SPEC suite, as any research compiler must now support four languages. From the perspective of evaluating system performance, it is even more interesting to examine the detailed SPEC reports, as they can reveal differences in optimization capabilities for specific languages in addition to hardware performance.

B.11. VIDEO-RELATED BENCHMARKING

A great and accurate way to determine what video cards/chipsets are hot and what's to avoid, and in which areas the former shine, is to run benchmarks. Bear in mind that it is very difficult to devise a benchmark which could test the video card singularly, without being influenced by other factors, most notably the CPU type and frequency, but also such factors as the bus speed. Nevertheless, benchmarks are the way to go if we want to gain insight into video card performance. Now-a-days, Windows performance has become a major preoccupation for many so that Windows-based benchmarking suites which incorporate video-related benchmarks have taken precedence over DOS-based benchmarks, which have all but disappeared today.

Two other areas of video card benchmarking worthy of notice are 3D benchmarks, rapidly becoming standard benchmarking fare, as well as game-related benchmarking, as illustrated by the popular Quake -based benchmarks, 3 of which you can find on this very site: the GlQuake Accelerated Benchmark List, the Quake TIMEDEMO Benchmark List and the Quake TIMEREFRESH Benchmark List.

Below is a breakdown of some of the most popular video-related benchmarks of old and new. Note that new versions of many of these diagnostics programs show up regularly.

B.11.1. DOS-Based Benchmarks

These benchmarks typically test video performance at low resolutions and low colour depths, since this simulates the typical DOS environment. Needless to say that these benchmarks are of special interest to gamers, since most people only use DOS for games now-a-days. Some of the more popular and effective benchmarks based in DOS are:

(1) 3DBench

This benchmark suite has been around for quite a while and was quite popular. I say "was" because it is next to useless today. While it works great for slower systems, today's upper-level Pentiums, Cyrix 6X86 and Pentium Pro machines are too powerful for this older benchmarking utility, resulting in inaccurate results. Indeed, it seems that for more powerful systems, 3Dbench yields benchmark scores which fall into a grove: for instance, the score "125" is very popular, while no system has been known to score something like "130" or "135" in the benchmark. "142.8" is another popular score...and the problem is that pretty much any P133 will end up with a score of 142.8....which tells us that 3Dbench is no good at measuring video card performance.

(2) Chris Dial's 3Dbenchmark

It was destined to be a replacement for the aging 3Dbench, and it is indeed a worthy replacement. It's a much more accurate benchmarking utility, works great on low and high-end systems, offers a benchmark in low resolution (320 x 200) and higher resolution (640 x 480), both in a 256 colour depth, and is good enough to register the difference between video cards, although perhaps not to the extent that a good low-level benchmark could. It is CPU dependent, and one of my gripes is that it also depends upon the FPU, although Chris claims that this was done since new games now make use of the FPU, such as the ever-so-popular Quake series. A good one nevertheless.

(3) Vidspeed 4.0

Vidspeed 4.0 is a good indicator of bus transfer speed. It measures the throughput of writing constant pixel data to video memory over the bus in graphics modes, and is more or less CPU-independent, which is what makes it nice in my eyes. It also offers a huge amount of modes to benchmark in, including some high resolution, high colour depth modes. Not sure how it behaves with high speed systems.

(4) PC Bench 9.0

This is a general benchmarking utility from Ziff Davis which runs in DOS and tests many aspects of your computer system. It has a video test, which isn't great but which you may want to try for the heck of it.

B.11.2. Windows-based Benchmarks

Benchmarks based in Windows will usually test your video card in higher resolutions and in a variety of colour depths. Ideally, resolutions should range from 640X480, though 800X600 and 1024X768, and perhaps even higher. Colour depths range from 16 colours (which is pretty useless in Windows these days) through 256 colours, 32K, 65K, and 16.7 million.

By far the most popular benchmarking software for Windows-based PCs are those released by Ziff-Davis. They've named themselves the Ziff-Davis Benchmark Operation , or ZDBOp for short, and are the world's largest independent developer of benchmark software. Many of you are probably already familiar with their Winbench, Winstone and 3DWinbench utilities.

The latest ZDBOp benchmarks are the 1998 revision. Rather than write about them here and use up precious space, read all about their 3 great benchmarking utilities right here.

B.12. OTHER BENCHMARKS

There's another approach to benchmarking any hardware which doesn't rely on "synthetic benchmarking" (i.e. programs specifically designed to measure or quantify the performance of hardware), and that's to use actual everyday programs to test just how fast your components really are. In the benchmarking world this concept is known as *real-world benchmarking* and it's definitely a process which I am 100% in favour of. The 3 Quake benchmark listed on this site are a prime example of using a program (in this case Quake) to test out video cards, motherboards and CPUs in the environment of interest to the benchmarker. After all, if you want the fastest video card for Quake, what better way to find this out than to test it right there in the game itself?

Well, here's a few more things you may want to look at in your quest for the ultimate video card:

GLQuake Accelerated Benchmark List: For those of you running the accelerated version of Quake, known as GLQuake, which uses the OpenGL API, don't miss this list on my site. Go submit your scores as well.

Quake TIMEREFRESH Benchmark List: The granddaddy of all Quake real-world benchmark lists, started right here in The Computer Hardware Site so long ago : It's a little obsolete today in terms of testing as the TIMEDEMO-based benchmarks above are better all-around benchmarks, but it's still a very useful resource.

DOOM FPS Version 1.9s: It's kind is old, but I thought I'd mention it....this measures the average frame rate of a DOOM 1 demo run, and can give some indication of video card performance, although it is dependent on other factors as well. Essentially it's a TIMEDEMO benchmark for DOOM.

B.13. PROCESSOR BENCHMARKING

Benchmarking is the process of evaluating and comparing devices or systems. Processors are frequently benchmarked, in the hope of coming up with a single number that can capture the value of the CPU and let it be easily compared to others. The value of benchmarking is that it allows this comparison. The problem with it is that there are so many ways to do it, and the answers different benchmarks give about the same processors aren't always consistent.

There is a difference between benchmarking processors and benchmarking whole systems. System benchmarking attempts to evaluate "real world" performance of the whole system, which involves much more than just the processor. Processor benchmarks show isolated performance of processors relative to one another.

Even in just looking at processors, there are many types of benchmarks, which can result in many different scores. Since the processor doesn't operate in a vacuum, the scores will always depend to some extent on the system that is used when performing the benchmark. Some benchmarks are more able to isolate the processor's performance from impact by other system components than others. In addition, benchmarks that are based on Windows or other multitasking operating systems, are subject to accuracy problems induced by how the software is set up, and what other programs might be in operation when they are run.

Some benchmarks can only be used by doing relative comparisons; since the absolute number that is produced depends on the configuration, it can realistically only be used to compare two processors when used in the same exact configuration (much the way the "P rating" calculation is done.) An absolute number from this sort of benchmark can only be given as an average of scores from many system configurations used to run it, or only given with also providing information about the configuration used. Other benchmarks provide roughly the same score for the processor across a wide array of system configurations.

In addition, the benchmark's score depends on what types of code it is executing to do the test. Is it 16-bit or 32-bit code? Many loops or just a few? There are many different types of code and some architectures are better than others at certain types. The power of some of the newer processors is understated when they are tested using older benchmarks (such as the original Norton SI index.) Also remember that benchmarks typically contain some floating-point code, which will mean older processors with no floating point unit will get lower scores.

B.14. THREE VECTORS OF MICROPROCESSOR PERFORMANCE

The microprocessor and the PC of today are designed to run a broad range of powerful software applications. Not every processor is equally capable of the same performance for each type of application. Benchmarks specifically designed for evaluating the performance of processors and systems running integer-, multimedia-, and floating-point-,intensive applications should be used to examine the complete performance of a processor or system.

Integer Benchmarks Typical productivity applications such as word processing, spreadsheets, presentation applications, and personal finance programs, to name a few, depend on integer performance. Popular, industry integer benchmarks include:

(1) Processor Level Benchmarks :

➡ SPECint95

➡ CPUmark32 Norton SI32

(2) System Level Benchmarks :

➡ SYSmark 32 for Windows 95

➡ SYSmarkNT

➡ Winstone98 Business Scores

➡ Winstone98 High End Scores

(3) Multimedia Benchmarks : Traditional benchmark tools were not designed to measure the performance of systems running today's applications rich in graphics, audio and video attributes. Multimedia benchmarks are designed specifically to simulate the activities of end users utilizing video, such as MPEG1 and MPEG2, Dolby Digital Sound, AVI, PC Imaging or Video Conferencing, and other similar media-rich applications. Some of the benchmarks that fall under this category are:

➡ Intel Media Benchmark Norton

➡ Multimedia Benchmark from Norton

➡ Utilities for Windows*95 Version 3.0

(4) Floating Point Benchmarks : Applications which use three-dimensional visualization techniques, such as games, are increasingly employing floating-point performance to support richer textures and enhanced lighting effects. Floating-point performance is also a critical factor for workstation applications such as Computer Aided Design (CAD). Benchmarks that measure floating-point performance include:

B.15. PENTIUM(R) II PROCESSOR PERFORMANCE INDICATORS

SPECint95 under Windows NT 4.0

Performance Comparison	SPECint95	SPECint_base95
Pentium II processor 233MHz	9.38	9.38
Pentium II processor 266MHz	10.70	10.70
Pentium II processor 300MHz	11.90	11.90
Pentium II processor 333MHz	13.00	13.00
Pentium II processor 350MHz	13.90	13.90
Pentium II processor 400MHz	15.80	15.80
Pentium II processor 450MHz	17.20	17.20

NB: All Pentium(r) II processors include 512KB L2 cache.

The SPECint95 benchmark test provides a comparison point for the performance of the microprocessor, memory architecture and compiler of a computer system on compute-intensive, 32-bit applications. SPEC benchmark test results for Intel microprocessors are determined using particular, well-configured systems. These results may or may not reflect the relative performance of Intel microprocessor in systems with different hardware or software designs or configurations (including compilers).

B.16. BENCHMARK LIMITATIONS

Benchmark and performance tests ("benchmarks") measure different aspects of processor and/or system performance. While no single numerical measurement can completely describe the performance of a complex device like a microprocessor or a personal computer, benchmarks can be useful tools for comparing different components and systems. The only totally accurate way to measure the performance of your system, however, is to test the software applications you use on your computer system. Benchmark results published by Intel are measured on specific systems or components using specific hardware and software configurations, and any differences between those configurations (including software) and your configuration may very well make those results inapplicable to your component or system.

Benchmarks can be divided into two kinds

- component
- system

B.16.1. Component V/s System Benchmarks

Component benchmarks measure the performance of specific parts of a computer system, such as a microprocessor or hard disk drive, while system benchmarks typically measure the performance of the entire computer system. In either case, the performance you see in day to day use will almost certainly vary from benchmark performance, for a number of reasons.

First, individual components must usually be tested in a complete computer system, and it is not always possible to eliminate the considerable effects that differences in system design and configuration will have on benchmark results. For instance, system vendors sell systems with a wide variety of disk capabilities and speeds, system memory, system bus features and video and graphics capabilities, all of which influence how the system components (such as the microprocessor) and the computer system perform in actual use and can dramatically affect benchmark results. Also, differences in software, including operating systems and compilers, will affect component and system performance.

Finally, benchmark tests are typically written to be exemplary of only a certain type of computer application, which may or may not be similar to your applications. Benchmarks are, at most, only one kind of information that you may use during the purchasing process. To get a true picture of the performance of a component or system you are considering purchasing, you must consult other sources of information (such as performance information on the exact system you are considering purchasing).

iCOMP

This is an acronym for the Intel COmparative Microprocessor Performance index. This is the "official" benchmark that Intel has chosen to use to state the relative performance of its processors. It is in fact a blending of the results from several other benchmarks. Intel used this rating to produce values for processors ranging from the 386SX-16 to the Pentium 166, before replacing it with iCOMP 2.0.

iCOMP 2.0

This is Intel's revised iCOMP benchmark, used for Pentium 75 and later processors. It too is an amalgam of several other benchmarks (including SI32 and CPUmark32). It focuses more on 32-bit performance than the original iCOMP index, and also partially incorporates a multimedia benchmark. The focus on 32-bit tasks and multimedia make this test give higher scores to Pentium Pro and MMX-capable processors, relative to the "classic" Pentium.

Norton SI

This is one of the earliest universal benchmarks, created by Peter Norton of Norton Utilities fame and used as part of the Norton System Information (sysinfo) utility. This benchmark has been around for years, and in fact is one of the few for which scores are available even for older processors. It is however, an outdated benchmark, and really doesn't tell the whole story when it comes to modern software being run on modern processors. Still though, its universality and the fact that it gives a quick-and-easy performance benchmark relative to the first IBM PC (which is defined to be 1.0) keeps it in use still today. It is sometimes called "SI 8.0" after the last DOS version of the Norton Utilities to use it, although earlier versions tend to return the same values.

Norton SI32

This is an updated version of the original Norton SI index and produces scores that are to a different scale than the old benchmark. It operates under Windows 95 and uses 32-bit code to perform its assessment. Scores are relative to the 386SX-16, which is defined as 1.0. Note that this benchmark does depend to some extent on the system that it is put into, so two systems with the same processor are likely to yield different results for this test.

CPUmark32

These is one of two processor benchmarks (the other being CPUmark16) that are part of the Ziff-Davis Benchmark Operation (ZDBOp) WinBench benchmark set. They test performance executing 32-bit and 16-bit code, respectively. They are widely used and are somewhat dependent on particular machine setup.

B.17. PERFORMANCE BENCHMARKS FOR SPECIFIC PROCESSORS

This section contains a summary table showing the benchmark values for all the processors covered by this web site. The benchmark values shown in table B.1 were obtained from many different sources. Where possible, I used official values from manufacturers, after cross-referencing with independent numbers. All numbers should be considered approximate, since system variations can cause a particular PC to score significantly above or below the benchmarks shown. This is more likely to happen with some benchmarks than others, as discussed here. Please do NOT send me email complaining about these numbers unless they are off by a large margin. This is not a benchmarking site and the numbers are for general comparison only.

Values with a "~" in front of them are extrapolated or approximated, or in general ones I feel less confident about.

The notation "!?" means "right now, I don't know".

The older ICOMP and Norton SI indexes are increasingly meaningless for the later CPUs and due to this fact, plus lack of availability of information, I am not using them for sixth-generation and later chips.

TABLE B.1. Benchmarks for various chips of different generations

Processor Family	Processor Version	ICOMP	ICOMP 2.0	Norton SI	Norton SI32	CP mark 32
8088	(regular)	—	—	1.0	—	—
	-8	—	—	1.7	—	—
8086	(regular)	—	—	!?	—	—
	-8	—	—	!?	—	—
	-10	—	—	!?	—	—
80286	-6	—	—	3.1	—	—
	-8	—	—	4.4	—	—
	-10	—	—	5.6	—	—
	-12	—	—	6.7	—	—
	-16	—	—	~9.0	—	—
	-20	—	—	~11.5	—	—
80386DX	-16	~29	—	~15	!?	—
	-20	~38	—	~20	!?	—
	-25	49	—	~25	!?	—
	-33	68	—	35	!?	—
	-40	~85	—	~43	!?	—
80386SX	-16	22	—	~11	!?	—
	-20	32	—	15	!?	—
	-25	39	—	~20	!?	—
	-33	56	—	~27	!?	—
80486DX	-25	122	—	54	!?	—
	-33	166	—	72	!?	—
	-50	249	—	109	!?	—
80486SX	-16	63	—	34	!?	
	-20	78	—	42	!?	—
	-25	100	—	54	!?	—
	-33	136	—	72	!?	—
80486DX2	-50	231	—	109	~7	~45
	-66	297	—	144	~8	~65
	-80	~340	—	173	~10	~105
80486DX4	-75	319	—	~155	~12	~100
	-100	435	—	198	15	~125
	-120	~530	—	259	17	~150
AMD 5x86	-133	~610	~67	288	18	~160
Cyrix 5x86	-100	~610	~67	264	~16	~150
	-120	~735	~81	316	19	~180

Processor Family	Processor Version	ICOMP	ICOMP 2.0	Norton SI	Norton SI32	CP mark 32
Pentium	60	510	51	190	~16	~120
	66	567	57	211	~18	~140
	75	610	67	237	23	181
	90	735	81	285	27	219
	100	815	90	317	30	243
	120	1000	100	380	32	270
	133	1110	111	421	36	300
	150	1176	114	476	35	308
	166	1308	127	529	40	343
	200	~1575	142	637	44	382
Pentium OverDrive	63	443	~42	~180	~15	~110
	83	581	~57	~240	~19	~140
	120/133	877/970	75/84	~340/	~370~27/ ~210/	~30 ~235
	125	1070	~105	~410	~32	~260
	150	1176	114	476	35	308
Pentium with MMX	166	~1575	160	~640	50.7	378
	200	~1900	182	~770	57.5	425
	233	~2210	203	~890	62	460
Pentium with MMX OverDrive	125	~1200	120	~480	~36	~290
	150	~1400	144	~570	~46	340
	166	~1575	160	~640	50.7	378
	180	~1700	~167	~670	~53	~390
	200	~1900	182	~770	57.5	425
	6x86					
	PR120+	—	—	676	32	—
	PR133+	—	—	750	~37	—
	PR150+	—	—	816	41	—
	PR166+	—	—	907	48	—
	PR200+	—	—	1020	~55	—
K5	PR75	—	—	286	~25	—
	PR90	—	—	359	30	—
	PR100	—	—	390	33	—
	PR120	—	—	~380	~37	—
	PR133	—	—	407	43	—
	PR166	—	—	~470	47	—
Pentium Pro	150	—	168	—	70	420
	166	—	186	—	78	465
	180	—	197	—	81	497
	200 (256 KB)	—	220	—	90	553
	200 (512 KB)	—	~240	—	98	611
	200 (1M)	—	!?	—	!?	!?
Pentium II	233	—	267	—	~115	~640
	266	—	303	—	~130	~700
	300	—	332	—	—	—
	333	—	366	—	—	—
K6	166	—	—	—	~73	~420
	200	—	—	—	~83	~490
	233	—	—	—	~91	!?
	266	—	—	—	~100	!?
6x86MX	PR166	—	—	—	—	~380
	PR200	—	—	—	—	—
	PR233	—	—	—	—	!?

B.18. PROCESSOR FAMILIES

This section provides detailed technical information about all the major processor families used on PC-compatible computers. It covers every major x86 processor on the market, from the first Intel 8088 used in the original IBM PC, to the latest released hot chips. The processors here are grouped by "families", where I consider a family of processors to be a group of processors that vary only in clock speed, not in architecture (with a couple of possible minor exceptions to keep the section from getting too large).

I organize this section by processor "generation". Which generation a processor belongs to, used to be clear when it was just Intel making processors and they were coming out at a slow rate and with very obvious performance differences. Now the situation is much cloudier, with many more chips on the market and many types of technology being used to implement them. I have made the decision of which processor to include in which generation based on technology and performance considerations, along with my interpretations of how the market perceives the chip. Pentium OverDrives use fifth generation technology so they are listed as fifth-generation chips.

Most of the information about each processor is presented in tabular format for easy reference. The rows of these tables summarize the statistics for each processor, and correlate to the other sections in the processor reference on this site where the terms and characteristics are explained in detail. The first section contains explanations for each of the rows in the tables, and links to where each of the characteristics is described in more detail.

Each section contains a brief discussion of the processor and its more interesting points. Included is an illustration of the processor's improvements and changes from the previous generation processor it most directly replaced.

B.19. EXPLANATION OF PROCESSOR SUMMARY TABLES

This section describes each of the characteristics contained in the processor summary tables included for each processor. The table below is identical to those used for each CPU, except that an explanation is given for what each statistic means.

General notes on the data in the summary tables:

The benchmark values were obtained from many different sources. All numbers should be considered approximate, since system variations can cause a particular PC to score significantly above or below the benchmarks shown. This is more likely to happen with some benchmarks than others; see here for more on these benchmarks. Values with a "~" in front of them are extrapolated or approximated, or in general ones I feel less confident about. Benchmark values are provided for general guidance purposes only. **Items with the value "!?" are unknown.**

(1) General Information Manufacturer: The company or companies that manufacture the processor.

- **Family Name:** The name of the processor family.
- **Code name:** The code name that was/is used to refer to this processor.
- **Processor Generation:** Processor generation, first to sixth.
- **Motherboard Generation:** The generation of the motherboard that this processor runs on; this is normally the same as the processor generation but not always.
- **Version:** The particular name of each version of the processor family.
- **Introduced:** The date that the processor was introduced. Note that this is the release date of the processor, which doesn't always equal the date that systems with the processor in them were first available.
- **Variants and Licensed Equivalents:** Some processors were manufactured by other companies under license from the original designer; these are listed here.

(2) Speed Specifications

➡ **Memory Bus Speed (MHz):** The speed of the main system memory bus in MHz.

➡ **Processor Clock Multiplier :** The multiplier value for the processor; the amount that is multiplied by the memory bus speed to get the processor speed.

➡ **Processor Speed (MHz) :** The speed of the processor in MHz.

➡ **"P" Rating :** For some processors, the "P rating" , which is the manufacturer's claim of the performance level of the processor regardless of its internal clock speed.

(3) Benchmarks

➡ **iCOMP Rating :** The processor's iCOMP benchmark . For Intel processors from the 80386 through the Pentium.

➡ **iCOMP 2.0 Rating :** The iCOMP 2.0 benchmark for the processor, for Intel processors from the Pentium on.

➡ **Norton SI:** The original Norton SI benchmark rating, also sometimes called Norton SI 8.0.

➡ **Norton SI32 :** The new Norton SI 32-bit benchmark rating for the processor.

➡ **CPUmark32 :** The processor's Ziff-Davis WinBench 32-bit CPU benchmark.

(4) Physical Characteristics

➡ **Process Technology :** The process technology used in manufacturing the chip; usually CMOS or Bipolar CMOS.

➡ **Circuit Size (microns) :** The circuit size in microns, reflecting the miniaturization level of the processor.

➡ **Die Size (mm^2):** The physical size of the actual silicon chip (not the package that it is placed in).

➡ **Transistors (millions):** The number of transistors used in making the chip; more complex processors use more transistors.

➡ Voltage, Power and Cooling

➡ External or I/O Voltage (V)

The external voltage level used by the processor to interface to the motherboard, also called I/O voltage.

➡ **Internal or Core Voltage (V) :** The internal voltage used by the core of the processor; lower than the external voltage on newer processors.

➡ **Power Management :** Power management protocol used by the processor, if any.

➡ **Cooling Requirements :** The generally-accepted cooling method required for the processor. Note that this is not really "cast in stone"; some manufacturers use a larger passive heat sink instead of an active one with no problems.

(5) Packaging Packaging Style: The packaging used for the processor, including the number of pins.

➡ **Motherboard Interface :** The socket or slot style used to connect the processor to the motherboard.

➡ **External Architecture :**

➡ **Data Bus Width (bits) :** The width of the data bus.

➡ **Address Bus Width (bits) :** The address bus width in bits.

(6) Maximum Addressable Memory: The maximum theoretical amount of memory that the processor can address (generally 2^n where n is the width of the address bus).

- **Level 2 Cache Type :** The type of secondary cache used, motherboard, integrated or processor card.
- **Level 2 Cache Size :** The secondary cache size. For processors that have the secondary cache on the motherboard, a typical range is provided, since the actual value in any system of course depends on the particular motherboard.
- **Level 2 Cache Bus Speed :** The speed of the secondary cache. This is usually either the same as the memory bus speed or the processor speed.
- **Multiprocessing :** Whether or not the processor can be used in a multiprocessing environment, and if so, how many can be used.
- **Internal Architecture Instruction Set :** The instruction set supported by the processor, including any extensions (other than MMX).
- **MMX Support :** Whether or not the processor supports the MMX instruction set extension.
- **Processor Modes :** The different processor modes supported.
- **x86 Execution Method:** The execution method that the processor uses, either native or emulation.

(7) Internal Components

- Register Size (bits) : The width of the processor's internal registers which indicates the processor's overall "size".
- Pipeline Depth (stages) : The number of steps in the processor's internal pipelines.
- **Level 1 Cache Size :** The size of the level 1 data and instruction caches .
- **Level 1 Cache Mapping :** The associativity of the level 1 cache.
- **Level 1 Cache Write Policy :** The write policy of the primary cache, either write-through or write-back (or both).
- **Integer Units :** The number of integer execution units in the processor. This includes specialized branch and load/store units, as well as dedicated MMX units.
- **Floating Point Unit / Math Coprocessor :** For processors that have a built-in FPU, say "Integrated". For those that use a separate coprocessor, indicates its name.
- **Instruction Decoders :** The number and type of decoders used in the processor.
- **Branch Prediction Buffer Size / Accuracy :** The number of entries in the branch target buffer and the claimed accuracy level of the branch prediction unit, for processors that use branch prediction.
- **Write Buffers :** The number of write buffers used to hold execution results.
- **Performance Enhancing Features :** Indicates any additional performance-enhancing features, such as out of order execution, speculative execution, register renaming, or superpipelining.

B.20. FIRST GENERATION PROCESSORS

First generation processors were used on the earliest of machines, the original IBM PC and XT, and the first clones. These machines were primitive in most ways compared to modern PCs, and the processors they used were of limited capabilities. First generation processors run at system bus speed and use the oldest processor technologies.

Table B.2 characteristics Information on Intel 8088

Characteristics	Characteristic Informaion	General Information
Manufacturer Family Name	8088	8088-8
Code name	—	
Processor Generation	First	
Motherboard Generation	First	
Version	8088	8088-8
Introduced	June 1979	!?
Variants and Licensed Equivalents	80C88, AMD 8088, NEC V20	
Speed Specifications		
Memory Bus Speed (MHz)	4.77	8
Processor Clock Multiplier	1.0	
Processor Speed (MHz)	4.77	8
"P" Rating	—	
Benchmarks		
iCOMP Rating	—	
iCOMP 2.0 Rating	—	
Norton SI	1.0	1.7
Norton SI32	—	
CPUmark32	—	
Physical Characteristics		
Process Technology	NMOS, CMOS	
Circuit Size (microns)	3.0	
Die Size (mm^2)	33	
Characteristics	8088	8088-8
Transistors (millions)	0.029	
Voltage, Power and Cooling		
External or I/O Voltage (V)	5	
Internal or Core Voltage (V)	5	
Power Management	None	
Cooling Requirements	None	
Packaging		
Packaging Style	40-Pin DIP	

(1) Intel 8088 : The first chip used in PCs was Intel's 8088. This was not, at the time it was chosen, the best available CPU, in fact Intel's own 8086 was more powerful and had been released earlier. The 8088 was chosen for reasons of economics: its 8-bit data bus required less costly motherboards than the 16-bit 8086. Also, at the time that the original PC was designed, most of the interface chips available were intended for use in 8-bit designs. It's ironic, isn't it, that Intel's first production chip was in a way, the "8086SX"? It originally shipped at 4.77 MHz and a "turbo" version was later produced that ran at 8 MHz.

This original chip used what would be considered today to be archaic technology. The 8088 offers performance less than one-

thousandth that of a modern processor, showing just how far we have come in 15 years.

Note: The NEC V20 is an 8088-compatible but actually delivers slightly better performance due to a slightly more efficient internal design.

Note: The original design of this chip was in NMOS, with newer ones in CMOS (80C88) for lower power consumption.Characteristics information for Intel 8088 family chip are shown in Table B.2.

Look here for an explanation of the categories in the processor summary table below <expl.htm>.

Characteristics	Characteristic Informaion	General Information
Motherboard Interface	DIP socket	
External Architecture		
Data Bus Width (bits)	8	
Maximum Data Bus Bandwidth	4.778 (Mbytes/sec)	
Address Bus Width (bits)	20	
Maximum Addressable Memory	1 MB	
Level 2 Cache Type	None	
Level 2 Cache Size	—	
Level 2 Cache Bus Speed	—	
Multiprocessing	No	
Internal Architecture		
Instruction Set	8088	
MMX Support	No	
Processor Modes	Real	
x86 Execution Method	Native	
Internal Components		
Register Size (bits)	16	
Pipeline Depth (stages)	1	
Level 1 Cache Size	None	
Level 1 Cache Mapping	—	
Level 1 Cache Write Policy	—	
Integer Units	1	
Floating Point Unit	Optional 8087 Coprocessor	
Math Coprocessor		
Instruction Decoders	1	
Branch Prediction	None	
Buffer Size / Accuracy		
Write Buffers	None	
Performance Enhancing Features	None	

(2) Intel 8086: The Intel 8086 chip is almost identical to the 8088, except that it uses a 16-bit-wide data bus instead of the 8-bit bus of the 8088. It was actually invented before the 8088 but was not chosen for the original PC because IBM wanted to keep costs down by going with an 8-bit bus design.

The 8086 was used in some PC clones and later on in early IBM PS/2 models, but was never a popular choice for PCs. This is probably because by the time it started to be used, the 80286 was introduced, offering much improved performance over the 8086. The 8086 is architecturally identical to the 8088 aside from the wider data bus.

Note: The NEC V30 is an 8086-compatible but actually delivers slightly better performance due to a slightly more efficient internal design.

Note: Original designs were made in NMOS, newer ones in CMOS (80C86) for lower power consumption. The characteristc information on 8086 chip are given in Table B.3.

Table B.3. Characteristic Information: Intel's 8086

Characteristics	**General Information**		
Manufacturer Detail Intel			
Family Name	8086		
Code name			
Processor Generation	First		
Motherboard Generation	First		
Version	**4.77 MHz**	**8 MHz**	**10 MHz**
Introduced	June 1978	!?	
Variants and Licensed Equivalents	80C86, AMD 8086	NEC V30	
Speed Specifications			
Memory Bus Speed (MHz)	8086	8086-8,	8086-10
Processor Clock Multiplier	1.0		
Processor Speed (MHz)	4.77	8	10
"P" Rating	—		
Benchmarks			
iCOMP Rating	—		
iCOMP 2.0 Rating			
Norton SI	!?	!?	!?
Norton SI32	—		
CPUmark32			
Physical Characteristics			
Process Technology	NMOS, CMOS		
Circuit Size (microns)	3.0		
Die Size (mm^2)	33		
Transistors (millions)	0.029		
Voltage, Power and Cooling			
External or I/O Voltage (V)	5		
Internal or Core Voltage (V)	5		
Power Management	None		
Cooling Requirements	None		
Packaging			
Packaging Style	40-Pin DIP		
Motherboard Interface	DIP socket		
External Architecture			
Data Bus Width (bits)	16		
Maximum Data Bus Bandwidth (Mbytes/sec)	9.54	16	20
Address Bus Width (bits)	20		
Characteristics	**General Information**		
Maximum Addressable Memory	1 MB		
Level 2 Cache Type	None		
Level 2 Cache Size	—		
Level 2 Cache Bus Speed	—		
Multiprocessing	No		
Internal Architecture			
Instruction Set	8088		
MMX Support	No		
Processor Modes	Real		
x86 Execution Method	Native		
Internal Components			
Register Size (bits)	16		
Pipeline Depth (stages)	1		
Level 1 Cache Size	None		
Level 1 Cache Mapping	—		
Level 1 Cache Write Policy	—		
Integer Units	1		
Floating Point Unit / Math Coprocessor	Optional 8087 Coprocessor		
Instruction Decoders	1		
Branch Prediction Buffer Size / Accuracy	None		
Write Buffers	None		
Performance Enhancing Features	None		

B.21. SECOND GENERATION PROCESSORS

The only chip generally considered second generation is the 80286. Intel also made a chip between the 8088 and the 80286, called the 80186. Largely due to a lack of performance enhancements compared to the 286, in addition to compatibility problems with support chips, the 80186 was never really used in PCs. It has however found a strong niche role in the design of embedded microcontrollers, even to this day.

(1) Intel 80286

The 80286 was the chip used in IBM's AT (advanced technology) system. The 286 was the first major step up in PC processors, providing significant performance increases over the 8088 and 8086—double or more performance at the same clock speed. The 286 also widened the address bus to allow access to 16 MB of memory, and introduced protected mode operation. It was originally available in 6 MHz and 8 MHz versions, but was later expanded to faster versions, all the way up to 20 MHz.

The 286 opened up the PC world to many users, but still was used mainly as the equivalent of a "turbo-charged 8088". At this time DOS was still the virtually exclusive operating system, and the protected mode the 286 offered was largely ignored. The 80286 is of course considered obsolete today. Information on Intel's 80286 microprocessor chip are given in Table B.4.

Look here for an explanation of the categories in the processor summary table below <expl.htm>, including links to more detailed explanations.

Table B.4. Characteristic Information : Intel's 80286 Microprocessor

Characteristics	General Details					
Manufacturer Detail	Intel					
Family Name	80286					
Code name	—					
Processor Generation	Second					
Motherboard Generation	Second					
Version	80286 -6	80286-8	80286-10	80286-12	80286-16	80286-20
Introduced	Feb. 1982	!?	!?	!?	!?	!?
General Information						
Variants and Licensed	AMD 286 Equivalents	286S				
Speed Specifications						
Memory Bus	6	8	10	12	16	20
Processor Clock Multiplier	1.0					
Processor Speed (MHz)	6	8	10	12	16	20
"P" Rating	—					
Benchmarks	—					
iCOMP Rating	—					
iCOMP 2.0 Rating	—					
Norton SI	3.1	4.4	5.6	6.7	~9.0	~11.5
Norton SI32	—					
CPUmark32	—					
Physical Characteristics						
Process Technology	CMOS Speed (MHz)					
Circuit Size (microns)	1.5					
Die Size (mm^2)	47					
Transistors (millions)	0.134					

Voltage, Power and Cooling						
External or I/O Voltage (V)	5					
Internal or Core Voltage (V)	5					
Power Management	None					
Cooling Requirements	None					
Packaging						
Packaging Style	68-Pin PGA					
Motherboard Interface	68-Pin Socket					
External Architecture						
Data Bus Width (bits)	16					
Maximum Data Bus Bandwidth (Mbytes/sec)	12	16	20	24	32	40
Address Bus Width (bits)	24					
Maximum Addressable Memory	16 MB					
Level 2 Cache Type	None					
Level 2 Cache Size	—					
General Information	80286 -6	80286-8	80286-10	80286-12	80286-16	80286-20
Level 2 Cache Bus Speed	—					
Characteristics	**General Details**					
Multiprocessing	No					
Internal Architecture						
Instruction Set	80286					
MMX Support	No					
Processor Modes	Real, Protected					
x86 Execution Method	Native					
Internal Components						
Register Size (bits)	16					
Pipeline Depth (stages)	1					
Level 1 Cache Size	None					
Level 1 Cache Mapping	—					
Level 1 Cache Write Policy	—					
Integer Units	1					
Floating Point Unit / Math Coprocessor	Optional 80287 Coprocessor					
Instruction Decoders	1					
Branch Prediction Buffer Size / Accuracy	None					
Write Buffers	None					
Performance Enhancing Features	None					

B.22. THIRD GENERATION PROCESSORS

The third generation processors (the 386 family) represent another step forward from the second generation class. Intel started with this generation of chips to create "subfamilies" of related chips with different capabilities, using the "DX" and "SX" designations.

(1) Intel 80386DX: The Intel 80386DX was the first true 32-bit processor used on the PC platform. Its internal register size was increased to 32 bits, and its data and address buses were as well, doubling data path width to the processor and increasing addressable memory to 4 GB theoretical. The 80386 family of chips offered more performance than the 80286s they replaced, largely through processor speed increases. The 386 did not offer the very large improvements over the 286 that the 286 did over the 8088.

The 386DX's increased power and the improved processor modes it offered (including full protected mode and virtual real modes) spurred the introduction of GUI-based operating systems on the PC, such as Microsoft Windows (although they are still quite slow on 386 chips). The instruction set of the 386 has set the standard for what is now called "x86" and hasn't changed very much since it was introduced. The 80386DX was the first to use pipelining to allow much improved processor performance through the use of much higher clock frequencies.

Invented by Intel, the 386 was also "cloned" by AMD and Cyrix. These are very good copies with no real compatibility problems. Intel only produced the 386DX up to 33 MHz (presumably to prevent overlapping into the 486's performance range) while AMD and Cyrix produced a 40 MHz version as well. This latter chip produced very good performance (for the time), comparable to many lower-end 486s. They are still however obsolete by today's standards, of course.

Table B.5. Characteristic Information of Intel's 80386 Family microprocessors

General Characteristic	**Other Details**				
Manufacturer's Detail	Intel	Intel, AMD, Cyrix		AMD,	
Family Name	80386DX				
Code name	—				
Processor Generation	Third				
Motherboard Generation	Third				
Version	**80386DX -16**	**80386DX -20**	**80386DX -25**	**80386DX -33**	**80386DX -40**
Introduced	Oct. 1985	Feb. 1987	April 1988	April 1989	!?
Speed Specifications					
Memory Bus Speed (MHz)	16	20	25	33	40
Processor Clock Multiplier	1.0				
General Characteristic	**Other Details**				
Processor Speed (MHz)	16	20	25	33	40
"P" Rating	—				
Benchmarks					
iCOMP Rating	~29	~38	49	68	~85
iCOMP 2.0 Rating	—				
Norton SI	~15	~20	~25	35	~43
Norton SI32	!?				
CPUmark32	—				
Physical Characteristics					
Process Technology	CMOS				
Circuit Size (microns)	originally 1.5	now 1.0			
Die Size (mm^2)	!?				
Transistors (millions)	0.275				
Voltage, Power and Cooling					
External or I/O Voltage (V)	5				
Internal or Core Voltage (V)	5				
Power Management	None				
Cooling Requirements	None				
Packaging	Packaging Style	132-Pin PGA			
Motherboard Interface	132-Pin Socket				

External Architecture					
Data Bus Width (bits)	32				
Maximum Data Bus Bandwidth (Mbytes/sec)	66	80	100	133	160
Address Bus Width (bits)	32				
Maximum Addressable Memory	4 GB				
Level 2 Cache Type	None				
Level 2 Cache Size	—				
Level 2 Cache Bus Speed	—				
Multiprocessing	No				
Internal Architecture					
Instruction Set	x86				
MMX Support	No				
Processor Modes	Real, Protected, Virtual Real				
x86 Execution Method	Native.				
Version	**80386DX -16**	**80386DX -20**	**80386DX -25**	**80386DX -33**	**80386DX -40**
Internal Components	Register Size (bits)	32			
Pipeline Depth (stages)	4				
Level 1 Cache Size	None				
Level 1 Cache Mapping	—				
Level 1 Cache Write Policy	—				
Integer Units	1				
Floating Point Unit / Math Coprocessor	Optional 80387	Coprocessor			
Instruction Decoders	1				
Branch Prediction Buffer Size / Accuracy	None				
Write Buffers	None				

B.23. FOURTH GENERATION PROCESSORS

The fourth generation of processors saw rapid growth in the CPUs' power and capabilities, and the introduction of several new technologies as well. It was here that AMD and Cyrix had their first real early successes in not just emulating Intel's designs but coming up with real value leaders of their own, at the top end of the fourth generation scale. Also, it was at this time that the new phenomenon of upgradable processors and standardized motherboard sockets began to become prevalent.

(1) Intel 80486DX: The Intel 80486DX was the first member of the 486 family (which has many more members than the 386 family did). It provided a very significant increase in power over the 80386DX processor, in fact, far more proportionately than the 386 did over the 286. A 486DX processor provides approximately 100 to 150% more performance than a 386DX of the same clock speed. The 80486 brought GUIs to the mainstream on PCs; it is the minimum processor that most people consider "usable" for running an operating system like Microsoft Windows.

Interestingly, the 486 does not provide its performance improvements by widening any of the buses, as had been the case in the previous two generations: it is still a 32-bit processor with 32-bit data and address buses, just like the 386DX. However, internally, the 486 incorporates several significant improvements over the 386:

Faster Overall Instruction Execution time: The core of the chip can execute instructions in less time than earlier processors.

Deeper Pipeline: The execution pipeline was increased by one step.

Primary Cache: The 486 processor was the first to incorporate level 1 cache on the chip, to reduce the number of required accesses to memory.

Integrated Floating Point Unit: The chip includes an integrated math coprocessor (not on the SX version however). In addition, the coprocessor provides much more performance than the optional 80387 used with 386 chips, in part because it is integrated into the chip.

Burst Mode: The 486 introduced the use of burst mode to reduce wait time on memory accesses.

Power Management: SL power management enhancements as an option (instead of a specialized SL chip as was the case for the 386SL).

Improved Support Architecture: In general, 486 motherboards were more efficient than 386 ones, and began to use secondary cache as well. This improves processor performance significantly.

Interestingly, the 80487SX coprocessor that is intended for use with the SX version of the 80486, is in fact a full-blown 80486DX processor. This is part of how Intel structured its fourth generation family's upgrade path. See the description of the 80486SX for more details on this.

AMD and Cyrix did produce clones of the original Intel 80486 processors, but they were not a big player in the 486 clone market until the higher-speed DX2 and DX4 processors.

The 486DX is considered obsolete, although the chip still has fairly good power for performing a wide variety of light tasks, such as word processing and some older games, and light Internet access. The 486DX-50, which runs on a 50 MHz system bus, provides performance comparable to the 486DX2-66 in many ways, because the latter uses only a 33 MHz system bus. The 486DX-50 was not used in nearly as many systems as the other processor speeds were. It should not be confused with the 486DX2-50, which runs at the same processor clock speed but is clock-doubled relative to the system bus (which runs at 25 MHz).

Note: The 486DX processor normally was purchased as part of a new system only, not as part of an upgrade. Most early 486 systems used a 168-pin socket for the chip, which predates the numbered standardized socket system that Intel created. The 486DX will fit into a Socket 1, Socket 2 or Socket 3 however.

Look here for an explanation of the categories in the processor summary table below <expl.htm>, including links to more detailed explanations.

Table B.6. Characteristic Information: Intel's 80486 Family Microprocessor

General Characterstic	General Information			
Manufacturer	Detail Intel, AMD, Cyrix			
Family Name	80486DX			
Code name	"P4"			
Processor Generation	Fourth			
Motherboard Generation	Fourth			
Version	**80486DX-25**	**80486DX-33**	**80486DX-50**	
Introduced	April 1989	May 1990	June 1991	
Variants and Licensed Equivalents	80487SX			
Speed Specifications	Memory Bus Speed (MHz)	25	33	50
Processor Clock Multiplier	1.0			

Processor Speed (MHz)	25	33	50
"P" Rating	—		
Benchmarks			
iCOMP Rating	122	166	249
iCOMP 2.0 Rating	—		
Norton SI	54	72	109
Norton SI32	!?		
CPUmark32	—		
Physical Characteristics			
Process Technology	CMOS		
Circuit Size (microns)	1.0	1.0	0.8
Die Size (mm^2)	81		81?
Version	**80486DX-25**	**80486DX-33**	**80486DX-50**
Transistors (millions)	1.2		
Voltage, Power and Cooling			
External or I/O Voltage (V)	5		
Internal or Core Voltage (V)	5		
Power Management	SMM in SL-enhanced versions		
Cooling Requirements	Generally none, some use passive heat sink.		
Packaging			
Packaging Style	168-Pin PGA		
Motherboard Interface	168-Pin Socket, Socket 1, Socket 2, Socket 3		
External Architecture			
Data Bus Width (bits)	32		
Maximum Data Bus Bandwidth (Mbytes/sec)	100	133	200
Address Bus Width (bits)	32		
Maximum Addressable Memory	4 GB		
Level 2 Cache Type	Motherboard		
Level 2 Cache Size	Usually 0 KB to 256 KB		
Level 2 Cache Bus Speed	Same as Memory Bus		
Multiprocessing	No		
Internal Architecture			
Instruction Set	x86		
MMX Support	No		
Processor Modes	Real, Protected, Virtual Real		
x86 Execution Method	Native		
Internal Components	Register Size (bits)	32	
Pipeline Depth (stages)	5		
Level 1 Cache Size	8 KB Unified		
Level 1 Cache Mapping	4-Way Set Associative		
Level 1 Cache Write Policy	Write-Through		
Integer Units	1		
Floating Point Unit /	Integrated Math Coprocessor		
Instruction Decoders	1		
Branch Prediction	None		
Buffer Size / Accuracy			
Write Buffers	None		
Performance Enhancing Features	None		

(2) Intel 80486SX: The Intel 80486SX is the same chip as the 80486DX with one exception: the lack of an integrated math coprocessor (floating point unit. Note that this is a different kind of difference between the SX and DX versions than is the case with the 386; the 386SX and 386DX both had no coprocessor and the 386SX had narrower data and address buses than the 386DX. The 486DX and 486SX have the same bus widths. Since it is the same chip except for the floating point processor, the 486SX has the same advantages over the 386 that the 486DX does. Note that the 486SX was made available in slower clock speeds than the 486DX; the SX comes in 16, 20, 25 and 33 MHz versions, while the DX is 25, 33 and 50 MHz.

Intel made some rather interesting (some would say bizarre) decisions about how to market the 486 line. The 486SX was in many ways a marketing gimmick only; it was in fact a 486DX with the math coprocessor disabled. Why did Intel bother to sell a full 486DX and a "crippled" one (the 486SX)? Mainly for market targeting reasons: they wanted to be able to fill the niche of low-end buyers without dropping the price on their "top of the line" 486DX. The 486SX was a popular choice for laptops due to its lower cost and lower power consumption.

The 80486SX uses a "math coprocessor" called the 80487SX (there is no 80487DX at all so don't be confused by that). The 80487SX is, in fact, a fully functional 80486DX chip. What Intel wanted people to think was that (like with its earlier coprocessors) you would put the 80487SX in and it would handle the math functions. In fact, when inserted, the 80487SX shuts down the 80486SX and handles both integer and floating point operations (since it is internally a 80486DX, which does both). This makes no difference from a performance standpoint but is kind of a technical curiosity.

With the large number of very cheap 486DX and faster chips and systems around, the 486SX is really quite obsolete. In addition, it does not come in a 50 MHz version. If you are not doing any floating point operations at all it is comparable to a 486DX of the same speed, but there is usually no need to have to worry about this given that even the 486DX is quite obsolete at this time anyway.

Note: The 486SX processor normally was purchased as part of a new system only, not as part of an upgrade. Most early 486 systems used a 168-pin socket for the chip, which predates the numbered standardized socket system that Intel created. The 486SX will fit into a Socket 1, Socket 2 or Socket 3 however.

TABLE B.7. Characteristic Information : Intel's 80486x Family Microprocessor

Characteristic Information	General Information			
Manufacturer	Intel, AMD, Cyrix			
Family Name	80486SX			
Code name	"P4S"			
Processor Generation	Fourth			
Motherboard Generation	Fourth			
Version	**80486SX-16**	**80486SX-20**	**80486SX-25**	**80486SX-3**
Introduced	Sept. 1991	April 1991	Sept. 1991	Sept. 1992
Variants and Licensed Equivalents	!?			
Speed Specifications				
Memory Bus Speed (MHz)	16	20	25	33
Processor Clock Multiplier	1.0			
Processor Speed (MHz)	16	20	25	33
"P" Rating	—			
Benchmarks				
iCOMP Rating	63	78	100	136
iCOMP 2.0 Rating	—			

Norton SI	34	42	54	72
Norton SI32	!?			
Version	**80486SX-16**	**80486SX-20**	**80486SX-25**	**80486SX-3**
CPUmark32	—			
Physical Characteristics				
Process Technology	CMOS			
Circuit Size (microns)	1.0			
Die Size (mm^2)	67			
Transistors (millions)	1.185			
Voltage, Power and Cooling				
External or I/O Voltage (V)	5			
Internal or Core Voltage (V)	5			
Power Management	SMM in SL-enhanced versions			
Cooling Requirements	None			
Packaging				
Packaging Style	168-Pin PGA			
Motherboard Interface	168-Pin Socket,	Socket 1	Socket 2	Socket 3
	External Architecture			
Data Bus Width (bits)	32			
Maximum Data Bus Bandwidth (Mbytes/sec)	66	80	100	133
Address Bus Width (bits)	32			
Maximum Addressable Memory	4 GB			
Level 2 Cache Type	Motherboard			
Level 2 Cache Size	Usually 0 KB to 256 KB			
Level 2 Cache Bus Speed	Same as Memory Bus			
Multiprocessing	No			
Internal Architecture	**Instruction Set x86**			
MMX Support	No			
Processor Modes	Real, Protected, Virtual Real			
x86 Execution Method	Native			
Internal Components	Register Size (bits)		32	
Pipeline Depth (stages)	5			
Level 1 Cache Size	8 KB Unified			
Level 1 Cache Mapping	4-Way Set Associative			
Level 1 Cache Write Policy	Write-Through			
Integer Units	1			
Coprocessor				

(3) Intel 80486DX2 and 80486DX2 OverDrive: The 80486DX2 was the first chip to use "clock doubling" technology, where the processor runs at a faster speed than the memory bus it talks to. This was done to allow the processor speed to be increased without having to deal with the much more difficult task of increasing motherboard speed. Chips that run at faster than memory bus speed improve performance but at a diminishing rate as the multiplier increases, due to the processor waiting for data from memory.

Intel produced 50 and 66 MHz DX2 chips, intended for use in 25 and 33 MHz system bus systems. These chips have been sold as regular chips intended for use in new systems, which generally come in 168 pin packages to go in the original 168 pin socket used in 486 systems. They have also been made in 169 pin OverDrive versions to go in Socket 1 (the original OverDrive socket). These can be used to upgrade older 486DX or 486SX systems.

AMD and Cyrix not only cloned the 66 MHz DX2 processor, they took Intel one step further with the 80486DX2-80, running at 80 MHz. This uses a 40 MHz system bus, which isn't a speed that is normally used by Intel systems but that became more popular late in the 486 life cycle due to the performance increase it gives over 33 MHz bus systems. In addition, the AMD (enhanced version) and Cyrix chips have several advantages over the Intel chips (they had the benefit of developing them well after Intel):

Power Management: They have SMM (power management) built in.

Write-Back Cache: They have write-back capability for the primary cache which provides a small boost in performance.

Lower Power: They use less power because they run at 3 volts, but will still work in standard 5 volt motherboard sockets without a converter or regulator (they are said to be "5 volt tolerant").

Other than clock speed, the 80486DX2 is virtually identical to the 80486DX. They are obsolete due to the availability of faster, very inexpensive processors such as the 5x86-133 that go in the same sockets, but 486DX2 systems are perfectly viable for many uses, including routine office word processing and spreadsheet work under DOS and Windows 3.x. The 80486DX2-66 is by far the most common version of this chip; a great number of these systems were produced and many are still in use today, especially in small businesses.

Note: The 486DX2 was the first processor to really require a heat sink in order to operate reliably. The increased speed of this chip means that it runs very hot (at least the Intel 5 volt versions).

Table B.8. Characteristic Information : Intel's 80486DX2 Microprocessor Family

Characteristics	**General Information**		
Manufacturer	Intel	Intel, AMD, Cyrix AMD, Cyrix	
Family Name	80486DX2		
Code name	"P24"		
Processor Generation	Fourth		
Motherboard Generation	Fourth		
Version	**80486DX2-50**	**80486DX2-66**	**80486DX2-80**
Introduced	March 1992	Aug. 1992	!?
Variants and Licensed Equivalents	—		
Speed Specifications			
Memory Bus Speed (MHz)	25	33	40
Processor Clock Multiplier	2.0		
Processor Speed (MHz)	50	66	80
"P" Rating	—		
Version	**80486DX2-50**	**80486DX2-66**	**80486DX2-80**
Benchmarks			
iCOMP Rating	231	297	~340
iCOMP 2.0 Rating	—		
Norton SI	109	144	173
Norton SI32	~7	~8	~10
CPUmark32	~45	~65	~105

Physical Characteristics	
Process Technology	CMOS
Circuit Size (microns)	0.8 (Intel), 0.5 (AMD), 0.65? (Cyrix)
Die Size (mm^2)	76 (Intel)
Transistors (millions)	1.2 (Intel)
Voltage, Power and Cooling	
External or I/O Voltage (V)	5 (Intel) 3.3 (AMD, Cyrix, 5V tolerant)
Internal or Core Voltage (V)	5 (Intel), 3.3 (AMD, Cyrix)
Power Management	SMM in AMD, Cyrix, and SL-enhanced Intel versions
Cooling Requirements	Passive or active heat sink
Packaging	
Packaging Style	168-Pin PGA (Regular), 169-Pin PGA (OverDrive)
Motherboard Interface	168-Pin Socket (Regular only), Socket 1, Socket 2, Socket 3 (Regular or OverDrive)
External Architecture	
Data Bus Width (bits)	32
Maximum Data Bus Bandwidth	(Mbytes/sec)100 133 160
Address Bus Width (bits)	32
Maximum Addressable Memory	4 GB
Level 2 Cache Type	Motherboard
Level 2 Cache Size	Usually 64 KB to 256 KB
Level 2 Cache Bus Speed	Same as Memory Bus
Multiprocessing	No
Internal Architecture	
Instruction Set	x86
MMX Support	No
Processor Modes	Real, Protected, Virtual Real
x86 Execution Method	Native
Internal Components	Register Size (bits) 32
Pipeline Depth (stages)	5
Level 1 Cache Size	8 KB Unified
Level 1 Cache Mapping	4-Way Set Associative
Level 1 Cache Write Policy	Write-Through, Write-Back (AMD, Cyrix)
Integer Units	1
Floating Point Unit / Math Coprocessor	Integrated
Instruction Decoders	1
Branch Prediction Buffer Size / Accuracy	None
Write Buffers	None
Performance Enhancing Features	None

(4) Intel 80486DX4 and 80486DX4 OverDrive: The 80486DX4 continued the trend started by the 80486DX2 towards faster clock speed processors on slower motherboards. The DX4 uses "clock tripling", where the processor runs at three times the speed of the memory bus (motherboard). Note that despite the name the DX4 does not run at four times the memory bus speed ("DX3" refers to 2.5 times the memory bus, but never became a shipping product for the 486.)

Intel's DX4 runs at two speeds: 75 MHz (for the 25 MHz bus) and 100 MHz (for the 33 MHz), with the 100 being by far the most popular. In order to keep power and heat to a manageable level the voltage of these chips is reduced to 3.3 volts. The AMD and Cyrix versions are very similar to their DX2 cousins, in that they are 5 volt tolerant, meaning they can handle being put in a 5 volt motherboard; the Intel chip is not 5 volt tolerant and cannot be put into a 5 volt board (well, you can physically put them into the board, but you shouldn't do it). AMD and Cyrix also make a 120 MHz version that runs on a 40 MHz bus; this is relatively rare. The Intel, AMD and Cyrix versions of the DX4 chip are quite similar. One difference is that the AMD and Cyrix support write-back cache while the Intel does not; however the Intel has its level 1 cache doubled to 16 KB. All three chips support power management.

The 486DX4 was most commonly put into systems as OverDrive processors for older, slower systems. Intel did not originally make the DX4 available as a stand-alone chip, presumably since it shipped after the Pentium had already been introduced. However, they did make the chip available stand-alone later on. New 486DX4-100 systems were often made with the AMD version of the chip. The OverDrive version is generally required for older motherboards that don't support 3.3 volt power.

The Like the 80486DX2, the 80486DX4 is technically obsolete, but the 100 MHz version especially has good processing power for routine applications. It is also acceptable for some other uses, but is not found in modern systems.

Table B.9. Characterstic Information: Intel's 80486DX4 Family Microprocessor.

Characteristic	General Information		
Manufacturer	Intel, AMD, Cyrix		AMD, Cyrix
Family Name	80486DX4		
Code name	"P24C"		
Processor Generation	Fourth		
Motherboard Generation	Fourth		
Version	**80486DX4-75**	**80486DX4-100**	**80486DX4-120**
Introduced	March 1994		!?
Variants and Licensed Equivalents	—		
Speed Specifications			
Memory Bus Speed (MHz)	25	33	40
Processor Clock Multiplier	3.0		
Processor Speed (MHz)	75	100	120
"P" Rating	—		
Benchmarks			
iCOMP Rating	319	435	~530
iCOMP 2.0 Rating	—		
Version	**80486DX4-75**	**80486DX4-100**	**80486DX4-120**
Norton SI	~155	198	259
Norton SI32	~12	15	17
CPUmark32	~100	~125	~150
Physical Characteristics			
Process Technology	CMOS		
Circuit Size (microns)	0.6 (Intel), 0.5 (AMD), 0.65? (Cyrix)		
Die Size (mm^2)	!?		
Transistors (millions)	1.6 (Intel)		
Voltage, Power and Cooling			
External or I/O Voltage (V)	3.3 (AMD, Cyrix, 5V tolerant)		
Internal or Core Voltage (V)	3.3		
Power Management	SMM		
Cooling Requirements	Active heat sink		
Packaging			
Packaging Style	168-Pin PGA (Regular), 169-Pin PGA (OverDrive)		

Motherboard Interface	168-Pin Socket (Regular only), Socket 1, Socket 2, Socket 3
External Architecture	
Data Bus Width (bits)	32
Maximum Data Bus Bandwidth (Mbytes/sec)	100 133 160
Address Bus Width (bits)	32
Maximum Addressable Memory	4 GB
Level 2 Cache Type	Motherboard
Level 2 Cache Size	Usually 64 KB to 256 KB
Level 2 Cache Bus Speed	Same as Memory Bus
Multiprocessing	No
Internal Architecture	Instruction Set x86
MMX Support	No
Processor Modes	Real, Protected, Virtual Real
x86 Execution Method	Native
Internal Components	Register Size (bits) 32
Pipeline Depth (stages)	5
Level 1 Cache Size	8 KB (AMD, Cyrix), 16 KB (Intel) Unified
Level 1 Cache Mapping	4-Way Set Associative
Level 1 Cache Write Policy	Write-Through, Write-Back (AMD, Cyrix)
Integer Units	1
Floating Point Unit / Math Coprocessor	Integrated
Instruction Decoders	1
Branch Prediction Buffer Size / Accuracy	None
Write Buffers	None

(5) AMD 5x86 (80486DX5): The 486DX4-100 was the fastest 486 processor made by Intel before they decided to leave the fourth generation and concentrate on the Pentium. AMD took clock multiplying one increment further with what it calls the "5x86" chip. Despite the implication that it is a fifth-generation chip, it is not—it is a high speed, clock-quadrupled 486 processor, that runs in 486 motherboards. It does use a rather advanced 0.35 micron CMOS process (advanced for a fourth-generation chip that is).

The AMD 5x86 was made available in one speed only, 133 MHz, for use in 33 MHz motherboards. The processor runs at four times the system clock and fits into a Socket 3. Since having a clock multiplier of four was not part of the original Socket 3 design, AMD made the 5x86 look for a two times setting from the motherboard and interpret that as four times instead. In other words, to use the 5x86 you want to set the motherboard to the 2x setting. This will actually cause the 5x86 to run at 4x. The chip will actually physically fit into an older 486 socket such as a socket 1 or 2 or the original 168-pin 486 socket, but doing this requires a voltage regulator since the AMD chip runs at 3.3 volts.

AMD calls this chip the "5x86-P75" because it offers performance comparable to low-end fifth-generation chips. In fact, it is comparable in integer performance to a 75 MHz Pentium. It is also called the 80486DX5-133, which is a more accurate reflection of what the chip really is. Architecturally, the chip is virtually identical to the 80486DX4, except that it matches Intel's 16 KB level 1 cache (the AMD 80486DX4 has only 8 KB). Note that this is not the case with Cyrix's 5x86 chip <g4_C5x86.htm>, which is quite different.

The 5x86-133 is the most powerful 486-class chip available (the Cyrix 5x86-120 is actually faster but was discontinued by Cyrix many months before AMD stopped mass-producing the 5x86-133). The 5x86-133 has traditionally been an excellent choice for an economy PC, especially for home-builders, because not only is the chip itself very inexpensive, so are the motherboards that use it. With 200+ MHz

Pentium and Pentium-compatible chips coming well down in price, the 5x86-133 is following the rest of the fourth-generation chips to obsolescence, quickly. However, since it is the "king of the hill" for 486 motherboards, it remains a good choice for those who want to upgrade their 486 systems without replacing the motherboard. The chips are still out there, though they are getting much more difficult to find now.

Table B.10. Characteristic Information : Intel's 80486 DX5 Microprocessor

Characteristic Information	General Information
Manufacturer Detail	AMD
Family Name	80486DX5, 5x86
Code name	"X5"
Processor Generation	Fourth
Motherboard Generation	Fourth
Version	80486DX5-133
Introduced	1995?
Variants and Licensed Equivalents	—
Speed Specifications	
Memory Bus Speed (MHz)	33
Processor Clock Multiplier	4.0
Processor Speed (MHz)	133
"P" Rating	P75
Benchmarks	
iCOMP Rating	~610
Characteristic	**Information**
iCOMP 2.0 Rating	~67
Norton SI	288
Norton SI32	18
CPUmark32	~160
Physical Characteristics	
Process Technology	CMOS
Circuit Size (microns)	0.35
Die Size (mm^2)	!?
Transistors (millions)	!?
Voltage, Power and Cooling	
External or I/O Voltage (V)	3.45
Internal or Core Voltage (V)	3.45
Power Management	SMM
Cooling Requirements	Active heat sink
Packaging	
Packaging Style	168-Pin PGA
Motherboard Interface	Socket 3; or 168-Pin Socket, Socket 1, Socket 2
External Architecture	
Data Bus Width (bits)	32
Maximum Data Bus Bandwidth (Mbytes/sec)	133
Address Bus Width (bits)	32
Maximum Addressable Memory	4 GB
Level 2 Cache Type	Motherboard
Level 2 Cache Size	Usually 256 KB
Level 2 Cache Bus Speed	Same as Memory Bus
Multiprocessing	No
Internal Architecture	
Instruction Set	x86
MMX Support	No

Processor Modes	Real, Protected, Virtual Real
x86 Execution Method	Native
Internal Components	
Register Size (bits)	32
Pipeline Depth (stages)	5
Level 1 Cache Size	16 KB Unified
Level 1 Cache Mapping	4-Way Set Associative
Level 1 Cache Write Policy	Write-Through, Write-Back
Integer Units	1
Floating Point Unit / Math Coprocessor	Integrated
Instruction Decoders	1
Branch Prediction Buffer Size / Accuracy	None
Write Buffers	None
Performance Enhancing Features	None

(6) Cyrix 5x86 ("Misc"): Despite having the same name as AMD's 5x86 processor, the Cyrix 5x86 is a totally different animal. While AMD designed its 5x86 by further increasing the clock on the 486DX4, Cyrix took the opposite approach by modifying its M1 processor core (used for the 6x86 processor) to make a "lite" version to work on 486 motherboards. As such, the Cyrix 5x86 in some ways resembles a Pentium OverDrive (which is a Pentium core modified to work in a 486 motherboard) internally more than it resembles the AMD 5x86. This chip is probably the hardest to classify as either fourth or fifth generation.

The 5x86 employs several architectural features that are normally found only in fifth-generation designs. The pipeline is extended to six stages, and the internal architecture is 64 bits wide. It has a larger (16 KB) primary cache than the 486DX4 chip. It uses branch prediction to improve performance.

The 5x86 was available in two speeds, 100 and 120 MHz. The 5x86-120 is the most powerful chip that will run in a 486 motherboard—it offers performance comparable to a Pentium 90 or 100. The 5x86 is still a clock-tripled design, so it runs in 33 and 40 MHz motherboards. (The 100 MHz version will actually run at 50x2 as well, but normally was run at 33 MHz.) It is a 3 volt design and is intended for a Socket 3 motherboard. It will run in an earlier 486 socket if a voltage regulator is used. I have heard that some motherboards will not run this chip properly so you may need to check with Cyrix if trying to use this chip in an older board. These chips have been discontinued by Cyrix but are still good performers, and for those with a compatible motherboard, as good as you can get. Unfortunately, they are extremely difficult to find now.

Look here for an explanation of the categories in the processor summary table below, including links to more detailed explanations.

Table B.11. Characteristic Information on Cyrix 5x86 Microprocessor Family

Characteristic Information	General Information	
Manufacturer	**Cyrix**	
Family Name	5x86	
Code name	"M1sc"	
Processor Generation	Fourth	
Motherboard Generation	Fourth	
Version	**5x86-100**	**5x86-120**
Introduced	1996?	
Variants and Licensed Equivalents	—	
Speed Specifications		

Memory Bus Speed (MHz)	33 / 50	40
Processor Clock Multiplier	3.0 / 2.0	3.0
Processor Speed (MHz)	100	120
"P" Rating	P75	P90
Benchmarks		
iCOMP Rating	~610	~735
iCOMP 2.0 Rating	~67	~81
Norton SI	264	316
Norton SI32	~16	19
CPUmark32	~150	~180
Physical Characteristics		
Process Technology	CMOS	
Circuit Size (microns)	0.65	
Die Size (mm^2)	144	
Transistors (millions)	2.0	
Version	**5x86-100**	**5x86-120**
Voltage, Power and Cooling		
External of I/O Voltage (V)	3.45	
Internal or Core Voltage (V)	3.45	
Power Management	SMM	
Cooling Requirements	Active heat sink	
Packaging		
Packaging Style	168-Pin PGA	
Motherboard Interface	Socket 3; or 168-Pin Socket	Socket 1, Socket 2
External Architecture		
Data Bus Width (bits)	32	
Maximum Data Bus Bandwidth (Mbytes/sec)	133	160
Address Bus Width (bits)	32	
Maximum Addressable Memory	4 GB	
Level 2 Cache Type	Motherboard	
Level 2 Cache Size	Usually 256 KB	
Level 2 Cache Bus Speed	Same as Memory Bus	
Multiprocessing	No	
Internal Architecture	Instruction Set	x86
MMX Support	No	
Processor Modes	Real, Protected, Virtual Real	
x86 Execution Method	Native	
Internal Components	Register Size (bits)	32
Pipeline Depth (stages)	6	
Level 1 Cache Size	16 KB Unified	
Level 1 Cache Mapping	4-Way Set Associative	
Level 1 Cache Write Policy	Write-Through, Write-Back	
Integer Units	1	
Floating Point Unit / Math Coprocessor	Integrated Instruction Decoders	1
Branch Prediction Buffer Size / Accuracy	!? entries /	!? %
Write Buffers	!?	
Performance Enhancing Features	—	

(7) Two Intel Pentium CPU's : All CPUs have model names/numbers and (clock) speeds. For example, the current generation of Intel CPUs are called Pentiums and Pentium Pros (these are actually two distinct, though related, CPU types). CPU speeds are measured in megahertz (MHz). Pentium CPUs come in 60, 66, 75, 90, 100, 120, 133, 150, and 166 MHz. Pentium Pro CPUs typically come in higher speeds. A higher clock speed does not necessarily mean better performance (Case in point, the 80486 generation of Intel processors came in 66Mhz, however a 60MHz Pentium would blow a 66MHz 486 out of the water), though within the same generation of processors, higher CPU speed is unquestionably better. When dealing with CPU speeds, higher numbers are better (i.e. 120 beats 100), though the actual chip generation (486? Pentium?) is most important of all.

All PCs have a CPU made by Intel or any of a number of companies that build intel-like CPUs. These include Cyrix, AMD, and NextGen. These companies' CPUs have generation names typically different from Intel's. For example, Cyrix calls its CPUs that are similar to Intel's Pentiums the "6x86" generation.

Therefore, If you were thinking of buying a CPU, there are three factors to consider: manufacturer, generation, and speed of the chip (clock speed). When you read ads such as "System contains a Pentium 100 CPU", Intel is generally assumed, though it never hurts to confirm this with the dealer. "100" in this case refers to 100MHz.

Modern CPUs, such as Pentiums and some fast 486s, need cooling fans or heat sinks on them so they won't overheat. This is not optional as excessive heat could ruin your CPU. These fans that you do need to buy just clip onto the top of the CPU and are really cheap. With some of the more expensive ones boasting teflon-coated ball-bearings for "super quiet" operation, though for the most part the sound of such fans is drowned out by the case.

(8) Intel Celeron Processor

The Intel(r) Celeron(tm) processor is designed for low cost, or "Basic PC", desktop systems. Based on Intel's P6 architecture, the same core on which the Pentium(r) II processor is based, the Celeron processor provides a solid foundation for the Basic PC.

Because it incorporates the P6 architecture and Intel MMX(tm) technology, the Celeron processor offers particularly good floating-point and multimedia performance. In addition, the Celeron 300A MHz and 333 MHz processors, which include an integrated L2 cache, offer Intel's highest level of integer, multimedia, and floating point performance for the Basic PC.

The Intel Celeron processor provides the capability to run today's most common PC applications on operating systems such as Windows* 98, Windows* 95, Windows NT* and UNIX*. The microprocessor and the PC of today are designed to run a broad range of powerful software applications. Not every processor is equally capable of the same performance for each type of application. Multimedia (such as video and audio), floating- point (such as 3D geometry calculations), and integer performance comprise three vectors of performance that should be considered when evaluating systems. Benchmarks designed for evaluating these vectors should be used to examine the complete performance of a processor or system.

The graph below highlights Celeron processor performance, relative to the Pentium processor with MMX technology, on popular and industry standard benchmarks that demonstrate the three vectors of performance mentioned above. Introduction Based on the Intel P6 architecture, the same core on which the Pentium II processor is based, the Intel Celeron(tm) processor is an exceptional value for the Basic PC. The Celeron processor opens the door to the internet and provides the capability to run today's most common PC applications.

The newest additions to the Celeron processor family are the Celeron 300A MHz and 333 MHz processors. These processors offer Intel's highest level of integer, multimedia, and floating point performance for the Basic PC. The Celeron processor family consists of the following products: 333 MHz Celeron processor 300A MHz Celeron processor 300 MHz Celeron processor 266 MHz Celeron processor. When evaluating the performance of a microprocessor, it is important to get a complete picture of how it executes various tasks. The increasing use of 3D and multimedia content in software today is placing new demands on the microprocessor. Applications such as video playback, 3D games, and PC imaging stress the multimedia and floating-point capabilities of the processor and the system.

Typical productivity applications, such as word processing, presentation applications, or personal finance programs, require the processor to have good integer performance. For the best all around computation, a system should deliver high performance in all three of these areas: multimedia, floating-point, and integer. This report provides benchmark results for Intel Celeron processor systems. Modern industry standard benchmarks were chosen to demonstrate the performance of the Celeron processor for the three vectors of performance. Multimedia performance can be compared with the Norton Multimedia Benchmark. Floating-point performance can be measured with 3D Winbench*98 as well as the FPU WinMark test.

Integer performance is covered by several compute-intensive 32-bit Windows*95 benchmarks as well as more system oriented benchmarks like BAPCo's SYSmark32 test. Intel is committed to using the most robust and relevant benchmarks in characterizing its products' performance and, over time, Intel will adapt this mix as newer benchmarks appear. Robust benchmark programs should be derived from how actual applications will execute. Performance is often the result of combined characteristics of a given computer architecture and many other tightly coupled system software/hardware constituents in addition to the microprocessor.

Operating system, compiler, library, memory design, and I/O subsystem characteristics may significantly impact the results and make comparisons difficult. This report illustrates Intel Celeron processor performance on a consistent configuration. The Intel(r) Celeron(tm) processor offers Intel's quality and dependability at an exceptional value, while providing the performance required to run today's applications on an operating system such as Windows 98. The Celeron processor is based on the Intel P6 architecture and includes Intel MMX(tm) technology. The processor core for the 266 and 300MHz parts has 7.5 million transistors. In addition, the Celeron 300A MHz and 333 MHz processors incorporate an integrated L2 cache. All Celeron processors are based on Intel's advanced 0.25 micron CMOS process technology The Celeron processor is backed by Intel's 25 years of experience in engineering and manufacturing high quality, reliable processors.

The Intel Celeron processor may contain design defects or errors known as errata. Current characterized errata are available upon request. The Intel Celeron(tm) Processor Product Feature Highlights: The Celeron(tm) processor is fully compatible with an entire library of PC software based on operating systems such as MS-DOS, Windows 3.1, Windows for Workgroups 3.11, Windows 98, Windows 95, OS/2, UnixWare, SCO UNIX, Windows NT, OPENSTEP, and Sun Solaris.

Architectural features of the Celeron processor include: Dynamic Execution Technology. Dynamic execution incorporates the concepts of out of order and speculative execution. The Celeron processor's implementation of these concepts removes the constraint of linear instruction sequencing between the traditional fetch and execute phases of instruction execution. Up to 3 instructions can be decoded per clock cycle. These decoded instructions are put into a buffer, which can hold up to 40 instructions. Instructions are executed from this buffer when their operands are available (versus instruction order). Up to 4 instructions can be executed per clock cycle.

Superpipelining: The pipeline of the P6 architecture consists of approximately 12 stages versus 5 for the Pentium(r) processor and 6 for the Pentium processor with MMX(tm) technology. This enables the Celeron processor to achieve about a 50% higher frequency than the Pentium processor on the same manufacturing technology.

The sophisticated, two-level, adaptive-training, branch prediction mechanism of the P6 microarchitecture is key to maintaining the efficiency of the Intel Celeron processor's superpipelined microarchitecture. High Performance Intel MMX Technology: Intel's MMX media enhancement technology is a major enhancement to the Intel Architecture which makes PCs richer multimedia and communications platforms. This technology introduces 57 instructions oriented to highly parallel operations with multimedia and communications data types. These instructions use a technique known as

SIMD (Single Instruction, Multiple Data) to deliver better performance for multimedia and communications computation. Intel processors that provide MMX technology support are fully compatible with previous generations of the Intel Architecture and the installed base of software. To further improve performance, the Celeron processor, like the Pentium II processor, can execute 2 Intel MMX instructions simultaneously.

Write Combining: The Write Combining technology of the P6 architecture can be utilized to achieve very high graphics I/O performance. This feature combines multiple writes to a region of memory (for example, a video controller's frame buffer) declared as WC type into a single burst write operation. This is well suited for the bus which is optimized for burst transfers. These writes are further combined by the chipset leading to high throughput for graphics I/O. This will further enhance multimedia performance and enable more realistic full motion video and realistic, fast graphics performance.

Caches: The Celeron processor has 32 KB of non-blocking L1 cache, which is divided into a 16 KB instruction cache and a 16 KB data cache. Each of these caches run at the processor frequency and provide fast access to heavily used data. The 333 MHz and 300A MHz Celeron processors have a 128 KB full-speed, integrated L2 cache which is unified for code and data, and is non-blocking.

□□□

INDEX

❑❑❑

Win Prizes

ATTENTION READERS

Win Prizes

We request for your frank assessment regarding various aspects of the book as given below :

A TEXTBOOK OF COMPUTER ARCHITECTURE (PTU)

By : AMIT KUMAR MISHRA

Please Fill up the Sheet in Neat and Clean Writing

(i) What topics of your syllabus, important for your examination point of view, are not covered in this book ?

....

....

....

....

(ii) What are the chapters wherein the treatment of the subject matter is not systematic or organised or updated ?

....

....

....

....

(iii) Have you come across misprints/mistakes/factual inaccuracies in the book ? Please specify the chapters and the page numbers.

....

....

....

....

(iv) What other books on the same subject have you found/heard better than the present book ? Please specify in terms of price and quality.

....

....

....

....

(v) Further suggestions and comments for the improvement of this book.

....

....

....

....

(PS : If need be please attach a separate sheet to write your views)

OTHER DETAILS

(i) Who recommended you this book ? (Please tick (✓) in the box near the option relevant to you.

☐ Teacher ☐ Friends ☐ Bookseller

....

....

....

....

(ii) Name of the recommending teacher, his designation and address:

....

....

....

....

(iii) Name and address of the bookseller you purchased the book from :

....

....

....

....

(iv) Name and address of your institution (Please mention the University or Board as the case may be)

....

....

....

....

(v) Your name and complete address :

....

....

....

....

(vi) Course you are enrolled in :

....

....

....

....

The best assessment will be awarded each month : The award will be in the form of our publications as decided by the Editorial Board.

Please mail the filled up coupon at the vearliest to :

S.K. KATARIA & SONS®

4760-61/23, Ansari Road,

Daryaganj, New Delhi-110002

Phone: +91-11-23243489, 23269324

TeleFax : +91-11-23243489

email : katariabooks@yahoo.com